IET ENERGY ENGINEERING 111

Energy Storage at Different Voltage Levels

Other volumes in this series:

Volume 1 **Power Circuit Breaker Theory and Design** C.H. Flurscheim (Editor)
Volume 4 **Industrial Microwave Heating** A.C. Metaxas and R.J. Meredith
Volume 7 **Insulators for High Voltages** J.S.T. Looms
Volume 8 **Variable Frequency AC Motor Drive Systems** D. Finney
Volume 10 **SF_6 Switchgear** H.M. Ryan and G.R. Jones
Volume 11 **Conduction and Induction Heating** E.J. Davies
Volume 13 **Statistical Techniques for High Voltage Engineering** W. Hauschild and W. Mosch
Volume 14 **Uninterruptible Power Supplies** J. Platts and J.D. St Aubyn (Editors)
Volume 15 **Digital Protection for Power Systems** A.T. Johns and S.K. Salman
Volume 16 **Electricity Economics and Planning** T.W. Berrie
Volume 18 **Vacuum Switchgear** A. Greenwood
Volume 19 **Electrical Safety: A guide to causes and prevention of hazards** J. Maxwell Adams
Volume 21 **Electricity Distribution Network Design, 2nd Edition** E. Lakervi and E.J. Holmes
Volume 22 **Artificial Intelligence Techniques in Power Systems** K. Warwick, A.O. Ekwue and R. Aggarwal (Editors)
Volume 24 **Power System Commissioning and Maintenance Practice** K. Harker
Volume 25 **Engineers' Handbook of Industrial Microwave Heating** R.J. Meredith
Volume 26 **Small Electric Motors** H. Moczala *et al.*
Volume 27 **AC–DC Power System Analysis** J. Arrillaga and B.C. Smith
Volume 29 **High Voltage Direct Current Transmission, 2nd Edition** J. Arrillaga
Volume 30 **Flexible AC Transmission Systems (FACTS)** Y.-H. Song (Editor)
Volume 31 **Embedded Generation** N. Jenkins *et al.*
Volume 32 **High Voltage Engineering and Testing, 2nd Edition** H.M. Ryan (Editor)
Volume 33 **Overvoltage Protection of Low-Voltage Systems, Revised Edition** P. Hasse
Volume 36 **Voltage Quality in Electrical Power Systems** J. Schlabbach *et al.*
Volume 37 **Electrical Steels for Rotating Machines** P. Beckley
Volume 38 **The Electric Car: Development and future of battery, hybrid and fuel-cell cars** M. Westbrook
Volume 39 **Power Systems Electromagnetic Transients Simulation** J. Arrillaga and N. Watson
Volume 40 **Advances in High Voltage Engineering** M. Haddad and D. Warne
Volume 41 **Electrical Operation of Electrostatic Precipitators** K. Parker
Volume 43 **Thermal Power Plant Simulation and Control** D. Flynn
Volume 44 **Economic Evaluation of Projects in the Electricity Supply Industry** H. Khatib
Volume 45 **Propulsion Systems for Hybrid Vehicles** J. Miller
Volume 46 **Distribution Switchgear** S. Stewart
Volume 47 **Protection of Electricity Distribution Networks, 2nd Edition** J. Gers and E. Holmes
Volume 48 **Wood Pole Overhead Lines** B. Wareing
Volume 49 **Electric Fuses, 3rd Edition** A. Wright and G. Newbery
Volume 50 **Wind Power Integration: Connection and system operational aspects** B. Fox *et al.*
Volume 51 **Short Circuit Currents** J. Schlabbach
Volume 52 **Nuclear Power** J. Wood
Volume 53 **Condition Assessment of High Voltage Insulation in Power System Equipment** R.E. James and Q. Su
Volume 55 **Local Energy: Distributed generation of heat and power** J. Wood
Volume 56 **Condition Monitoring of Rotating Electrical Machines** P. Tavner, L. Ran, J. Penman and H. Sedding
Volume 57 **The Control Techniques Drives and Controls Handbook, 2nd Edition** B. Drury
Volume 58 **Lightning Protection** V. Cooray (Editor)
Volume 59 **Ultracapacitor Applications** J.M. Miller
Volume 62 **Lightning Electromagnetics** V. Cooray
Volume 63 **Energy Storage for Power Systems, 2nd Edition** A. Ter-Gazarian
Volume 65 **Protection of Electricity Distribution Networks, 3rd Edition** J. Gers
Volume 66 **High Voltage Engineering Testing, 3rd Edition** H. Ryan (Editor)
Volume 67 **Multicore Simulation of Power System Transients** F.M. Uriate
Volume 68 **Distribution System Analysis and Automation** J. Gers
Volume 69 **The Lightening Flash, 2nd Edition** V. Cooray (Editor)
Volume 70 **Economic Evaluation of Projects in the Electricity Supply Industry, 3rd Edition** H. Khatib
Volume 72 **Control Circuits in Power Electronics: Practical issues in design and implementation** M. Castilla (Editor)
Volume 73 **Wide Area Monitoring, Protection and Control Systems: The enabler for Smarter Grids** A. Vaccaro and A. Zobaa (Editors)
Volume 74 **Power Electronic Converters and Systems: Frontiers and applications** A.M. Trzynadlowski (Editor)
Volume 75 **Power Distribution Automation** B. Das (Editor)
Volume 76 **Power System Stability: Modelling, analysis and control** B. Om P. Malik
Volume 78 **Numerical Analysis of Power System Transients and Dynamics** A. Ametani (Editor)
Volume 79 **Vehicle-to-Grid: Linking electric vehicles to the smart grid** J. Lu and J. Hossain (Editors)
Volume 81 **Cyber-Physical-Social Systems and Constructs in Electric Power Engineering** S. Suryanarayanan, R. Roche and T.M. Hansen (Editors)
Volume 82 **Periodic Control of Power Electronic Converters** F. Blaabjerg, K.Zhou, D. Wang and Y. Yang
Volume 86 **Advances in Power System Modelling, Control and Stability Analysis** F. Milano (Editor)
Volume 87 **Cogeneration: Technologies, optimisation and implementation** C.A. Frangopoulos (Editor)
Volume 88 **Smarter Energy: From smart metering to the smart grid** H. Sun, N. Hatziargyriou, H.V. Poor, L. Carpanini and M.A. Sánchez Fornié (Editors)
Volume 89 **Hydrogen Production, Separation and Purification for Energy** A. Basile, F. Dalena, J. Tong and T.N. Veziroğlu (Editors)
Volume 90 **Clean Energy Microgrids** S. Obara and J. Morel (Editors)
Volume 91 **Fuzzy Logic Control in Energy Systems with Design Applications in MATLAB®/ Simulink®** İ.H. Altaş
Volume 92 **Power Quality in Future Electrical Power Systems** A.F. Zobaa and S.H.E.A. Aleem (Editors)
Volume 93 **Cogeneration and District Energy Systems: Modelling, analysis and optimization** M.A. Rosen and S. Koohi-Fayegh
Volume 94 **Introduction to the Smart Grid: Concepts, technologies and evolution** S.K. Salman
Volume 95 **Communication, Control and Security Challenges for the Smart Grid** S.M. Muyeen and S. Rahman (Editors)
Volume 97 **Synchronized Phasor Measurements for Smart Grids** M.J.B. Reddy and D.K. Mohanta (Editors)
Volume 98 **Large Scale Grid Integration of Renewable Energy Sources** A. Moreno-Munoz (Editor)
Volume 100 **Modeling and Dynamic Behaviour of Hydropower Plants** N. Kishor and J. Fraile-Ardanuy (Editors)
Volume 101 **Methane and Hydrogen for Energy Storage** R. Carriveau and D.S.-K. Ting
Volume 104 **Power Transformer Condition Monitoring and Diagnosis** A. Abu-Siada (Editor)
Volume 108 **Fault Diagnosis of Induction Motors** J. Faiz, V. Ghorbanian and G. Joksimović
Volume 110 **High Voltage Power Network Construction** K. Harker
Volume 112 **Wireless Power Transfer: Theory, technology and application** N.Shinohara
Volume 119 **Thermal Power Plant Control and Instrumentation: The control of boilers and HRSGs, 2nd Edition** D. Lindsley, J. Grist and D. Parker
Volume 123 **Power Systems Electromagnetic Transients Simulation, 2nd Edition** N. Watson and J. Arrillaga
Volume 124 **Power Market Transformation** B. Murray
Volume 130 **Wind and Solar Based Energy Systems for Communities** R. Carriveau and D.S.-K. Ting (Editors)
Volume 131 **Metaheuristic Optimization in Power Engineering** J. Radosavljević
Volume 905 **Power System Protection, 4 volumes**

Energy Storage at Different Voltage Levels

Technology, integration, and market aspects

Edited by
Ahmed F. Zobaa, Paulo F. Ribeiro,
Shady H.E. Abdel Aleem and Sara N. Afifi

The Institution of Engineering and Technology

Published by The Institution of Engineering and Technology, London, United Kingdom

The Institution of Engineering and Technology is registered as a Charity in England & Wales (no. 211014) and Scotland (no. SC038698).

First published 2018

The Institution of Engineering and Technology
Michael Faraday House
Six Hills Way, Stevenage
Herts, SG1 2AY, United Kingdom

www.theiet.org

British Library Cataloguing in Publication Data
A catalogue record for this product is available from the British Library

ISBN 978-1-78561-349-4 (hardback)
ISBN 978-1-78561-350-0 (PDF)

Typeset in India by MPS Limited

Contents

About the editors

Ahmed Faheem Zobaa is a senior lecturer in electrical and power engineering, an MSc Course Director and a full member of the Institute of Energy Futures at Brunel University London, UK. His main areas of expertise include power quality, renewable energy and smart grids. Dr. Zobaa is a senior fellow of the Higher Education Academy. He is a fellow of the Institution of Engineering and Technology, the Energy Institute, the Chartered Institution of Building Services Engineers, the Institution of Mechanical Engineers and the African Academy of Science.

Paulo F. Ribeiro is a full professor at the Federal University of Itajubá, Brazil. His research interests include power systems, power electronics, power quality engineering, active transmission and distribution systems, signal processing applied to power systems and interfaces for energy storage systems. He has conducted research at EPRI, NASA and the Center for Advanced Power Systems (CAPS), and taught engineering in the United States, New Zealand and the Netherlands. He is active within CIGRE, IEEE, IET and IEC Working Groups. He has published over 300 papers, several book chapters and three books. He is a fellow of the IEEE and IET.

Sara Nader Afifi received her BSc (Hons) in Electrical Power & Machines from Cairo University, Egypt, in 2005. She received her MSc in Energy Policy and Sustainability from the University of Exeter, UK, in 2009. Also, she received her PhD in Electrical Engineering & Electronics Research from Brunel University London, UK, in 2017. She has several years of industrial experience in building services consultancy and electric distribution utilities. Currently, she is a network planning and investment engineer at Scottish and Southern Electricity Networks, UK. Her main areas of expertise include power distribution systems, distributed generation and energy policy. Dr. Afifi is an associate fellow of the Higher Education Academy of UK. Also, she is a member of the Institution of Engineering and Technology.

Shady H.E. Abdel Aleem received his B.Sc., M.Sc., and Ph.D. degrees in Electrical Power and Machines from the Faculty of Engineering, Helwan University, Egypt, in 2002, and the Faculty of Engineering, Cairo University, Egypt, in 2010 and 2013, respectively. Starting from September 2013, he is an assistant professor at 15th of May Higher Institute of Engineering. His research interests include harmonic problems in power systems, power quality, renewable energy, smart grid, energy efficiency, decision-making, optimization, green energy, and economics. He is the author or co-author of many refereed journal and conference papers. He has published over 70 plus journal and conference papers, six book chapters, and four edited books with the Institution of Engineering and Technology (IET), Elsevier, and InTechOpen publishers. He has been awarded the Albert Nelson Marquis Lifetime Achievement Prestige Award in 2017 from the US and the State Encouragement Award in Engineering Sciences in 2018 from Egypt. Dr. Shady is a member of the Institute of Electrical and Electronics Engineers (IEEE). He is also a member of the Institution of Engineering and Technology (IET).

List of acronyms

ADNs	Active distribution networks
AFC	Advanced frequency control
AGC	Automatic generation control
ALPs	Annual load profiles
ARRA	American Reinvestment and Recovery Act of 2009
ASP	Ancillary Services Provider
BATT	Batteries for Advanced Transportation Technology
BESS	Battery energy storage systems
BESSM	Battery Energy Storage System Management
BFS	Backward forward sweep
CAES	Compressed air energy storage
CAESS	Compressed air ESS
CCM	Centre for Concepts in Mechatronics
CDLPs	Characteristic daily load profiles
CERL	Consortium of European Research Libraries
CERTS	Consortium for Electrical Reliability Technology Solutions
CES	Chemical energy storage
CETEEM	Clean Energy Technology Economic and Emissions Model
CFC	Conventional frequency control
CM	Capacity market
CPP	Critical peak pricing
CPR	Critical peak rebate
CSIRO	Commonwealth Scientific and Industrial Research organization
CU	Classical unit
CVR	Conservation voltage regulation
CVaR	Conditional-value-at-risk
DAEs	Differential algebraic equations
DERs	Distributed energy resources
DERMS	Distributed Energy Resources Management System

DGs	Distributed generators
DISPERSE	DIStributed Power Economic Rationale Selection
DLP	Daily load profile
DoD	Depth of discharge
DOE	Department of Energy
DP	Dynamic programming
DPG	Decentralized Power Generation
DRPs	Demand response programmes
DSM	Demand Side Management
EES	Electrical energy storage
EENS	Expected energy not supplied
EFR	Enhanced frequency response
ELDCs	Electric double-layer capacitors
EMALS	Electromagnetic Aircraft Launch System
EMRA	Energy Markets Regulator Authority
EMRP	International European Metrology Research Program
EMS	Energy Management Systems
EPA	Environmental Protection Agency
ER	Engineering Recommendation
ES	Energy storage
ESS	Energy storage system
EV	Expected value
EV	Electric vehicle
FCDM	Frequency control by demand management
FERC	Federal Energy Regulatory Commission
FES	Flywheel energy storage
FESS	Flywheel energy storage system
FFR	Firm frequency response
FMOO	Fuzzy multi-objective optimization
FOSS	Forecasting, Optimisation, Scheduling System
FR	Frequency response
GHG	Greenhouse Gases
GIS	Geographic İnformation System
GPS	Global Positioning System
HESS	Hydrogen ESS
HH	Half-hour

HOMER	Hybrid optimization of multiple energy resources
ICT	Information and Communication Technology
IEEE	Institute of Electrical and Electronics Engineers
KKT	Karush–Kuhn–Tucker
LCA	Life Cycle Analysis
LMP	Locational marginal price
M2M	Machine To Machine
MAE	Mean absolute error
MEGs	Micro-Energy Grids
MEMS	Micro-Electro-Mechanical Systems
MENR	The Ministry of Energy and Natural Resources
MES	Mechanical energy storage
MG	MicroGrid
MGCC	MG Central Controller
MILP	Mixed-integer linear program
MLR	Multiple linear regression
MPEC	Mathematical problem with equilibrium constraints
NG	National Grid
NG	Natural Gas
NIST	National Institute Of Standards and Technology
NR	Newton–Raphson
OCD	Optimality condition decomposition
PEM	Point estimate method
PEVs	Plug-in electric vehicles
PfG	Power-from-grid
PHS	Pumped hydroelectric storage
PI	Proportional-integral
PMFs	Probability mass functions
PMSMs	Permanent magnet synchronous machines
PMUs	Phasor measurement units
PPAs	Power-purchase agreements
PS	Peak shaving
PtG	Power-to-grid
PV	Photovoltaic
RE	Renewable Energy
REIF	Renewable Energy Integration Facility

RES	Renewable energy sources
RLC	Resistor-inductor-capacitor
RTDS	Digital real-time simulator
RTP	Real-time pricing
RTTRs	Real-time thermal ratings
SCED	Security-constrained economic dispatch
SEP	Solar energy profiler
SEN	Semantic Energy Network
SESSs	Smart energy storage systems
SG	Smart Grid
SGE	Smart Grid Energy
SMES	Superconducting magnetic energy storage
SMPC	Stochastic model predictive control
SNS	Smarter Network Storage
SoC	State of charge
SOCS	Smart Optimisation and Control System
STOR	Short-term operating reserve
STS	Sequential time-simulation
TCOs	Tap-changing operations
TESS	Thermal ESS
THD	Total harmonic distortion
ToD	Time of day
ToU	Time-of-use
TSOs	Transmission system operators
TSPF	Time-series power flow
TUoS	Transmission use of system charge
T&D	Transmission and distribution
UPS	Uninterruptable power supply
UVF	Unbalance voltage factor
V2G	Vehicle to grid
VPP	Variable peak pricing
VRs	Voltage regulators
VRMs	Variable reluctance machines
VSG	Virtual synchronous generator
WACC	Weighted average cost of capital
WinDS	Wind Deployment Systems
WTs	Wind turbines

Preface

Currently, global warming, pollution and fossil fuel shortages have forced all stakeholders to focus more on renewable energy sources as environmental-friendly sources of energy. However, power generated from renewable resources does not have the ability to provide immediate response to demand as these resources suffer from daily intermittency and seasonal variations. For renewables, energy storage is the energy capture that may allow firming of the capacity of the output power to provide a reasonably constant output from them or their combination with other conventional energy generation plants, in addition to the primary function of the energy storage technologies in storing energy for later use. Energy storage is the way of enabling renewables to power our lives in a green manner with almost no to little global warming emissions.

For electric grids, energy storage makes them reliable and improves their stability despite the challenges of the growth of decentralized production. For small-scale applications, different electrically powered devices have to carry their energy sources with them. Briefly, we do not need to look any further than energy storage to see how directly the energy scene, renewables and green power applications are currently changing. This book presents the essential aspects of the current and future roles of energy storage, prospects and challenges in the various generation, transmission, distribution and customer levels. It introduces a comprehensive overview of energy storage technologies. It discloses different scenarios to expect what will the next years bring for these storage technologies, and demonstrates the risks and mitigation solutions for integration problems of these techniques while illustrating the significance of storing energy for the sustainable development of modern power system grids.

Impacts of cost-effective energy storage by enabling low-carbon energy technologies will be discussed while demonstrating various mixed electricity scenarios at different levels and the problems of finding the optimal coordination between generation and storage under uncertainties. Additionally, management aspects of future planning uncertainty and the role of storage in transmission investment deferral will be addressed. Energy storage in micro-grids as an integral part of modern power system networks will be presented taking into account the new technologies and their advanced control methods.

Energy storage sizing problem will be presented while addressing markets, policies and time of use pricing, and also economic risks will be discussed using various existing case studies from different countries.

The book is principally focused on applications, but each of the book's chapters begins with the fundamental structure of the problem required for a rudimentary understanding of the methods described. The book is sorted out and organized in 12 chapters.

Chapter 1: This chapter presents an overview of energy storage technologies. Both the technical and economic advantages of these technologies are addressed. A comparative analysis between these technologies is presented.
Chapter 2: The aim of this chapter is to discuss technologies, applications, tendencies and challenges for engineers who want to employ energy storage devices at different voltage levels in distribution and transmission systems. This is particularly important in face of smart grid technologies that are about to shape power systems in the near future. It summarizes and evaluates some of the available storage technologies. For this sake, international trends and potential applications and recommendations on how to consider the energy storage along with other pilot projects on technological routes are addressed.
Chapter 3: There are still several technical challenges regarding the control, planning and operation of grids and micro-grids due to the intermittent nature of renewable energy resources, uncertainties in load variations and lack of inertia proposed by inverter-connected distributed energy resources. Technical concerns, associated with control and operation of micro-grids, include voltage regulation, system stability and reliability, and power quality. Energy storage systems are considered as a solution to improve power quality, dynamic stability, reliability and controllability, of micro-grids in the presence of renewable energy resources. They can act as an energy buffer to compensate renewable intermittency and load variations. They benefit electricity consumers by maximizing renewable energy utilization and minimizing electricity bills through optimal energy management. This chapter investigates the various microgrid challenges, energy storage technologies and applications of energy storage within a microgrid in detail.
Chapter 4: Unlike in vertically integrated power systems, where energy storage is coordinated with the rest of the system to minimize the overall generation cost, operations of energy storage in systems with electricity markets are driven by preferences of their owners. The owners are typically profit-seeking entities that aim to collect the maximum profit possible for given system conditions. These preferences of the storage owners do not necessarily lead to the minimization of the overall generation cost in the system. Generally, merchant storage devices will provide services that can be monetized and that are most profitable. This generally includes energy arbitrage, frequency regulation, voltage support, etc. This chapter discusses energy storage in energy markets as price takers and makers. Numerical examples are presented. In addition, policies to enhance energy storage flexibility, generalized payment schemes and the dilemma of trade energy or frequency regulation are addressed in detail.
Chapter 5: This chapter presents the role of storage in transmission investment deferral and the management of future planning uncertainty. Full mathematical formulation of the stochastic transmission expansion planning with storage

problem is presented. Besides, a decomposition scheme that renders the analysis of marge systems tractable is presented and discussed. A case study using the IEEE-24 bus system is investigated where the planner identifies the optimal investment strategy while facing uncertainty with respect to future generation connections and demand growth. Our focus is computing the benefit of deploying energy storage to defer premature commitment to conventional reinforcement projects and manage interim uncertainty.

Chapter 6: This chapter focuses on the optimal sizing of battery systems in end-user's applications aimed at minimizing customer costs under time of use tariffs. Thanks to the presence of the battery system, in fact, variations of off- and on-peak prices result in significant benefits obtainable by the modification of consumers' energy-spending patterns without impacting their comfort and/or manufacturing process. More specifically, a method for the optimal sizing of a battery system used for the reduction of the end-user's electricity bill is described taking into account the technical and economic uncertainties involved in the sizing procedure. The probabilistic optimal sizing is obtained applying the point estimate method that guarantees highly accurate results. The point estimate method has the advantage of less computational efforts than the Monte Carlo simulation procedures. Numerical applications are reported with reference to industrial, commercial and residential customers.

Chapter 7: This chapter focuses on the analysis and assessment of multiple technical benefits of battery energy storage systems in improving daily network operations of distribution systems. While this kind of analysis needs to be based on modelling and simulation of distribution networks including battery energy storage systems, the time-series power flow simulations are also required at first in the development of dispatch schedule for battery energy storage systems and at later stages to assess the efficacy of the scheduling algorithm over longer periods including different case scenarios. A multi-objective optimization problem using fuzzy systems followed by simulation results and discussion for two distribution feeders are presented.

Chapter 8: The analysis of various case studies with storage systems in Italy and Romania was conducted. In Italy, the results from collaboration with Enel Distribution for black start function with a storage system with lithium batteries are presented. A dynamic model for presenting the entire system was elaborated and through RTDS simulations considering the protection and control systems was achieved. An optimization model for minimizing operational costs of a smart grid supplying cortical and interruptible loads in Romania was also presented. The case study is based on a real facility equipped with renewable sources and storage systems for clean and reliable power supply. The analysis is based on perfectly known photovoltaic production data obtained from 1 year-recorded measurements of the facility. The characteristic electrical data of the sources and loads within the smart grid are based on real data obtained from the producers.

Chapter 9: This chapter presents different case studies of energy storage systems in the United States, as well as the segmentation of markets using these energy systems.

Chapter 10: This chapter reviews flywheel energy storage systems, considerations and their applications. This chapter also extends recent reviews on flywheel systems through discussions on aerospace, energy harvesting and other applications of flywheel systems. Additionally, companies and research institutions active in flywheel systems research are discussed, along with recent developments of facilities using these systems for power quality studies.
Chapter 11: To reach the current levels of confidence in the coming wide-scale adoption of energy storage, the real-world operation of large-scale energy storage systems is needed to prove that effective operation can be achieved. This was the purpose of the Smarter Network Storage (SNS) project delivered by UK Power Networks from 2013 to 2016. This chapter discusses significantly on the development, implementation and operation of the systems required to support the technical and commercial management of the SNS project. Forecasting and optimization for multi-purpose application of energy storage systems to deliver grid services of the SNS project are discussed in detail.
Chapter 12: This chapter focuses on one of the most significant applications in electric power systems – real-power balancing (active power balancing) as an example to demonstrate how energy storage devices can effectively work with conventional generation to address the variation and intermittency introduced by renewable energy sources. This chapter presents an innovative and systematic approach to tackle the real-power balancing problem by means of optimal coordination between conventional generation and energy storage.

Finally, this book aims to introduce good practice with new research outcomes, programmes and ideas that connect the past, current and future roles of energy storage technologies from different perspectives. It is a tool for the planners, designers, operators and practising engineers of electrical power systems that are concerned with energy storage technologies, reliability and security. Likewise, it is a key resource for advanced students, postgraduates, academics and researchers who had some background in electrical power systems.

The editors also hope that a reader will both use and receive this book as a useful engineering tool and thought-provoking guide, as any scientific or engineering work such as this can be either 'received' or 'used.' When 'used', it assists our own tasks and activities. When 'received', it will help to exercise our senses and imagination according to the patterns suggested by the authors.

The editors
Ahmed, Paulo, Shady and Sara

Chapter 1

Overview of energy storage technologies

Navid Rezaei[1], *Abdollah Ahmadi*[2], *Sara N. Afifi*[3], *Ahmed F. Zobaa*[4] *and Shady H.E. Abdel Aleem*[5]

1.1 Introduction

Electric power systems are gradually maturing in the operational and management architecture. The eventual goal of the system operators is to provide more reliable and high-quality energy services in a cost-effective and environmental framework. To that end, new applications and technologies should be innovated and integrated into the system infrastructure sustainably. Energy storage system (ESS) is an essential part of the power system. ESS can be used for several reasons such as [1–3]:

- *Providing a high sustainable penetration of renewable energy sources (RESs)*: Joint energy management of ESSs and RESs, by mitigating the intermittent nature of renewable resources, assures the operators to rely on higher RES penetration levels.
- *Guaranteeing the system reserve requirement in a more eco-friendly portfolio*: A coordinated integration of ESSs with hydro-thermal generators gets an extended spinning and non-spinning reserve capacity promising the system reliability. Besides, owing to the cleaner production of the ESS technologies, through effective lowering the share of the thermal generating units, system reserve procurement constraint is obtained in a more environment-friendly form.
- *Realisation of electric vehicle (EV) epidemic-applicability vision:* By developing the energy storage technologies, the ever-increasing utilisation of EV

[1]Department of Electrical Engineering, University of Kurdistan, Sanandaj, Iran
[2]Department of Electrical Engineering and Telecommunications, University of New South Wales, Sydney, Australia
[3]Network Planning and Investment Engineering, Scottish & Southern Electricity Networks, Berkshir, United Kingdom
[4]Department of Electronic & Computer Engineering, College of Engineering, Design & Physical Sciences, Brunel University London, Uxbridge, United Kingdom
[5]Department of Mathematical and Physical Sciences, 15th of May Higher Institute of Engineering, Cairo, Egypt

can be vindicated. Furthermore, in the light of vehicle to grid (V2G) capability, Plug-in EVs can play a vital role in power system operation and management through optimal energy storage discharging.
- *Development in the concept of smart/microgrids*: Efficient integration of ESS technologies allows the idea of smart grid construction to be visualised reliably. Besides, the customers can also participate in demand response programmes (DRPs) more dependable with least discomfort. Moreover, microgrids (MGs) as the smart grid flexible prosumers rely on the ESSs operative energy management ensuring a secure operation and high controllability, particularly in islanded mode.

Various types of ESS technologies, the associated characteristics and benefits are overviewed in this section. First, a statistical analysis of the related literature is performed. Section 1.3 presents the bibliometric networks. Next, the techno-economic-environmental effects of ESSs on the power system operational aspects are discussed concisely in Section 1.4. The main technical features of ESSs will be characterised. Then the primary classification of the ESS technologies is presented and explained one-by-one in Section 1.6. Section 1.7 presents the comparison of ESS technologies. In Section 1.8, some relevant conclusions and future trends are derived out.

1.2 The statistical analysis

This section presents statistics for documents in the field of energy storage technologies which have been published until 2017. The term 'energy storage technologies' has been searched in Scopus database in the article title, abstract and keywords, which resulted in 1,192 documents.

Figure 1.1 illustrates that the number of publications increased from only 3 documents in 1976 to 205 documents in 2017. From the graph, it is quite clear that the number of documents has been increasing over the period.

Figure 1.2 illustrates the numbers of documents by type; it is clear that 553 and 427 documents have been published as articles and conference papers, respectively.

Figure 1.3 illustrates that 628 and 606 documents have been published in energy and engineering subject areas, respectively.

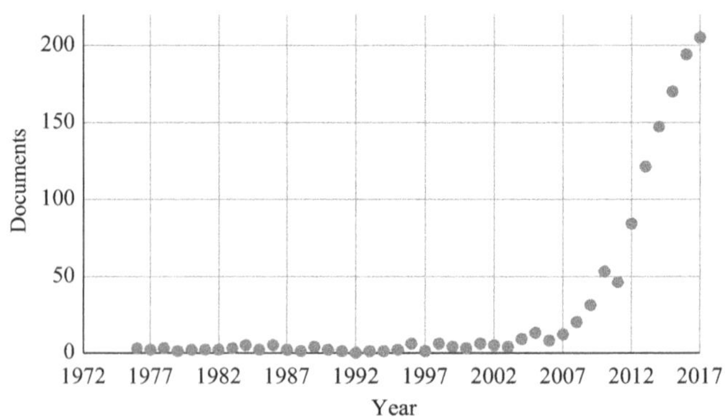

Figure 1.1 Documents by year

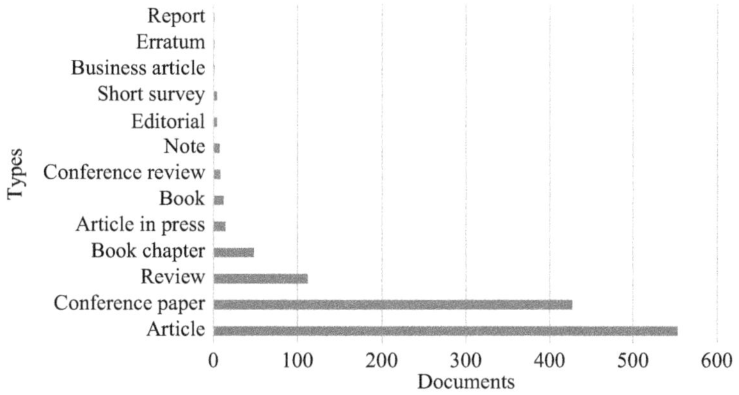

Figure 1.2 Documents by type

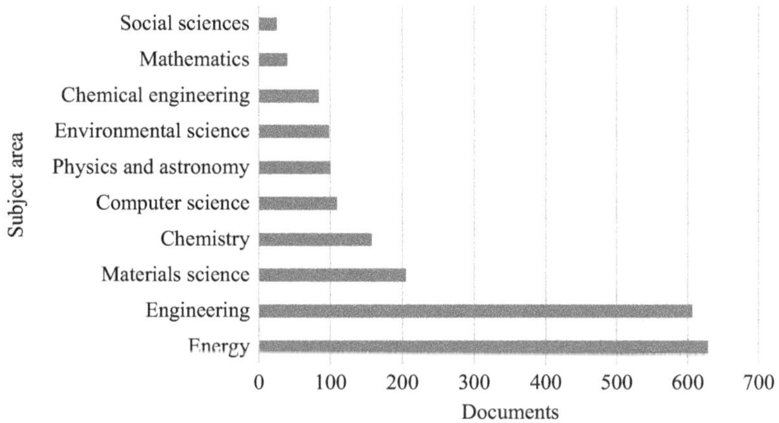

Figure 1.3 Documents by subject area

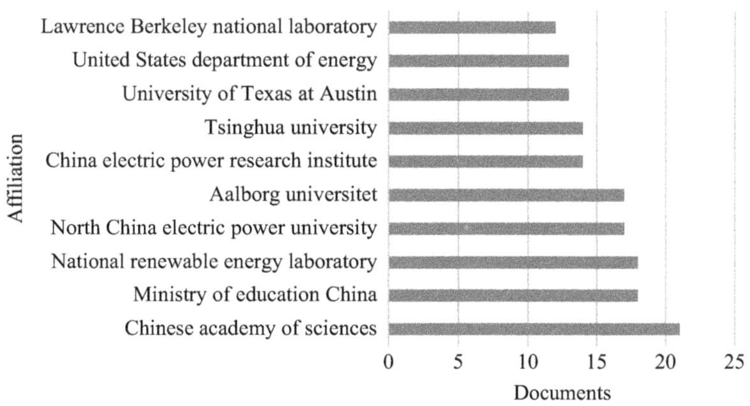

Figure 1.4 Documents by affiliation

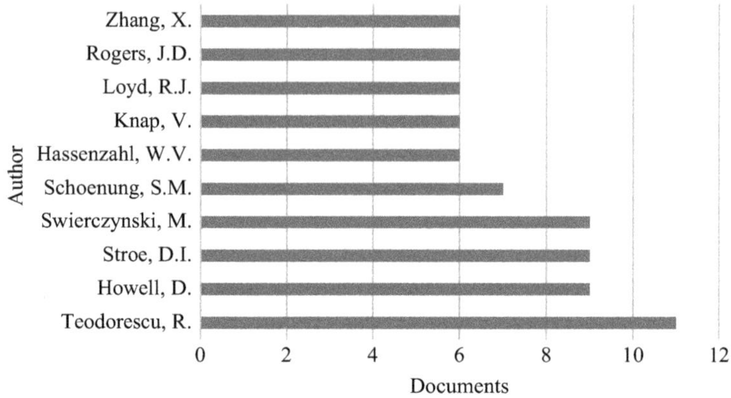

Figure 1.5 Documents by author

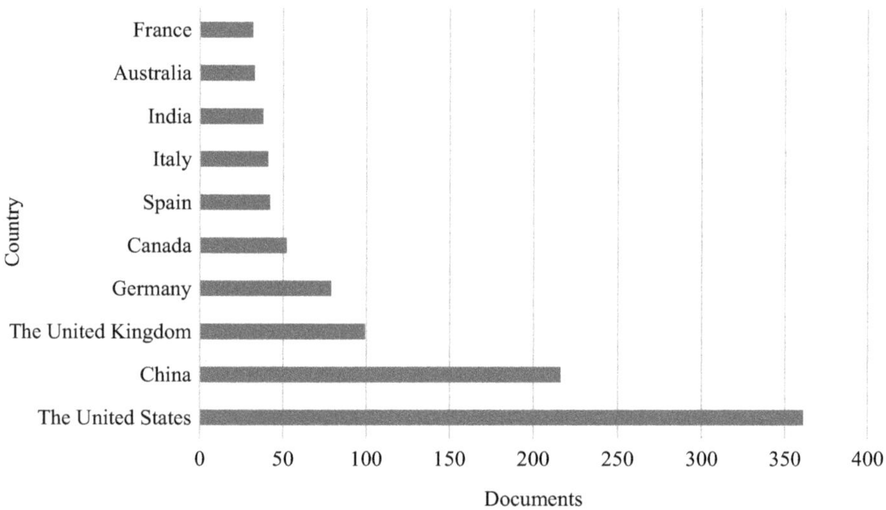

Figure 1.6 Documents by country

Figure 1.4 shows that 21 and 18 documents have been published with the affiliation of the Chinese Academy of Sciences and Ministry of Education China, respectively.

Figure 1.5 shows that Teodorescu and Howell have published 11 and 9 documents, respectively.

Figure 1.6 shows that researchers from the United States and China have published 361 and 216 documents, respectively.

1.3 The bibliometric networks

This section presents the bibliometric networks using VOSVIEWER software [4].

1.3.1 The co-occurrence analysis

Figure 1.7 shows the most common keywords used in the documents by full counting method [5, 6].

The keywords such as energy storage and energy storage technologies in the middle of Figure 1.7 are the most common keywords, while the keywords such as cathode and anode in the outer parts are the less commonly used keywords.

Table 1.1 shows that energy storage and energy storage technologies with 679 and 599 times occurrence, respectively, are the two top popular keywords in energy storage technologies field.

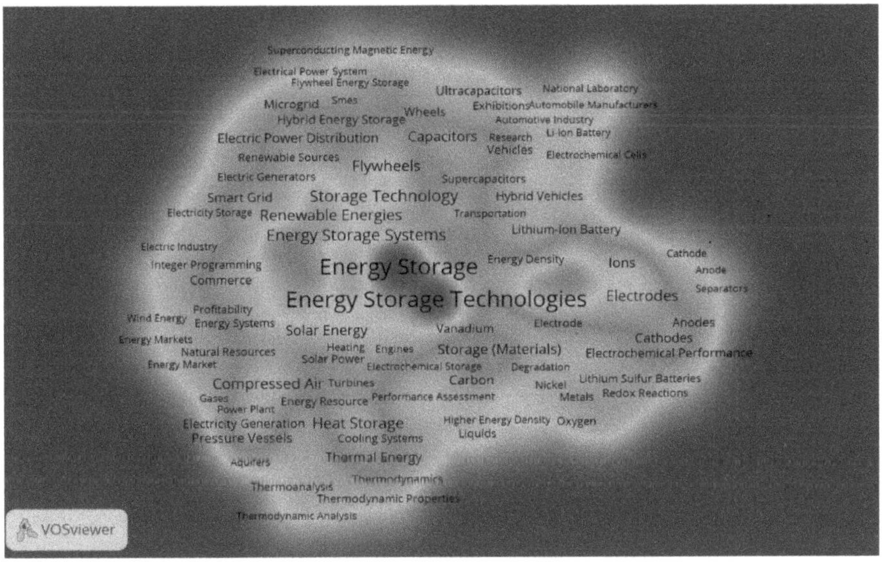

Figure 1.7 The most common keywords used in the documents

Table 1.1 The ten most commonly used keywords

Number	Keywords	Occurrences
1	Energy storage	679
2	Energy storage technologies	599
3	Electric batteries	212
4	Electric energy storage	181
5	Renewable energy resources	153
6	Secondary batteries	140
7	Energy storage systems	130
8	Wind power	117
9	Flywheels	113
10	Electric power transmission networks	108

1.3.2 The co-authorship analysis

Table 1.2 illustrates the co-authorship analysis for the top ten countries with the highest number of citations. Table 1.2 shows that the United States published 365 documents and with 8,230 citations is the most top country based on the co-authorship analysis.

1.3.3 The citation analysis

Table 1.3 shows that Renewable and Sustainable Energy Reviews with 35 documents and 2,880 citations is the most relevant sources in the field of energy storage technologies.

Table 1.4 shows the citation analysis of the top ten documents with the highest number of citations. From Table 1.4, it is clear that [7] received the highest number of citation in the field of energy storage technologies.

Table 1.2 The co-authorship analysis based on countries

Number	Country	Documents	Citations
1	The United States	365	8,230
2	China	219	3,886
3	The United Kingdom	98	4,064
4	Germany	79	1,815
5	Canada	52	567
6	Spain	42	1,913
7	Italy	41	306
8	India	38	1,130
9	Australia	34	550
10	France	32	699

Table 1.3 The co-citation analysis based on sources

Number	Source	Documents	Citations
1	*Renewable and Sustainable Energy Reviews*	35	2,880
2	*Energy Conversion and Management*	16	1,270
3	*Journal of Power Sources*	24	1,232
4	*Proceedings of the IEEE*	10	1,218
5	*Science*	1	1,201
6	*Applied Energy*	34	973
7	*Advanced Functional Materials*	7	933
8	*Progress in Natural Science*	1	928
9	*IEEE Transactions on Industrial Electronics*	1	463
10	*Nature*	1	402

Table 1.4 The citation analysis for the top ten documents

Number	Authors	Reference	Journal/Conference	Citations	Year
1	J. Untivich *et al.*	[7]	*Science*	1,201	2011
2	H. Chen *et al.*	[1]	*Progress in Natural Science*	928	2009
3	C. Ponce de León *et al.*	[8]	*Journal of Power Sources*	580	2006
4	S. M. Hasnain	[2]	*Energy Conversion and Management*	574	1998
5	P. F. Ribeiro *et al.*	[9]	*Proceedings of the IEEE*	548	2001
6	S. Vazquez *et al.*	[10]	*IEEE Transactions on Industrial Electronics*	463	2010
7	A. F. Burke	[11]	*Proceedings of the IEEE*	462	2007
8	L. Hadjipaschalis *et al.*	[3]	*Renewable and Sustainable Energy Reviews*	457	2009
9	Díaz-González *et al.*	[12]	*Renewable and Sustainable Energy Reviews*	452	2012
10	E. J. W. Crossland *et al.*	[13]	*Nature*	402	2013

1.4 ESS benefits

Power system reliability is at risk of insecurity violation due to sudden changes in the load consumption and occurrence 'N-1' contingencies. Furthermore, deployment of the RESs in energy generation causes the system behave more uncertain which leads to higher risks [14–16]. A promising tool that covers the short-term uncertainties without imposing involuntary load-shedding to the end-users or unnecessary starting-up other generating units is the utilisation of energy storage technologies [17–19]. Storage devices can be employed at different voltage levels from the generation to the distribution systems [3, 20–22]. Obviously, ESSs have significant impacts on system stability, security, power quality and peak load serving. In the following, the main techno-economic advantages of ESS integration are discussed.

- **System reliability improvement**: ESSs can enhance the reliability index of the power system by procuring a ratio of system reserve requirements. Scheduled reserve resources are deployed to mitigate the system imbalances. Spinning reserves are the fastest reserve types which provided from the free capacity of committed generators [23]. Thermal unit scheduled reserves can be replaced by ESSs. By this way, not only emission of the pollutants are reduced but also overall reliability of power system is improved substantially. Furthermore, ESSs can decrease the frequency and number of energy interruptions in the distribution system and help the restoration procedures become faster and more sustainable.

  One of the eventual targets of the smart grid is to construct a 'self-healing' system through which faulted elements of the system can be isolated and restored automatically and without human interposition. ESSs can perform a vital role in vindicating a self-heal power system.

- **Facilitating peak load energy serving**: ESSs can store energy during off-peak periods and discharge during peak hours. In other words, in off-peak hours, surplus generation of wind turbines (WTs) and photovoltaic panels, less-pollutant or inexpensive generators can be stored by charging the ESSs. In peak hours, instead of using high pollutant or expensive generators, ESSs can be discharged. Consequently, economic and environmental indices are considered. Furthermore, power system energy efficiency is augmented due to the more uniform load factor which is provided by arbitrating the energy from off-peak to peak hours [3].
- **Frequency control**: Imbalances in power system generation and consumption yield frequency excursions. Frequency deviations should be managed securely employing adequate reserve resources. Based on the fast and reliable response of the ESSs, they can be charged or discharge during frequency excursion scenarios to control the positive or negative frequency excursions. Analogously, voltage fluctuations can be regulated using the stored energy in the ESSs. Conclusively, ESSs can play a critical role in enhancing the frequency/voltage stability of the power systems.
- **Power quality improvement:** Variations in magnitude and shape of voltage and current waveforms referred to power quality phenomena [24]. The main issues related to the power quality phenomena are harmonics, voltage sag and swell, flickers and transients, which are dominantly the resultants of the non-linear loads. ESSs can mitigate the power quality disturbances and provide qualified services without any distortions.
- **Cost reduction of system operational planning:** ESSs can be charged when the energy price is low (usually during off-peak periods) and discharge when the market price is high. Thus, the cost of energy provision is reduced considerably. As a result, social welfare of the system is improved which is one of the eventual goals of the operators.
- **Generation expansion postponement:** ESS utilisation, particularly during peak hours, reduces the requirements to the power-plant generations. Consequently, based on strengthening the system generation capacity, the need for installing new generation capacities is deferred and economic system indices are elevated by saving capital costs [25].
- **Customer cost-effective bill management:** ESSs can by discharging the stored energy in peak hours when most of the expensive generating units are committed, can lower the market prices. As a result, the consumption bill of the end-user customers reduced. This is when the market price signal can be transferred to the customers through real-time pricing mechanism and using Advanced Metering Infrastructure. Dynamic time-of-use (TOU) programmes can also be affected by market prices. In other words, when the price signal is accessible to the customers, they can charge ESS in hours with low prices and supply their consumption demands during peak hours when energy price is high. By this way, the consumption bill is reduced. Moreover, employing ESS technologies prevents financial losses face to the consumers due to service interruptions. The financial losses are more bolded for the industrial and commercial customers.

- **RES generation profile levelling:** Coordinated management of RES and ESS technologies can help the intermittent nature of the RES covered. The spillage of the RES output power can be stored in the ESS, and it can be discharged at hours while the output power is lower than the forecasted values. Hence, the generation profile of the RES becomes uniform. In other words, ESS motivates the operators to utilise the RES technologies with higher assurance.

1.5 ESS general characteristics

Some of the performance characteristics common among all the ESS technologies are described in this section.

- *Power capacity* is defined as the amount of power (kW or MW) that an ESS can procure in its maximum instantaneous output.
- *Energy storage capacity* is the amount of kWh or MWh of electrical energy an ESS can store in each period.
- *Efficiency* is the ratio of recovered electricity to input electricity used to charge the ESS.
- *Response time* is the time duration required for releasing full power.
- *The depth of discharge (DoD)* indicates how much power can be discharged during a cycle.
- *Discharge duration* is the amount of time that an ESS can discharge at its rated power output.
- *Specific power/Energy density* is the power output/energy content per-unit ESS volume in W/l or J/l.
- *ESS lifetime* is the degradation time of ESS that can be measured as calendar life and cycle life. The number of years an ESS is expected to last is its calendar life, while cycle life is predicated on how many charge–discharge cycles an ESS can be expected to procure before failure [26]. ESS lifetime is affected by the type of ESS, environmental conditions, DoD and other operational parameters.

The mathematical formulation usually used to explain the general performance of the ESSs can be deduced as follows:

$$\text{Energy}(t) = Energy(t-1) + \eta_{charge} \cdot Power_{charge}(t) \cdot \Delta t - \frac{1}{\eta_{discharge}} Power_{discharge}(t) \cdot \Delta t \quad (1.1)$$

where $Energy(t)$, $Power_{charge}(t)$ and $Power_{discharge}(t)$ indicate stored energy, charged power and discharged power of the ESS in period t, respectively. Charge and discharge efficiency of the ESS are shown by η_{charge} and $\eta_{discharge}$, respectively. Δt states the response time of the ESS which depends on its type.

The essential characteristics of various ESSs may be different owing to the application. Table 1.5 summarises some of the features versus ESSs applications. In Figure 1.8, the applications of the ESSs in different voltage levels according to their power ratings are schematically depicted.

Table 1.5 Key characteristics of ESS according to various applications [27]

Application	ESS output	Power capacity (MW)	Discharge duration	Duty cycle	Response time
Load following	Electricity, heat	1 to 2,000	15 min to 1 day	1 to 29 times per day	Less than 15 min
Frequency control	Electricity	1 to 2,000	1 to 15 min	20 to 40 times per day	1 min
Arbitrage	Electricity	100 to 2,000	8 to 24 h	0.25 to 1 times per day	More than 1 h
Voltage support	Electricity	1 to 40	1 s to 1 min	10 to 100 times per day	ms to s
Congestion management	Electricity, heat	10 to 500	2 to 4 h	0.14 to 1.25 times per day	More than 1 h
Asset deferral	Electricity, heat	1 to 500	2 to 5 h	0.75 to 1.25 times per day	More than 1 h
Load levelling	Electricity, heat	0.001 to 1	Min to hours	1 to 29 times per day	Less than 15 min
RES integration	Electricity, heat	1 to 400	1 min to hours	0.5 to 2 times per day	Less than 15 min
Operational reserves	Electricity	10 to 2,000	15 min to 2 h	0.5 to 2 times per day	Less than 15 min

1.6 ESS technologies

To store electrical energy, first, it should be converted to DC or other forms of the energy such as thermal, chemical, electrical and mechanical energies. When the stored energy is required, it should reversely convert to the AC electrical energy. From a technical point of view, ESSs have various technologies. Although the ESS technologies are updated daily, in the following, the main ESS technologies integrated into power system different voltage levels are explained. The primary classification of ESS technologies is represented in Figure 1.9.

In the above classification, pumped hydro ESS (PHESS) and compressed air ESS (CAESS) have both large power, greater than 50 MW, and energy storage capacities, greater than 100 MWh. Batteries are categorised into 1–50 MW power rating and 5–100 MWh energy capacities, which are known as medium power and energy ESSs. Hydrogen ESS (HESS) and thermal ESS (TESS) are small power and energy ESSs. Flywheel ESS (FESS), superconducting magnetic ESS (SMESS) and capacitor ESS can have large power or large energy capacity but not both [29]. Obviously, large-scale ESSs can be used as power plants, while the medium and small ones are useful for remote communities and EV and satellite applications [30].

The ESS technology applications should be adapted to the network where it is utilised and to the type of production it is required to be coordinated. Figure 1.10 shows the relation between the power and energy capacity of various ESS technologies and associated timescale in which they have higher performance and adaptability. For instance, PHESSs, CAESSs, fuel cells, TESSs and large-scale

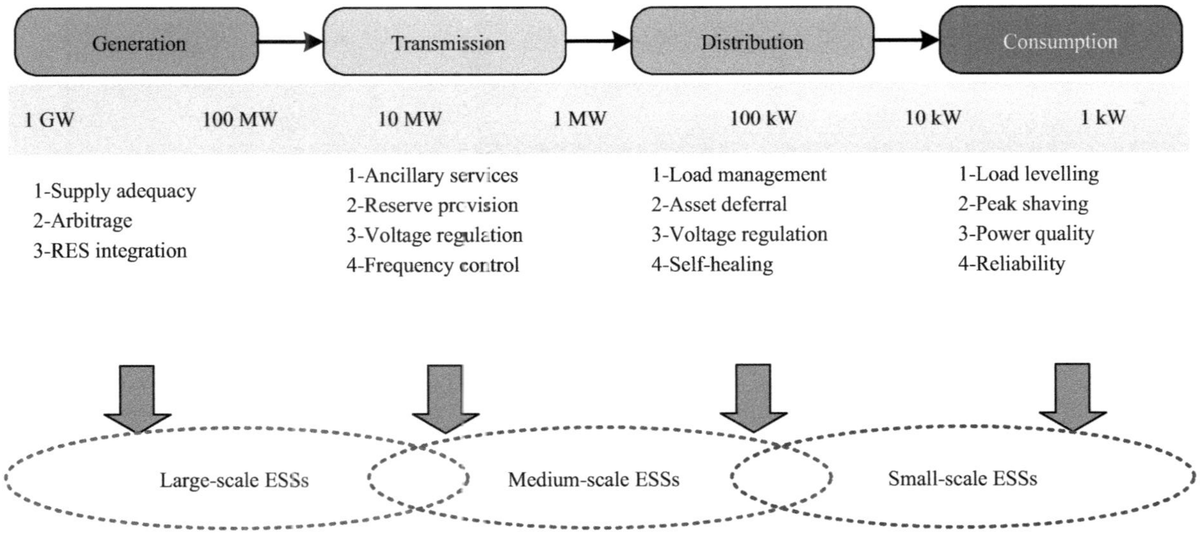

Figure 1.8 ESS various applications according to power ratings [28]

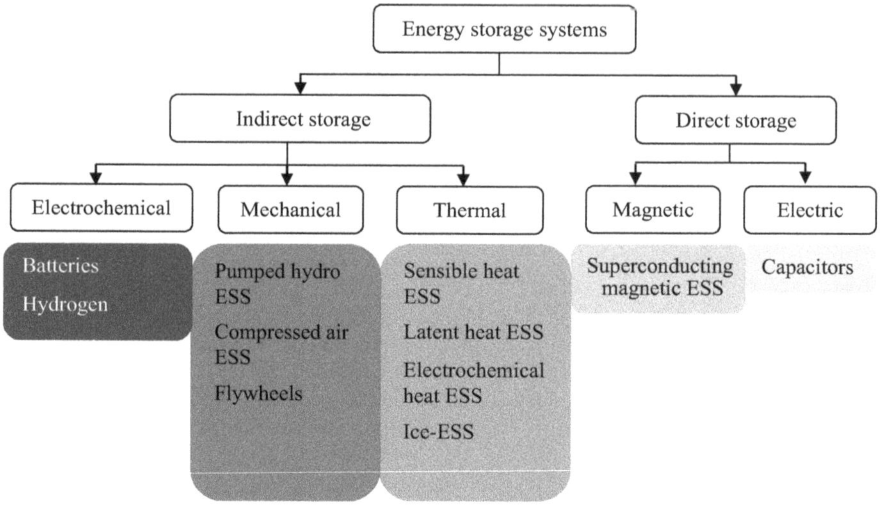

Figure 1.9 Classification of ESS technologies [9]

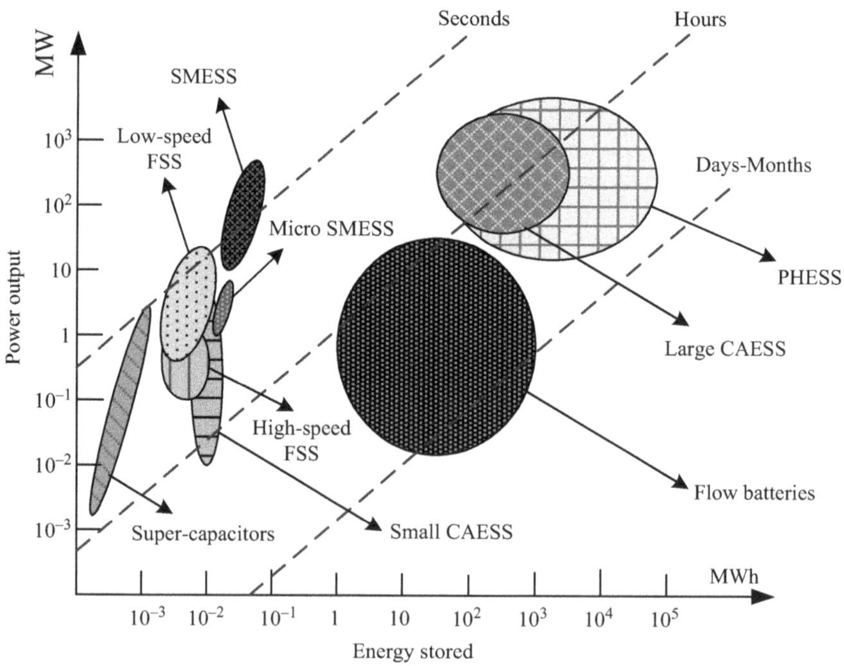

Figure 1.10 ESS technologies' power and energy outputs [31]

batteries are functionally most appropriated to the energy management objectives, while power quality and reliability-based targets are more adapted to the SMESSs, FESSs, super-capacitors and batteries [30].

A preliminary comparison among the ESS technologies is listed in Table 1.6. The primary applications, characteristics and operational challenges are listed.

- **PHESSs:** PHESSs are the oldest and largest ESSs in power systems. As depicted in Figure 1.11, pumped hydro storage units usually have two lavers, one in high altitude and the other in low height. In peak hours, the stored water in the upper pool is released, and its height dependent potential energy is converted to kinetic energy using turbines at lower pool. The coupled turbines and synchronous generators convert the mechanical energy to the electrical one [32, 33].

  During off-peak hours, the pumps are activated and the reserved water is transferred from the lower pool to the upper one. Accordingly, it avoids the system experiencing over-frequency states. Moreover, since valley filling and peak shaving procedures are executed in a 24 h operational period, PHESSs help the consumption profile being flatter. The performance of a typical PHESS is depicted over a daily operating period as is presented in Figure 1.11. It is worth mentioning that the typical PHESS capacity rating is 1,000 MW with 75 per cent to 82 per cent efficiency. The height difference between the upper and lower reservoir should be at least 300 m in such a way that the PHESS construction is justified. The calendar life of PHESSs is 80 years.

  The PHESSs are good candidates to be coordinated with base-load generation units such as nuclear power plants. In off-peak hours, when it is not techno-economic to OFF the nuclear units, PHESSs can go to pumping mode and enhance the efficiency of the nuclear units. Vice versa, in peak hours the nuclear units can supply the required energy to run the generating mode.

  Coordinated energy management between PHESSs and WTs plays a vital role in RES higher integration and power system economical operational planning.
- **Compressed air ESS:** CAESS main components usually include compressor, air storage cavern, combustion chamber and turbine generator set [34]. In this system, air is compressed up to 75 bars during off-peak periods in a hypogeal reservoir (cavern). By this way, electric energy is converted to pressurised potential energy and stored in the reservoir. In peak periods, the converted energy is released from the cavern and combined with fuel in the combustion chamber and used to generate electrical energy in the turbine-generator set. The cavern can be constructed artificially which is very costly or can be located in salt caverns, underground water reservoirs or hard rock mines. The compressed air ESSs can be constructed in small size to large-scale units. They usually have more than 100 MW power rating with 70 per cent to 80 per cent efficiency and 12 kWh/m^3 energy density [31]. Two 290 MW and 110 MW CAESS units are currently built in Germany and the USA, respectively. Figure 1.12 shows the structure and necessary components of a typical CAESS.

Table 1.6 ESS technologies' applications, characteristics and challenges [1]

ESS technology	Primary application	Characteristics	Operational challenges
PHESS	1-Energy management 2-Reserve provision	1-Fast ramp rate 2-Mature technology 3-Most economical	1-Geographical restriction 2-High investment cost 3-Environmental effects
CAESS	1-Energy management 2-Reserve provision 3-RES integration	1-New technology 2-Fast ramp rate (faster than gas turbines)	1-Geographical restriction 2- Environmental effects 3-Low response time versus to FESS and batteries
FESS	1-Load levelling 2-Peak shaving 3-Transient stability 4-Frequency regulation	1-Modular technology 2-Free size 3-Long lifetime 4-Fast response time 5-High efficiency 6-High power without excessive heating	1-Constructional limitations 2-Long recharging time due to high frictional losses
SMESS	1-Power quality 2-Frequency regulation	1-Highest efficiency	1-Low energy density 2-Expensive materials 3-Manufacturing complexity
Super-capacitor	1-Power quality 2-Frequency regulation	1-Long lifetime 2-Fast discharge	1-High operational costs
TESS	1-Load levelling 2-Load management 3-Grid stability	1-Very high energy density	1-High operational costs
Batteries	1-Power quality 2-Congestion relief 3-Grid stability 4-Load levelling 5-RES integration	1-Long lifetime 2-Mature technology 3-Fast response time 4-Good scaling potential	1-Discharge restriction 2-Lower energy density 3-Emission footprint 4-Sensitive to heat
Flow batteries	1-Fast ramp rate 2-Peak shaving 3-Frequency regulation 4-Power quality	1-Very long lifetime 2-Lower efficiency	1-Developing technology 2-Design complexity 3-Lower energy density

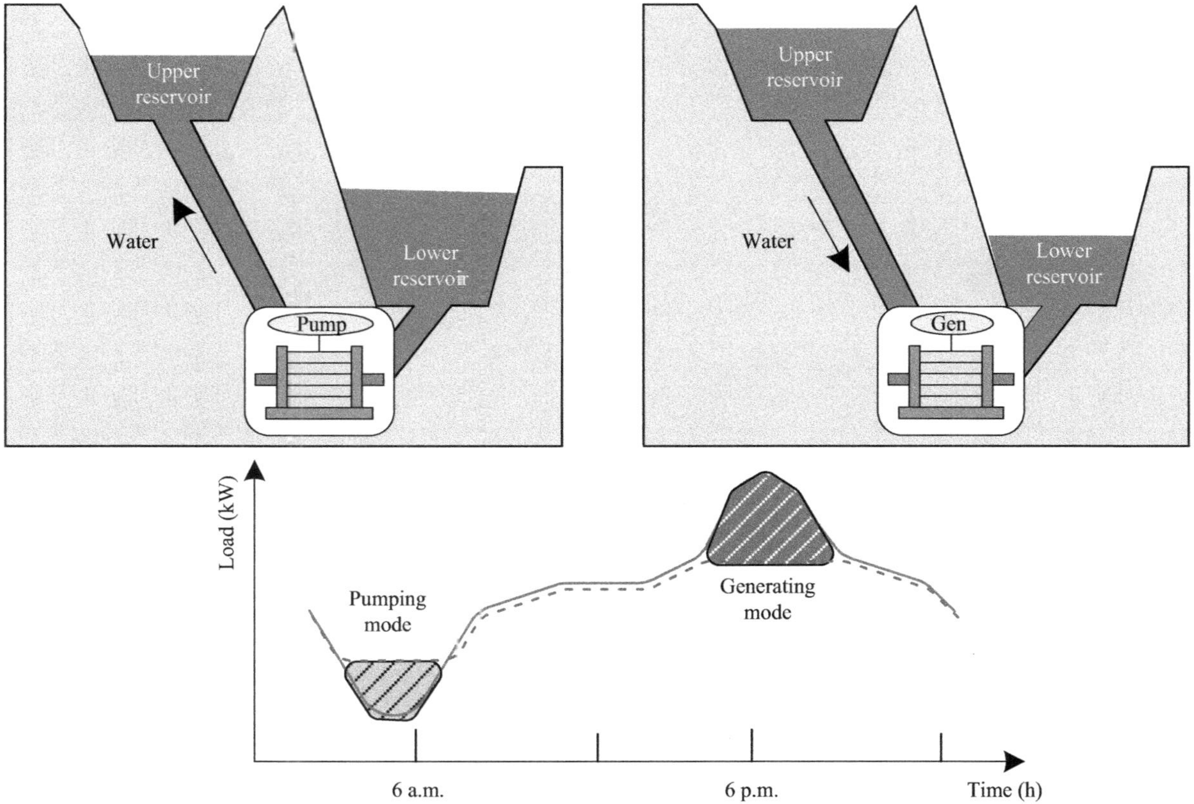

Figure 1.11 The PHESS daily performance in pumping and generating modes

- **Flywheel ESS:** Electrical energy is stored in the rotor inertia through accelerating the rotor to very high speeds [35]. The stored kinetic energy is dependable to the squared speed. Light rotors have higher speed and energy storage capability. Therefore, it is tried to use light materials in the structure of the rotor such new composites. In the charging mode (during off-peak), the flywheel plays as a motor, while it is likewise a generator in the discharging mode (in peak periods). In the generating mode, the stored mechanical energy is depleted gradually by the reduction of the rotor speed. The maintenance costs of the flywheels are relatively low, and this makes them more attractive. The typical efficiency of FESS state-of-the-art technology is higher than 90 per cent.

  FESSs suffer from frictional losses due to the bearings. To eliminate the losses, the vacuum enclosure and magnetic bearings are utilised. Therefore, although the safety and economic aspects may be degraded, the new FESSs become more efficient and compact with no noises.

  One of the exciting applications of the FESS technology is providing virtual inertia, which is mainly, very critical in islanded MGs with dominantly inertia-less inverter-interfaced distributed generations. In this way, the MG frequency stability and damping margins can be ensured.

  In Figure 1.13, a typical FESS model with its power conditioning inverters is depicted.

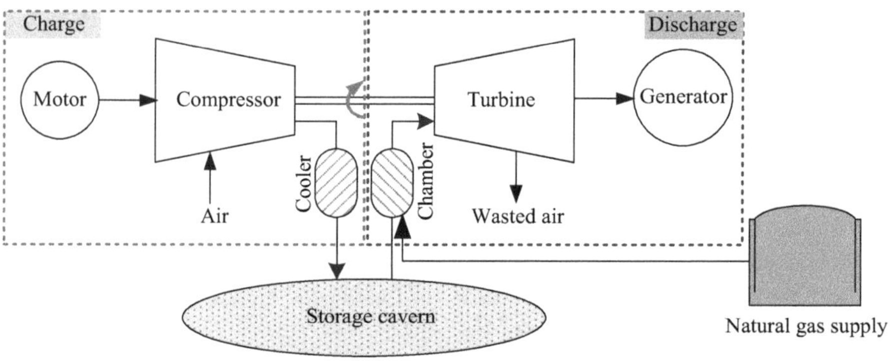

Figure 1.12 A typical CAESS structure and essential components

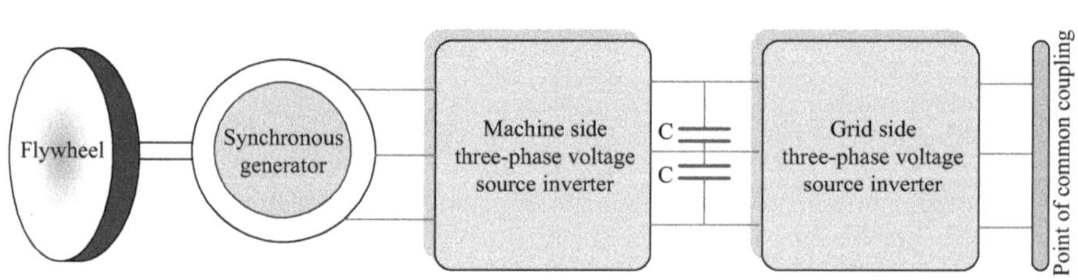

Figure 1.13 FESS model including power regulators

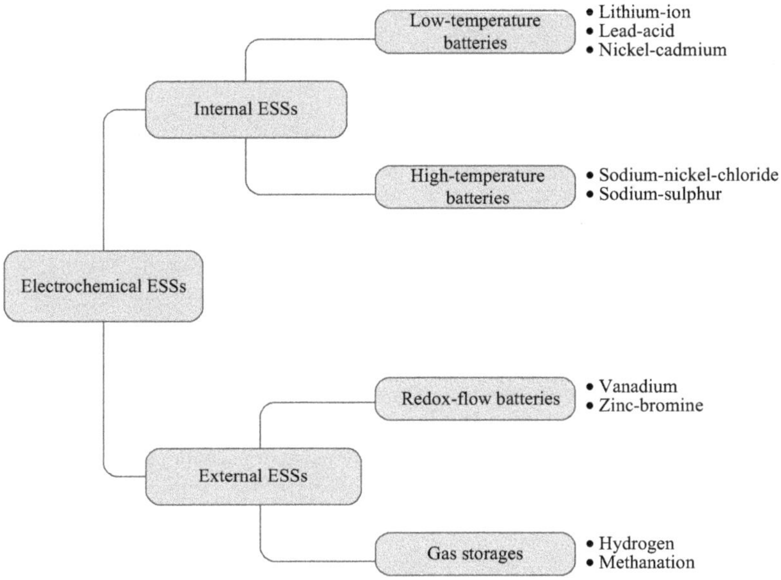

Figure 1.14 Electrochemical ESS classification

- **Batteries:** Substantially, electrochemical storages based on the energy conveying material storage method are divided into two main categories: internal and external. In external storages, the reactant material section and electric power generating section are designed independently. The hydrogen and methane gas storages are included in external storage category. In internal storages, both the two reactant material and power generating sections are located in one common volume. In Figure 1.14, classification of electrochemical ESSs is illustrated [36]. Although *batteries* are reversible electrochemical cells, there have been various technologies to produce them as ESSs.

  Currently, batteries are best options for energy storing in the medium and small sizes. These ESSs have two electrodes and betwixt electrolyte. In charging mode, the ions move from the anode to cathode. A reverse moving is performed during the discharging mode. Indeed, energy is stored in batteries electrochemically through creating electrically charged ions. A wide range of battery technologies is fabricated in the industry such as lead-acid, lithium-ion, nickel-cadmium, sodium-sulphur, sodium-nickel-chloride and zinc-air. Batteries are relatively inexpensive, and most of them have a matured technology. They can be applied in either portable or permanent application. Large-scale batteries can be utilised in system emergency control actions and providing back-up for higher RES integration [31]. When the power system experiences a contingency or disturbance, it may transfer to the emergency state. Batteries can preserve the system stability limitations in the acceptable ranges and evoke it to the normal state. In an economic point of view, batteries can buy energy in hours with low market prices and sell the stored energy at hours when the price is high. Through this arbitrage, they can provide considerable revenues for the operators. Besides, they have also an advantageous role in providing higher load factor. Medium-size batteries are beneficial for supplying hybrid EV and electric ship requirements.

In lithium-ion batteries, the cathode is made from a combination of lithium such as lithium-cobalt-oxide and anode is made from carbon. The electrolyte is made of the lithium salt in an organic solvent. The efficiency of lithium-ion batteries is about 85 per cent with 300 Wh/l energy density. The cycle life is about 1,000 to 5,000 cycles. The DoD is up to 100 per cent. Recently, nanomaterials have considerable impacts on the development of lithium-ion batteries.

In lead-acid batteries, electrodes are lead, and antipodean metals dipped in sulphuric acid as the electrolyte. Exchange of negative and positive ions between anode and cathode in the electrolyte is the basis of the battery charge and discharge. The efficiency of lead-acid batteries is about 80 per cent with 100 Wh/l energy density. The cycle life is about 500 to 2,000 cycles. Their DoD is 70 per cent.

In sodium-sulphur batteries, the electrolyte is made of liquid sodium and liquid sulphur with solid beta alumina. The charge and discharge process is similar to the lithium-ion batteries. Energy density in sodium-sulphur batteries is three times that of the lead-acid storages. Sodium-sulphur batteries can efficiently work under high-temperature conditions about 600 °F. The efficiency is about 80 per cent with 250 Wh/l energy density and 5,000 to 10,000 cycles lifecycle. DoD is 100 per cent.

Flow batteries are rechargeable, low cost and useful technologies for sustainable energy storage. In flow batteries, the electrolyte is kept in an external vessel. Ion exchange between two distinctive electrolytes results in electricity generation. The power output of these batteries is in the range of kW to MW. Despite low energy density and high cost due to structure complexity, the main advantages of the flow batteries are very fast charge/discharge process, high power and energy ratings, a long lifetime by proper maintenance, the use of non-toxic materials and controllable electricity generation. The efficiency of typical flow batteries is about 75 per cent. Figure 1.15 represents the basic structure of the flow batteries.

Different types of the flow batteries are mainly developed based on bromine combinations such as zinc (ZnBr), sodium (NaBr), vanadium (VBr) and sodium polysulphide. Charge and discharge procedures can be reversed using electrochemical reactions through a cell membrane.

- **Superconducting magnetic ESSs:** Superconducting materials extend the storage capacity and at low temperatures have no resistance against the current. In SMESSs, the goal is to store electric energy in a magnetic field which is produced using a superconducting coil conveying a DC. The response rate of SMESSs is lower than 100 ms, which is relatively fast. This makes them appropriate candidates for regulating distribution power system stability and utilisation as power quality conditioners. Their capacity rating is up to 3 MW with more than 90 per cent efficiency.

SMESSs consist of three essential parts: superconducting coil, cooling system and power conditioner system. The superconductivity feature of the coil leads to negligible losses and the passing DC remained unchanged for extended times. To sustain the coil superconductivity feature, the cooling system reduces the temperature to 50–77 K. The power conditioner system is employed to convert the AC to DC and vice versa. The switching process of the internal inverters is controlled to have the required output. Figure 1.16 shows a typical structure of an SMESS.

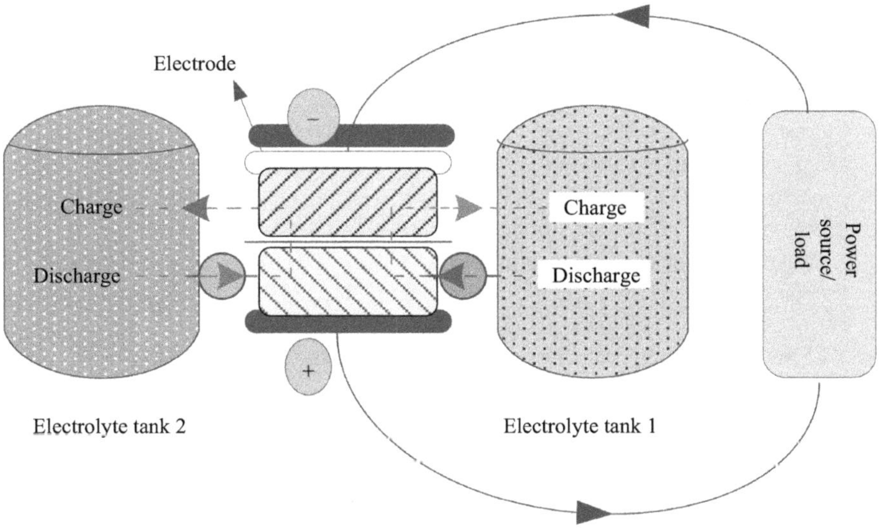

Figure 1.15 A typical flow battery illustration

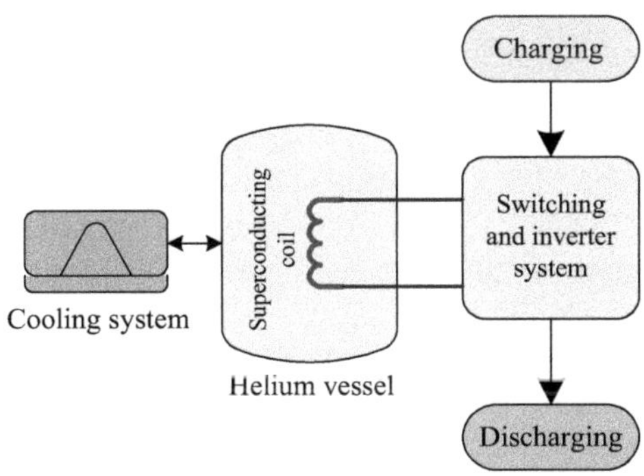

Figure 1.16 Schematic structure of a SMESS

Due to high required energy for the cooling system and high costs of the superconducting coils, currently, SMESSs are used for short-term energy storing and regarding the power quality targets. Coil inductance and its passing current affect the amount of stored energy. The higher inductance means higher capital costs. Usually, in the SMESSs, 70 per cent of the overall costs belong to power conditioner system. In 2018, an SMESS has 35,000$/kW cost which is relatively high.

- **Super-capacitors:** In super-capacitors, the energy is stored in the electric field between two electrodes. They may have a volume thousand times larger than a capacitor, and their energy density is much higher. The largest super-capacitor has an energy density of 0.03 kWh/kg. They have high permeability and very adjacent electrodes with a longer lifetime. Electric double-layer capacitors (ELDCs) with very high power density, low internal resistance and high efficiency higher than 95 per cent are the most common super-capacitors. The super-capacitors can be utilised in military lasers, medical equipment, WTs, solar systems, electromagnetic drivers, microwaves, large LED drivers and hybrid EV. Numerically, cycle life and calendar life of the super-capacitors are 1 million and 15 years, respectively. DoD index is 75 per cent. Super-capacitors have 2 Wh/l to 10 Wh/l energy density and up to 15 kW/l power density [36].

  In small scales, super-capacitors are operated for energy saving. If the energy density has been developed, it is expected that electrochemical batteries could be replaced by super-capacitors.
- **TESSs:** Energy can be stored in thermal ESSs using the following three methodologies:
  (a) Energy storage through heating a liquid or solid without state change. It is a **sensible heat TESS** which is relying on the temperature of the underlying material. Insensible TESSs stored energy is utilised for both heating and cooling purposes. This system allows peak shifting and leads a more uniform load factor. During off-peak hours, the interfacing material may be heated or cooled, and in peak hours, the appropriate heating or cooling can be reached through blowing the air on the interfacing material. The interfacing material can be water, underground layers and building a concrete foundation. The energy density of the sensible TESS is typically 25 kWh/m^3. The main advantages of this ESS are clean and reliable heating and cooling, low power output oscillations [31, 36].
  (b) Energy storage through heating a material whose state is altered. It is a **latent heat TESS.** The amount of stored energy depends on the latent heat of utilised material and associated mass. The interfacing material used in latent TESSs should have the capability of the phase changing during charges and discharges such paraffin and water. Since there is no temperature change in the interfacing material during the phase change, the constant temperature during discharging is the main advantage of latent TESSs [31].

(c) Energy storage through providing required heat to activate a physical or chemical process. It is known as **electrochemical heat TESS**. The process is performed in a thermally insulated container such that the thermal energy can be stored efficiently.

Hydrogen can be used in electrochemical TESS as HESS. Based on the electrochemical process of the electrolyte, the electrical energy can be stored in the form of the chemical energy and discharged vice versa. The conversion is by ion transfer between electrochemical cells using water electrolysis. The produced hydrogen is compressed and stored in pressurised buffer tanks. The stored hydrogen can be used in fuel cells, residential consumers, expansive turbines and hydrogen-based hybrid EV. HESSs are good candidates for isolated and remote power systems, in particular, to mitigate intermittencies of RESs. Figure 1.17 illustrates the process of generating hydrogen and its applications.

The low density of gaseous hydrogen restricts its application. Consequently, the liquid hydrogen due to its higher energy density is considered. Storage of liquid hydrogen requires insulated cryogenic freezing processes which are very costly. Moreover, the life expectancy is a constraint which affects power network applications of the HESSs.

(d) Recently, ice-storage technology is introduced as an innovative ESS. In off-peak, electric energy is used to freeze the water in an insular tank. In peak hours, the stored ice is utilised as a cooling system. Ice storage has excellent effects on comfort index of the consumers and hence can be used as heater/chiller systems in various seasons.

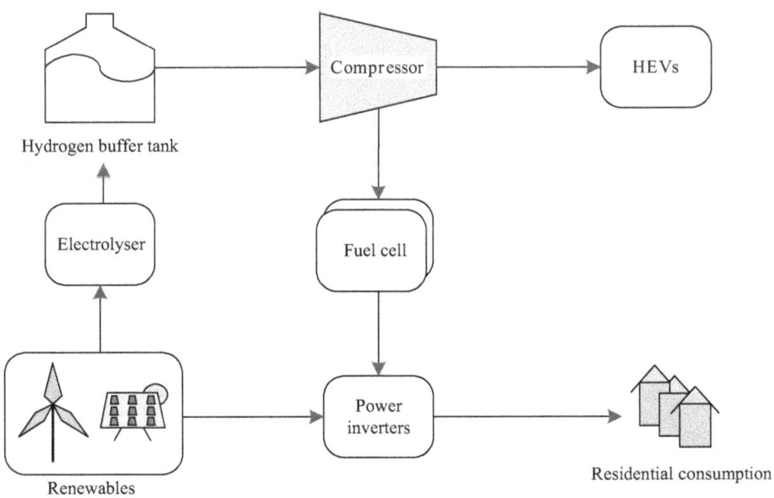

Figure 1.17 HESS process and applications

1.7 Comparison of ESS technologies

In this section, various technologies of ESS are compared subject to different techno-economic indices. It is worth mentioning that there is no ideal ESS technology which can meet all the requirements of power system applications. An ideal ESS has a significant capacity, high density, high efficiency, low cost and long lifetime. However, so far none of the ESS technologies have these features simultaneously [37]. Evidently, ESSs can be utilised in all parts of the power system infrastructure from generation to consumption. Accordingly, to select a more appropriate technology, power system operators should consider some key aspects such as economic, environmental, security, stability, optimal size and location, the system voltage level and ESS response time concurrently. Where the high storage capacity is required, the PHESS, CAESS and TESS can be selected by operators, while the battery and ice-storage technologies are more effective in low-voltage power systems and particularly in islanded MGs. For instance, the PHESS and TESS technologies are suited for application in the generation side. In the transmission and distribution levels, the battery and FESS technologies are good candidates. In the consumption level, battery, ice storage and some of the TESS technologies can be considered.

1.7.1 Technology maturity level

In Figure 1.18, ESS technologies are assessed from the maturity view point concerning capital requirements and technology risk indices. The ESSs are classified into developing, developed and mature categories [27]. PHESS, CAESS and lead-acid have mature technologies and low investment and technical risks. FESS, ice storage, sodium-sulphur and lithium-ion batteries are developed while the super-capacitors, SMESS, HESS, TESS and flow batteries are in the process of development.

ESS technology commercialisation versus the power capacity is demonstrated in Figure 1.19. Obviously, PHESS has the fully commercialised while the SMESS and super-capacitor technologies are brand-new small-scale technologies.

Indeed, PHESSs are the best options for the bulk power system optimisation issues due to high capacity and owing to discharge time at rated power [38].

1.7.2 ESS costs

To compare the operational costs of the ESS technologies, device efficiency, energy/power investment cost and calendar/cycle lifetime should be considered simultaneously. Table 1.7 summarises the ESS technology substantial cost parameters [39].

1.8 Conclusions and future trends

In this section, the critical ESS technologies have been reviewed. With the increase of the RES penetration, ESSs can provide a sustainable integration of renewable in a techno-economic portfolio. Both technical and economic advantages of various

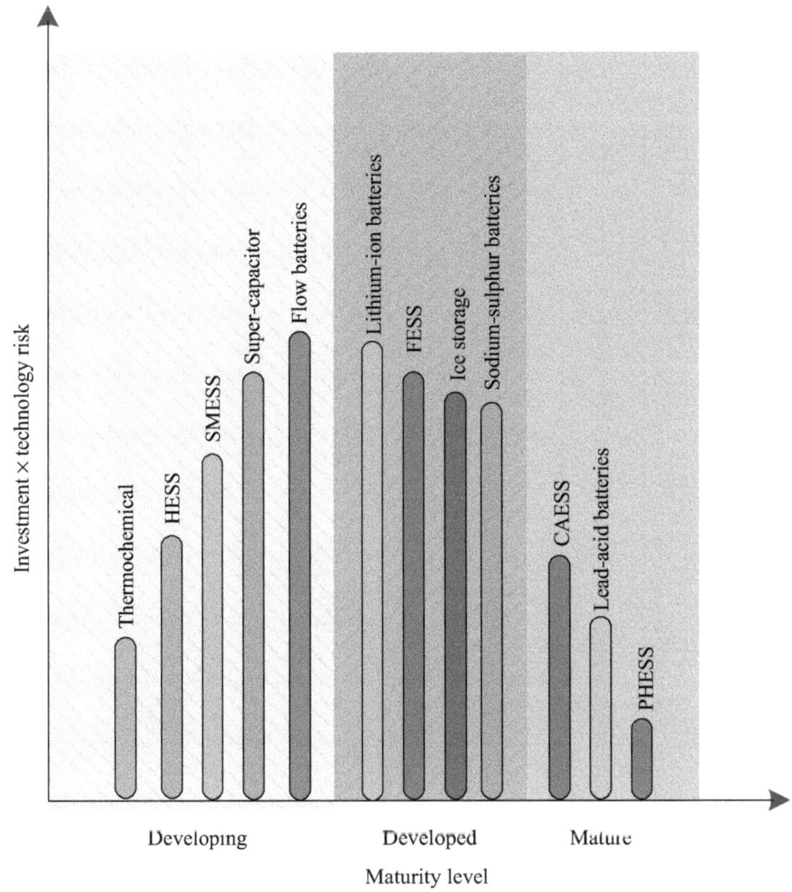

Figure 1.18 ESS technology maturity assessment [27]

ESSs have been reviewed. The details of each technology are discussed concisely and compared with other ones. Based on the review, the following concluding remarks could be derived out [31]:

- There is no ESS technology which can meet the ideal storage requirements of the power system. Some of the technologies are mature and cost-effective such as PHESSs.
- The ESS can be classified as high-efficiency technologies with higher than 90 per cent. Some of them are low efficient below the 50 per cent like the HESS which motivates to higher research and development.
- In the selection of optimal ESS technologies, the capital costs, the lifetimes, the energy and power densities and environmental footprints should be considered concurrently.

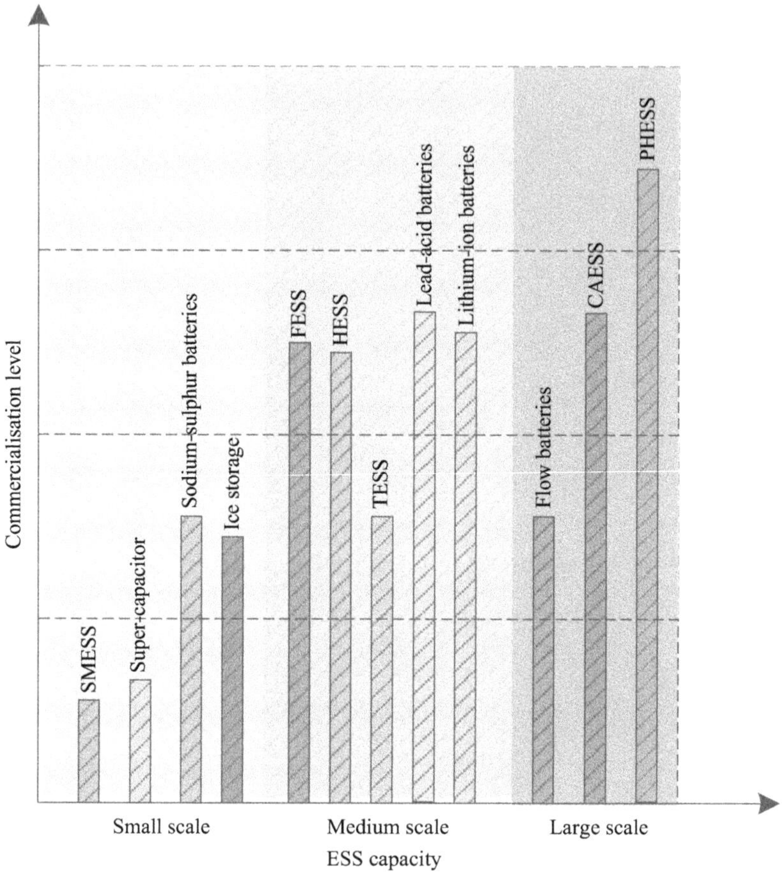

Figure 1.19 Commercialisation level of different ESS technologies [38]

As future trends, smart energy storage systems (SESSs) with intelligent, controllable and flexible charging/discharging procedure are innovative levels of technological progress in the field of the ESSs. SESSs have considerable impacts on the reliable integration of plug-in hybrid EV with the capability of providing V2G and grid-to-vehicle (G2V) systems. Furthermore, SESSs can help the government around the world to develop in smart grid evolution. One of the imperative tools is realising the MG concept at the distribution level. SESSs can ensure sustainable operation of the MGs, particularly in the islanded mode. Coordinated energy management strategy of distributed generation and ESSs is a critical function in the MG Central Controller (MGCC) control panel.

As one of the innovations in the SESS category, we can refer to the Electric Springs. The principle of electric springs is dual to mechanical springs and based on Hooke's law, i.e., $F = -kx$, where x is the physical displacement of spring through which potential energy ($PE = \frac{1}{2}kx^2$) can be stored or released by spring.

Table 1.7 Comparison of various ESS technologies considering operational cost parameters

Technology	Investment cost			Calendar and cycle lifetime		Efficiency (%)	Energy and power density			
	$/kW	$/kWh	Cent/kWh -per cycle	Calendar lifetime (years)	Cycle lifetime (cycles)		Wh/kg	W/kg	Wh/L	W/L
PHESS	600–2,000	5–100	0.1–1.4	40–60	-	75–85	0.5–1.5	-	-	0.5–1.5
CAESS	400–800	2–50	2–4	20–40	-	40–55	30–60	-	3–6	0.5–2
Lead-acid	300–600	200–400	20–100	5–15	500–1,000	60–94	30–50	75–300	50–80	10–400
NaS	1,000–3,000	300–500	8–20	10–15	2,500	85–90	150–240	150–230	150–250	-
Li-ion	1,200–4,000	600–2,500	15–100	5–15	1,000–10,000+	85–100	75–200	150–315	200–500	-
Flow batteries	700–2,500	150–1,000	5–80	5–15	2,000–12,000+	70–75	10–50	-	16–60	-
SMESS	200–300	1,000–10,000	-	20+	100,000+	95	0.5–5	500–2,000	0.2–2.5	1,000–4,000
FESS	250–350	1,000 – 5,000	3–25	15	20,000+	85–92	10–30	400–1,500	20–80	1,000–2,000
Super-capacitor	100–300	300–2,000	2–20	5	50,000+	85–97	0.05–5	10,000	2–10	100,000 +
TESS	200–300	20–60	2–4	5–20	-	20–50	80–200	-	80–500	-
HESS	10,000 +	-	6,000–20,000	5–15	10,000+	20–50	800–10,000	500+	500–3,000	500+

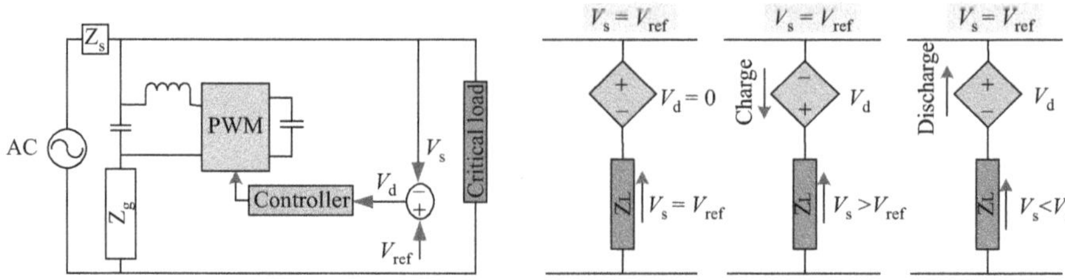

Figure 1.20 Electric spring equivalent circuit and charge/discharge modes

Analogously, using the power-electronic static switches, the concept of electric springs is identified [40], which follows the $q = -CV$ relation. Through charging or discharging, the electric springs are transferred to capacitive or inductive modes and the electric energy can be stored or released, respectively. For example, when the system voltage is deviated from the set-point reference value, the spring is changed to charge or discharge modes to regulate the power system voltage level. In Figure 1.20, the structure of an electric spring, equivalent circuit and operating modes is represented.

To reach to a near ideal ESS, the operators can combine the ESS technologies in an integrated and controllable hybrid ESS. It also needs insight into both power system and ESS technical aspects to provide efficient hybrid ESS technology. ESS can also combine with distribution flexible AC transmission systems devices to enhance the reliable performance and strengthen the compensation targets. Another exciting research topic is optimisation of distributed ESSs type, size and location in MGs.

References

[1] H. Chen, T. N. Cong, W. Yang, C. Tan, Y. Li, and Y. Ding, 'Progress in electrical energy storage system: A critical review'. *Progress in Natural Science.* 2009;19(3):291–312.

[2] S. M. Hasnain, 'Review on sustainable thermal energy storage technologies, Part I: Heat storage materials and techniques'. *Energy Conversion and Management.* 1998; 39(11):1127–38.

[3] I. Hadjipaschalis, A. Poullikkas, and V. Efthimiou, 'Overview of current and future energy storage technologies for electric power applications'. *Renewable and Sustainable Energy Reviews.* 2009;13(6):1513–22.

[4] N. J. van Eck, and L. Waltman, 'Software survey: VOSviewer, a computer program for bibliometric mapping'. *Scientometrics Scientometrics.* 2010;84 (2):523–38.

[5] A. Perianes-Rodriguez, L. Waltman, and N. J. van Eck, 'Constructing bibliometric networks: A comparison between full and fractional counting'. *Journal of Informetrics.* 2016;10(4):1178–95.

[6] L. Leydesdorff and H. W. Park, 'Full and fractional counting in bibliometric networks'. *Journal of Informetrics.* 2017;11(1):117–20.

[7] J. Suntivich, K. J. May, H. A. Gasteiger, J. B. Goodenough, and Y. Shao-Horn, 'A perovskite oxide optimized for oxygen evolution catalysis from molecular orbital principles'. *Science.* 2011;334(6061):1383–85.

[8] C. Ponce de León, A. Frías-Ferrer, J. González-García, D. A. Szánto, and F. C. Walsh, 'Redox flow cells for energy conversion'. *Journal of Power Sources.* 2006;160(1):716–32.

[9] P. F. Ribeiro, B. K. Johnson, M. L. Crow, A. Arsoy, and Y. Liu, 'Energy storage systems for advanced power applications'. *Proceedings of the IEEE.* 2001; 89(12):1744–56.

[10] S. Vazquez, S. M. Lukic, E. Galvan, L. G. Franquelo, and J. M. Carrasco, 'Energy storage systems for transport and grid applications'. *IEEE Transactions on Industrial Electronics.* 2010;57(12):3881–95.

[11] A. F. Burke, 'Batteries and ultracapacitors for electric, hybrid, and fuel cell vehicles'. *Proceedings of the IEEE.* 2007;95(4):806–20.

[12] F. Díaz-González, A. Sumper, O. Gomis-Bellmunt, and R. Villafáfila-Robles, 'A review of energy storage technologies for wind power applications'. *Renewable and Sustainable Energy Reviews.* 2012;16(4):2154–71.

[13] E. J. W. Crossland, N. Noel, V. Sivaram, T. Leijtens, J. A. Alexander-Webber, and H. J. Snaith, 'Mesoporous TiO_2 single crystals delivering enhanced mobility and optoelectronic device performance'. *Nature.* 2013;495:215.

[14] M. Hasheminamin, V. G. Agelidis, A. Ahmadi, P. Siano, and R. Teodorescu, 'Single-point reactive power control method on voltage rise mitigation in residential networks with high PV penetration'. *Renewable Energy.* 2018;119:504–12.

[15] M. Charwand, A. Ahmadi, P. Siano, V. Dargahi, and D. Sarno, 'Exploring the tradeoff between competing objectives for electricity energy retailers through a novel multi-objective framework'. *Energy Conversion and Management.* 2015;91:12–18.

[16] H. Moghimi, A. Ahmadi, J. Aghaei, and A. Rabiee, 'Stochastic techno-economic operation of power systems in the presence of distributed energy resources'. *International Journal of Electrical Power & Energy Systems.* 2013;45(1):477–88.

[17] A. Rezvani, M. Gandomkar, M. Izadbakhsh, and A. Ahmadi, 'Environmental/economic scheduling of a micro-grid with renewable energy resources'. *Journal of Cleaner Production.* 2015;87:216–26.

[18] M. Izadbakhsh, M. Gandomkar, A. Rezvani, and A. Ahmadi, 'Short-term resource scheduling of a renewable energy based micro grid'. *Renewable Energy.* 2015;75:598–606.

[19] M. Khalid, A. Ahmadi, A. V. Savkin, and V. G. Agelidis, 'Minimising the energy cost for microgrids integrated with renewable energy resources and conventional generation using controlled battery energy storage'. *Renewable Energy.* 2016;97:646–55.

[20] S. Koohi-Kamali, V. V. Tyagi, N. A. Rahim, N. L. Panwar, and H. Mokhlis, 'Emergence of energy storage technologies as the solution for reliable

operation of smart power systems: A review'. *Renewable and Sustainable Energy Reviews.* 2013;25:135–65.

[21] X. Tan, Q. Li, and H. Wang, 'Advances and trends of energy storage technology in microgrid'. *International Journal of Electrical Power & Energy Systems.* 2013;44(1):179–91.

[22] D. O. Akinyele and R. K. Rayudu, 'Review of energy storage technologies for sustainable power networks'. *Sustainable Energy Technologies and Assessments.* 2014;8:74–91.

[23] A. Ahmadi, J. Aghaei, and H. A. Shayanfar, 'Stochastic self-scheduling of hydro units in joint energy and reserves markets'. in *2011 19th Iranian Conference on Electrical Engineering*, 2011; pp. 1–5.

[24] F. H. Gandoman, A. Ahmadi, A. M. Sharaf, *et al.*, 'Review of FACTS technologies and applications for power quality in smart grids with renewable energy systems'. *Renewable and Sustainable Energy Reviews.* 2018;82:502–14.

[25] H. Mavalizadeh, A. Ahmadi, F. H. Gandoman, P. Siano, and H. A. Shayanfar, 'Multiobjective robust power system expansion planning considering generation units retirement'. *IEEE Systems Journal.* 2017; PP(99):1–12.

[26] H. L. Ferreira, R. Garde, G. Fulli, W. Kling, and J. P. Lopes, 'Characterisation of electrical energy storage technologies'. *Energy.* 2013;53:288–98.

[27] M. C. Lott, S.-I. Kim, C. Tam, D. Houssin, and J. Gagné, 'Technology roadmap: Energy storage'. International Energy Agency; *Paris.* 2014.

[28] J. Eyer, G. Corey, and S. N. Laboratories, '*Energy Storage for the Electricity Grid: Benefits and Market Potential Assessment Guide : A Study for the DOE Energy Storage Systems Program*'. United States: Sandia National Laboratories; 2010.

[29] D. Connolly, *'A review of energy storage technologies'.* Ireland: University of Limerick, 2009.

[30] S. Sabihuddin, A. Kiprakis, and M. Mueller, 'A numerical and graphical review of energy storage technologies'. *Energies.* 2015;8(1):172–216.

[31] *'Grid Energy Storage'.* U.S. Department of Energy;2013.

[32] A. Esmaeily, A. Ahmadi, F. Raeisi, M. R. Ahmadi, A. Esmaeel Nezhad, and M. Janghorbani, 'Evaluating the effectiveness of mixed-integer linear programming for day-ahead hydro-thermal self-scheduling considering price uncertainty and forced outage rate'. *Energy.* 2017;122:182–93.

[33] A. Ahmadi, J. Aghaei, H. A. Shayanfar, and A. Rabiee, 'Mixed integer programming of multiobjective hydro-thermal self scheduling'. *Applied Soft Computing.* 2012;12(8):2137–46.

[34] S. Shafiee, H. Zareipour, A. M. Knight, N. Amjady, and B. Mohammadi-Ivatloo, 'Risk-constrained bidding and offering strategy for a merchant compressed air energy storage plant'. *IEEE Transactions on Power Systems.* 2017;32(2):946–57.

[35] B. Bolund, H. Bernhoff, and M. Leijon, 'Flywheel energy and power storage systems'. *Renewable and Sustainable Energy Reviews.* 2007;11 (2):235–58.

[36] A. F. Zobaa. *Energy storage-technologies and applications*. Croatia: InTech Publisher; 2013.
[37] S. C. Smith, P. K. Sen, and B. Kroposki, 'Advancement of energy storage devices and applications in electrical power system'. in *2008 IEEE Power and Energy Society General Meeting – Conversion and Delivery of Electrical Energy in the 21st Century*. 2008; pp. 1–8.
[38] H. Akbari and O. Sezgen, 'Performance evaluation of thermal energy storage systems'. *Energy and Buildings*. 1995;22(1):15–24.
[39] www.electricitystorage.org [Accessed 12 March 2018]
[40] S. C. Tan, C. K. Lee, and S. Y. Hui, 'General steady-state analysis and control principle of electric springs with active and reactive power compensations'. *IEEE Transactions on Power Electronics*. 2013;28(3):3958–69.

Chapter 2

Energy storage systems: technology, integration and market

Paulo F. Ribeiro[1], *Antonio C. Zambroni Souza*[1] *and Yuri R. Rodrigues*[2]

2.1 Introduction

The vast growth of new power generation technologies, especially renewables, and the modifications in consumption profile are causing significant changes in the operation of generation, transmission, and distribution systems. This is taking place on a global scale, with particular characteristics to each location, but still a worldwide phenomenon.

Photovoltaic panels, wind farms, electronic converters, and storage systems with batteries are some examples of the addition of new values with respect to both generation and loads of power systems [1].

The integration of renewable power generation devices suggests a more distributed generation approach. Consequently, the use of equipment associated with an energy storage system opens new possibilities for power systems.

Energy storage can be done directly, through the electricity accumulation in super capacitors, or indirectly, by either an electrochemical process such as batteries, or mechanically, as flywheels. All these technologies are constantly evolving and are the subject of research aimed primarily to reduce production costs, increase their storage capacity, and reduce the environmental impacts.

Associated to the possibilities enabled by smart grids, batteries of various technologies have been tested in order to allow the creation of networks having their own reconfiguration skills, in which intelligent feeders and protection devices are linked in a real-time connection to eliminate possible faults and reestablish the demand supply quickly, when necessary, without the need to wait for a decision from the control center. This would be the model of island systems that, even when connected to the main grid, would be able to remain active in case of a disconnection.

Figure 2.1 depicts a summary of several energy storage technologies that may be employed in power systems.

[1]Institute of Electrical Systems and Energy, Federal University of Itajuba (UNIFEI), Brazil
[2]School of Engineering, The University of British Columbia (UBC), Canada

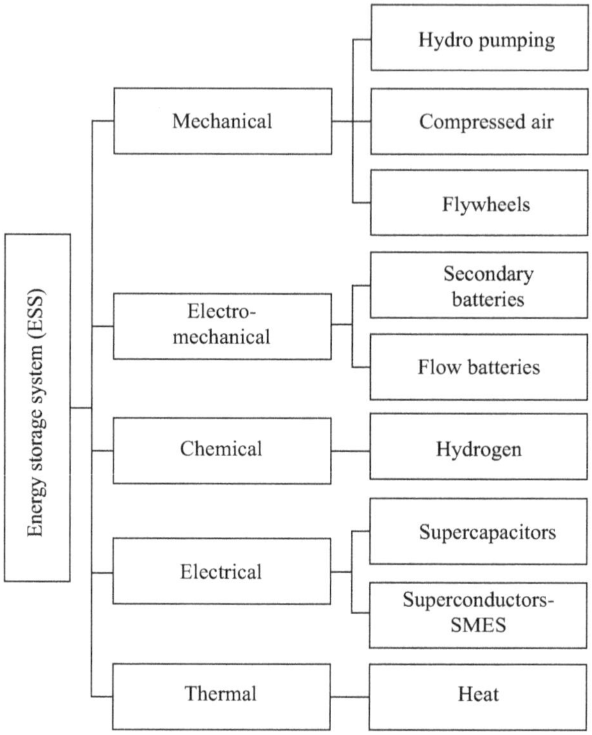

Figure 2.1 Energy storage systems (ESS) technologies

2.2 Benefits

The following benefits may be listed as likely applications of energy storage devices in transmission systems:

- Voltage support;
- Grid stabilizer;
- Loss reduction;
- Increase in reliability;
- Postponement of investment in transmission lines;
- Improvement of renewable sources generation;
- Smoothing of renewable generation;
- Shifting of renewable generation;
- Frequency regulation;
- Virtual inertia;
- Replacement of spinning reserves;
- Ramp management;
- Reduction of energy cost;
- Load leveling.

The technological portfolio of electric energy storage solutions covers a wide range of services required by the power system, which can be sorted based on the relationship between stored energy and dispatchable power.

In order to an ESS be successful, the storage technology must:

- Be well designed for applications with high capacity and long discharge periods. This yields larger profitability for utilities since they could take advantage of the fluctuations in energy prices throughout the day.
- Have an economically viable lifespan.
- Be efficient (low losses) and have a low maintenance cost.

A comprehensive perspective of the adequate position of storage technologies regarding different applications and discharge times is shown in Figure 2.2.

In this chapter, storage applications are primarily focused on the transmission and distribution (T&D) systems by considering the significant increase in renewable sources insertion.

2.3 State of the art

Electric energy storage is not a novel phenomenon or application. In 1799, the modern battery was developed by Volta. Following, in 1780, Galvani proved "animal electricity". In 1836, batteries were demonstrated in telegraphic systems. Around 1880, some New York City private systems were nightly supplied by the lead-acid batteries connected to a DC system. This application anticipates the demand-response concept, absorbing generation surplus during periods of low demand and supplying power to the grid in high-demand scenarios.

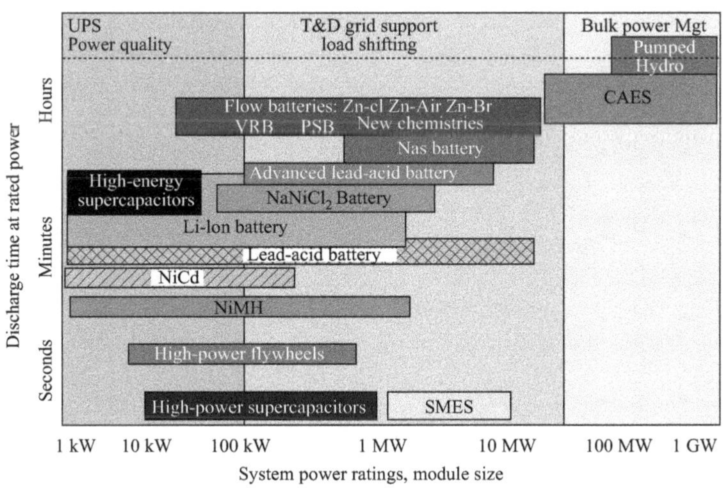

Figure 2.2 Position of energy storage technologies [2]

The initial electricity storage system on a large scale in the United States was developed in 1929. It was a pumped storage of 31 MW located at Connecticut Light & Power Plant Rocky River. Until the mid-1980s, the energy storage in the United States was used only for coal transportation to replace natural gas in peak load. These large energy storage systems can store excess of electricity production in low-energy demand periods. This practice not only allows the time changing but also reduces the need for combustion turbine operation for peak demand. These operative and economic benefits justified the development of a large number of pumping projects. Since the mid-1920s up to the 1980s, over 22 GW of power plants were constructed for pumping in the United States.

In the 1980s, the impetus to develop storage technologies through batteries was very strong. However, these technologies could not meet the large storage capacity provided by the hydro pumping. In this period, researchers from the Department of Energy/Sandia National Lab and Electric Power Research Institute identified other operative requirements of the power network that required supply for shorter periods, from 1 to 6 h. This was a huge incentive to keep research in this area as appealing.

2.4 International scenario

In view of smart grids development and the use of renewable generation, California recently adopted the first-term energy storage, demanding that the three major state energy companies have 1,325 MW of peak power electricity storage capacity by the end of 2020 and 200 MW peak by the end of 2014. This requirement stems from the state's strategy to set a goal that 33% of electricity should come from renewable generation by 2020. Although this innovative rule has been unanimously adopted by five commissions, two concerns arise: the cost to consumers and the fact that the pumped facilities (hydraulic) are not considered within this mandate.

There is a wide variety of technologies, including batteries, flywheels, but the costs are high. In the United States, there may be up to 14 GW (peak) electric storage in 2022, but this will only be possible if the costs are reduced to about \$700–\$750 per kWh. The wind and solar energy can only reach their full potential if storage is available as a reserve for times when the wind is not blowing and the solar irradiation is not faced. Many experts believe that recent advances will greatly influence the development of energy storage industry.

In California, several studies on storage use were considered, being divided into different categories based on the level of application, as shown in Table 2.1.

Hence, one of the challenges is the proper choice of adequate storage technology(ies) for various applications in the T&D systems.

As an example, one can consider a transmission system with peak electrical load approaching the maximum capacity of the system. If there is interest in the postponement of investments in transmission lines, in some cases, the installation of a small volume of energy storage somewhere in the transmission grid may delay systems expansion for some years.

Table 2.1 Storage studies in California [3]

Case study applications	Categories
Transmission level	Large storage systems Ancillary services Storage with local generation
Distribution level	End of feeder Substation Community
Customer level	Energy management Load shifting Electric vehicles

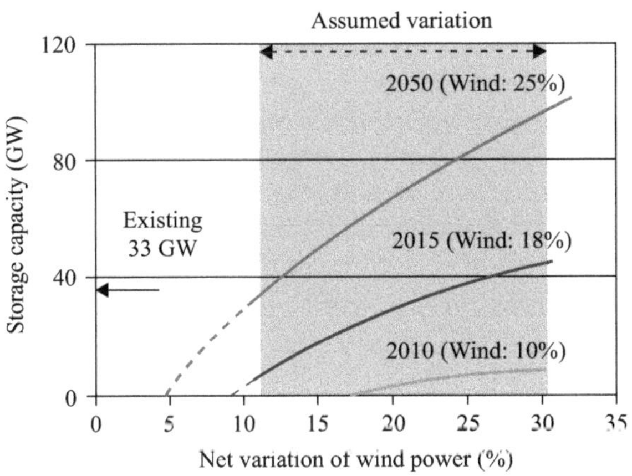

Figure 2.3 Storage capacity for Europe versus variation in wind power generation [4]

These several applications for small storages, such as deferral of investment in transmission equipment, can provide significant reduction in costs, improvement of assets usage, as well as enhancement of the grid development, since this capital can now be employed for other projects. This is possible because usually in a transmission system, the highest demands arise in very especial scenarios representing a few days or hours per year.

Although one of the basic benefits of storage for transmission is the deferral of works, it is possible to extend the benefit to the increase of the transmission equipment operating life. This can be a particularly attractive solution for transmission equipment with difficult access for maintenance and replacement work.

Furthermore, the increasing importance of renewable generation makes ESS even more significant. In Figure 2.3, the estimated storage required to compensate the variability of wind power generation in Western Europe is depicted. For example, if the change in wind energy is 30% of its nominal power, the required amount of storage would have to increase from 3 GW to 90 GW over a period of 40 years in order to support the expected development in wind power generation. The required storage capacity in 2050 is expected to be within 50 GW to 90 GW for a net variation in the rate of wind energy generation between 15% and 30%.

The European Commission concluded recently a report about energy storage that states:

> Therefore, there should be a stronger focus on energy storage and environmental regulation in the European Community, and better coordination about the issue of storage. Energy storage should be integrated and supported by all relevant bodies of energy and climate legislation within the EC as well as energy infrastructure strategies [5].

The three most innovative and economically powerful regions of the world, Asia, Europe, and the US, have adopted distinct strategies for energy storage. In general, the governments of the Asian countries are betting on advanced technologies for large batteries. In Europe, each country is adopting its own investment strategy to foment technology companies, aiming to use energy in the most efficient way by employing the best technology available. The United States has chosen to fund various types of innovative technologies, not just a specific one.

Thus, the most technologically diversified region for energy storage is North America where companies have deployed about 15 different technologies, reflecting a market driven by innovation. Meanwhile, Europe has proposed about 11 different technologies, compared to 8 in Asia.

Moreover, the power system of the future will have to accommodate bidirectional power flows and there will be an increasing transmission of information. An increasing number of the wind and photovoltaic generators will be integrated into the medium and low voltage networks. Thus, the conventional energy market should be transformed into a new model in which generation is evaluated in a control center, which performs a general supervisory function, according to market forces.

In addition, some new applications for storage are emerging to support the expansion of renewable energy generation and smart grids. A recent study [6] shows that by 2030, there will be a price reduction in energy storage technologies. The most promising market will be the use of energy in a time-shifted way. Another attractive market will be the energy balance, which aims to adjust the power supply of the substation through storage systems to meet fluctuations in demand for short periods of time. Large storage technologies such as hydraulic pumping and compressed air are already economically viable for this application. The need for energy balance will tend to increase as the generation of renewable energy causes more fluctuations on the utility side.

Some markets are experiencing substantial incentives for the development of renewable energy, being the implementation of storage capacity a crucial factor to accomplish these ambitious goals. To do so, several studies have being conducted looking for the determination of the exact quantity of storage required to guarantee the proper operation of the planned renewable insertion.

Another study on the potential market for storage regarding the increase in renewable energy worldwide was performed by the authors in [4]. The study has shown that by 2050, the increase in wind power will tend to increase the amount of storage in the market.

It is important to emphasize that the introduction of new storage technologies is a key factor in the future design of the network and will not be an easy task. Major changes will require an adjustment in business practices, as well as motivating acts (economic incentives) from governments.

Several factors drive attention to possible energy storage applications:

- Requirements for a quick and efficient response to changes in peak demand, due to transmission and generation limits;
- Renewable generation integration into the bulk system, given the distributed placement and intermittency of supply;
- The increasing congestion of T&D systems;
- Storage support in ancillary services for the efficient and reliable operation of the power grid;
- The growing necessity for consistent power due to the increasing participation of power electronics, and information and communication systems that are very sensitive to variations in the power supply.

Electricity storage can be deployed in any of the five major subsystems of electric power: generation, transmission, substations, distribution, and end consumers. Storage can play a role at the generator level by providing additional power when the generators are operating at full capacity and by storing power when the excess capacity for low-cost generation is available.

In transmission systems, the storage can provide additional energy at delivery points and alleviate the system in cases of congestion.

In terms of technological maturity, hydraulic pumping storage, acid, and sodium lead batteries and sulfur batteries can be classified as mature technologies. Other technologies also deployed, or under demonstration, include lithium ion, lead-acid advanced batteries, sodium chloride batteries, nickel, sodium ion batteries, inertia flywheels, and compressed air storage systems.

Although the market for distributed thermal storage is relatively large, the copper intensity is relatively limited, which limits the overall copper demand associated with this segment. Other types of distributed energy storage have higher copper intensities but are likely to have limited market penetration over the next five years.

For transmission systems, water pumping and compressed air seem to be the most suitable.

2.5 Technologies and applications

There are several energy storage solutions able to provide a wide range of services for the power system depending on the power and energy available. Solutions such as hydro pumps and compressed air are able to supply power to the grid for tens of hours at high demand rates that can reach 1,000 MW, thus being more applicable to transmission systems. In contrast, battery technologies, flywheels, electrochemical processes, etc., with smaller powers and energies have reached maturity and are being used in several projects. In this section, some characteristics of several of these technologies and applications are discussed.

2.5.1 Flywheel

Flywheel energy storage (FES) is classified as a mechanic storage device. It stores energy in the form of rotational kinetic energy. The stored energy depends on the moment of inertia and the speed of rotation of the flywheel. Its main components are the rotating cylinder which is coupled to the shaft, bearings, and the transmission device (motor/generator). When the steering wheel rotation is constant, regardless of the losses, there is no storage or power supply. To accelerate the flywheel, the transmission device supplies power to the rotating cylinder, operating as a motor. Increasing the speed of the rotation cylinder results in an increase in the amount of stored energy. If the speed of the rotating cylinder increases, it means that energy has been withdrawn from the system and the transmission device operates as a motor. The fast response time of the steering wheel makes it suitable for different applications in power systems.

More advanced systems have rotors made of carbon filaments of high mechanical strength, suspended by magnetic bearings with rotation speeds of more than 50,000 rpm with the rotating cylinder in the vacuum, which helps in reducing losses.

Among the main characteristics of the FES, there are excellent stability, long service life, low maintenance and high power density. Currently, FES is commercially employed in energy quality and in power sources for supply without interruption of energy. Efforts have been made to optimize the flywheel for long-lasting operations of up to several hours, as an energy storing element. Figure 2.4 illustrates a flywheel structure.

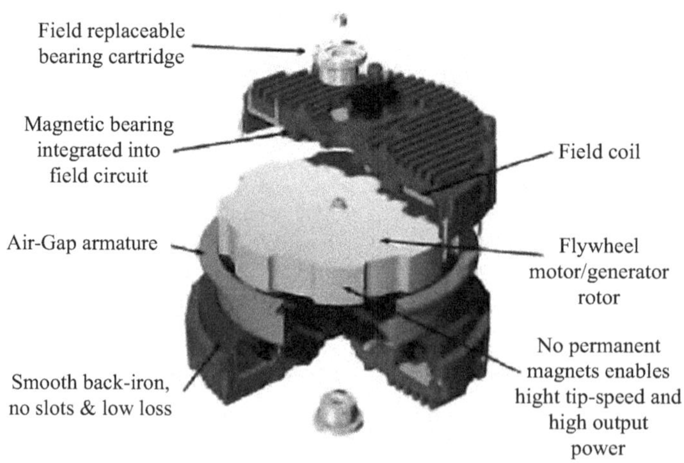

Figure 2.4 A flywheel [7]

2.5.2 Batteries

The battery is an electrochemical device, composed of stacked cells, where the chemical energy is converted into electrical energy and vice versa. The desired voltage and current levels are obtained by connecting the cells in series and in parallel. Features such as conversion efficiency, life (in a number of cycles), operating temperature, discharge depth (usually the batteries are not completely discharged), self-discharge rate and energy density should be considered when choosing which technology to use.

Currently, several types of battery are under development. Some of them are already commercially available, while others are in experimental stage.

Batteries used in electrical systems applications have so far been of the intensive cycle (similar to those used in electric vehicles), with power capacity ranging from 17 to 40 MWh and efficiency between 0.7 and 0.8.

Some of the battery technologies are more suitable for applications in power systems:

Acid lead: Lead-acid battery uses lead dioxide as the active material of the positive electrode and metallic lead as negative active material in a porous structure with a high surface area. The electrolyte used is a solution of sulfuric acid. Commercial units used by power utilities have capacities between 300 kW and 40 MWh, discharge depth between 200 and 2,000 cycles, and operating temperature between −5 and 40 °C [8].

Sodium-sulfur (NaS) batteries: This type of battery is based on the process of oxidation of sodium ions. The electrochemical cells are composed of a positive sulfur electrode and a negative sodium electrode in contact with a β-alumina electrolyte. During the discharge, the positive ions of sodium flow through the electrolyte and the electrons flow through the external circuit of the battery, producing a voltage of 2 V. The battery temperature should be kept at about 300 °C for this process to occur. This technology is in a mature phase of development, being produced since 2003 by NGK in Japan and being applied for energy storage of photovoltaic and wind power plants and in substations with powers between 500 kW up to 80 MW and 60 MWh [9].

Lithium ion: This battery possesses a high level of energy density, as well as a long service life, supporting around 3,000 cycles at 80% capacity and with efficiency, generally above 90% [10]. Due to its cost of installation, its application in large-scale systems is limited, being applied in small portable devices, such as cellular, computers, electric bicycle and more recently to electric vehicles. Its installation also requires additional security systems, since overheating scenarios can occur due to the thermally unstable characteristic and the possibility of decomposition at elevated temperatures of most metal oxide electrodes. To reduce this risk, lithium-ion batteries should be associated with a monitoring device to prevent over-charging and over-discharge. Normally, a voltage monitoring system is also added to control each cell voltage in order to avoid voltage deviations between them [7].

2.5.3 Supercapacitor

Dual-layer electrochemical capacitors are also known as supercapacitors or ultracapacitors. Supercapacitors present a high capacitance due to the enormous surface area of the electrodes and the small distance between the charges. Thus, supercapacitors may store a much larger amount of energy than any conventional capacitor of equivalent mass or volume.

Supercapacitors have been known for a long time. The first patent dates back to 1957, where a high surface area carbon capacitor was described by [11]. Later, in the 1970s, the first attempt was made to commercialize this device [12].

Because its charge and discharge process are due to the movement of the ions in the electrolyte (without chemical reactions), supercapacitors have a very small equivalent series resistance, causing the load and discharge current limit to be quite high.

Supercapacitors operate below 100 kW rate, having a life cycle of 10,000 cycles and efficiency of 0.95. Additionally, they have a fast response time and discharge time of less than 1 min.

2.5.4 Hydraulic pumping plant

Hydraulic pumping is widely adopted as an electricity storage technology. There are hundreds of pumping stations operating with a total capacity of 127 GW worldwide.

The pumped storage facility has two reservoirs. In times of high energy demand, the pumped storage plant releases water from the upper reservoir to the lower reservoir, rotating the turbines along the path. When power demand decreases, the plant pumps water from the lower reservoir back into the upper reservoir. Part of the power plant itself is used during pumping. Droughts do not affect pumped storage plants because water can be continuously pumped into the upper reservoir to keep electricity production stable. However, the construction of these plants has high costs. In addition, it is difficult to find the right location to accommodate two reservoirs.

2.5.5 Solar thermal power plant

The solar thermal plants do not directly use the solar rays in the production of energy, that is, its principle differs from the photovoltaic panels. The energy conversion follows the same cycle as a thermoelectric plant generation, but the required heat is generated by the sun. It is an interesting alternative since this process does not generate harmful gasses to the environment.

In the design of these plants, mirrors are placed reflecting light into a receiving tower, where the fluid (usually liquefied salt) present is heated to high temperatures and remains in that stored state. When there is demand for electricity, the fluid is driven by the generator and the steam moves the turbines of the primary machine, thus generating electricity. Similar to a cycle of a conventional thermoelectric plant, the fluid is reused for the same process later.

2.5.6 Superconducting magnetic energy storage (SMES)

Though superconductivity knowledge dates back to 1911, it took until the 1970s to the proposition of an SMES energy storage solution for power systems [13].

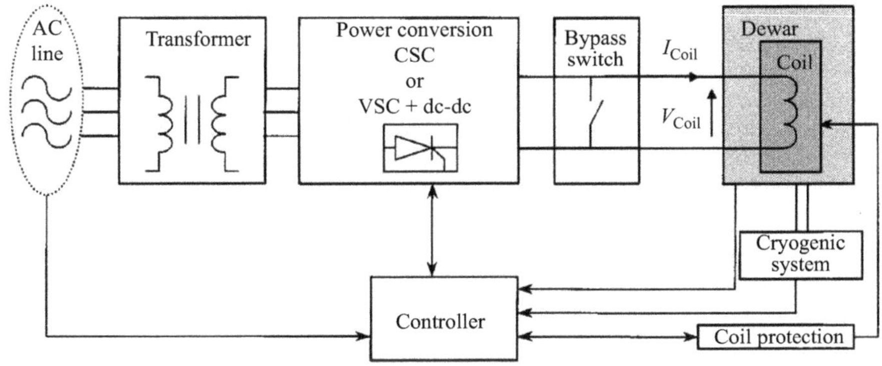

Figure 2.5 SMES components

The SMES have driven applications in electrical and military systems because of its rapid response and great efficiency (over 95%). Among its applications, there are dynamic, transient and voltage stability, load leveling, frequency regulation, transmission enhancement, and power supply quality.

Compared to other energy storage, SMES is still very expensive. Although the combination with FACTS equipment can reduce the cost, it condenses the needs in terms of electronic conversion.

A SMES unit is a magnetic field-based energy storage device that uses the current flowing through a superconductor coil. Figure 2.5 illustrates an SMES system [14].

2.5.7 Decoupling in time between generation and consumption

In an electrical system integrated with renewable sources, which usually present intermittent generation characteristics, it is interesting to enable the consumption of energy outside the time when it is generated. For this, the energy storage elements are used, which absorb the surplus energy generated and deliver it according to the demand. The main requirement for this type of application is a sufficient storage capacity, giving the volume of intermittent generation source.

The integration of grid-connected photovoltaic systems is possible by the application of energy storage systems, particularly batteries. They are used to enable energy consumption at different times from those at which it is generated, such as consuming at night (when photovoltaic generation is negligible) the energy that was generated during the day with plenty of sunlight.

Considering the possibility of tariff variation throughout the day, the consumers may choose when to consume and/or store energy from the electrical system. The stores can be used to stock up on energy at times of reduced fares to be consumed when fares are higher.

In 1987, the Crescent Electric Membership Cooperative in the United States, a 500 kW lead-acid battery was implanted to store energy for part of the day and to provide that accumulated portion at a later time. Because of this deployment, it was able to reduce the payload demand for generation and transmission to the North Carolina Electric Membership Cooperative [15].

2.5.8 Emergency energy sources

The spinning reserve is a portion of the unused generation, but always available to provide, during a given time interval, active power to the electric power system [16]. A fast response and large storage capacity to supply the load for a few minutes (while reserve generators are activated) are the main requirements for energy storage of this type of application.

2.6 Standards for storage devices

For mature storage systems, there are already several standards covering technical characteristics and systems incorporation aspects. In contrast, nonmature technologies have some standards addressing at special issues, but none deals with the complete integration of technology. Such standards must include:

- Terminology;
- Basic characteristics of components and systems;
- Definitions and measurement methods for comparison;
- Technical evaluation such as capacity, power, lifetime, discharge time, and communication between components and safety protocols;
- Interconnection requirements as acceptable voltage deviations, power quality, frequency synchronization and measurement;
- Tests;
- Implementation guides.

2.7 The role of virtual inertia

The growing penetration of intermittent sources raises some important issues, such as the interaction of the components connected to power electronic-based equipment, the management of intermittent power availability, and the problem of virtual inertia. These aspects are particularly important when the number of photovoltaic panels connected to the system escalates. Traditionally in electrical networks, synchronous generators are responsible to inject and absorb kinetic energy to compensate the frequency deviation in the grid due to abrupt changes between generation and load. This brings a problem to systems with high penetration of photovoltaic panels, which worsens if islanded microgrids are considered. In this case, such an operation state is possible thanks to the intermittent sources with low, or none, inertia. To overcome this problem, it is suggested that a short-term energy storage is added to the system. Therefore, when combined with the power electronic components, it behaves as a synchronous generator, yielding the name VSG, an acronym for virtual synchronous generator. This structure is illustrated in Figure 2.6.

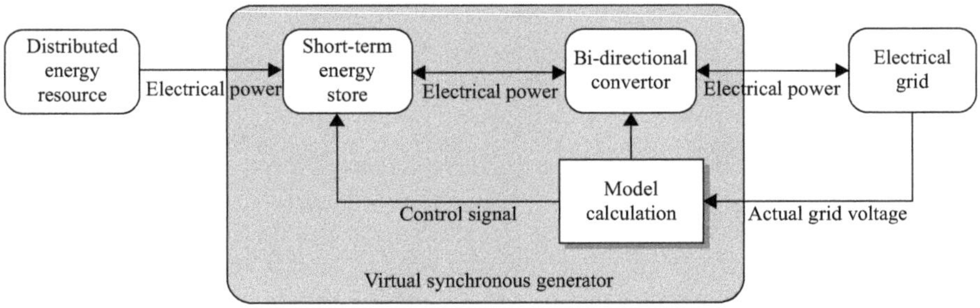

Figure 2.6 VSG structure

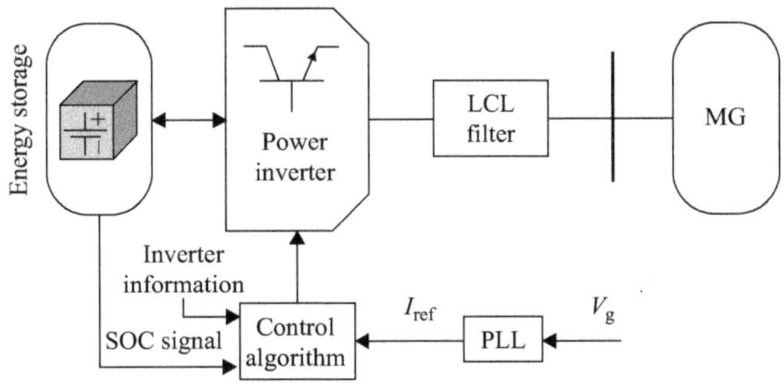

Figure 2.7 VSYNC project diagram

The structure above must be carefully analyzed. For wind farms, the combination of the momentum of inertia and the electronic devices plays the VSG. For distributed generators with no means of energy storage and photovoltaic panels, short-term storage devices are necessary.

- VSG topologies

The idea is to make the inverter work as a synchronous generator. The power provided by the VSG is given by Eq. (2.1).

$$P_{\text{VSG}} = P_0 + K_I \frac{d\Delta\omega}{dt} + K_p \Delta\omega \tag{2.1}$$

where K_I is a scalar related to virtual mechanic inertia and K_p is the mechanic damping, while P_0 is the power provided.

In the following sections, few topologies available in the literature addressing this topic are shown.

2.7.1 VSYNC's project

This topology consists of a storage battery connected to the network through a frequency inverter and an LCL filter, where L and C respectively denote inductor and capacitor element, as depicted in Figure 2.7.

The phase-locked loop block, when correctly adjusted, works as a synchronous machine, so its output is the reference for the inverter. The control block provides information about the batteries loading level and the virtual angular velocity establishes the active power to be generated. As for the reactive power, it is calculated from the voltage level.

2.7.2 IEPE's topology

In this topology, the virtual machine is responsible to specify the reference current for the inverter. The associated diagram is depicted in Figure 2.8. In this figure, E_p stands for the induced *fem,* whereas M_{mech} is the mechanical torque.

The elements in the lower part of the diagram model the mechanical aspect, where J is the inertia moment and K_{d} is the damping coefficient. The accelerating torque of the VSG is obtained as a combination of M_{mech}, M_{e} (electrical torque) and M_{d} (damping torque). The upper components are related to the electrical part. L_{s} is the stator inductance and R_{s} is the electrical resistance.

2.7.3 ISE lab's topology

In this topology, the variable of utmost interest is the mechanical angle θ_{m}. Its value is given to the pulse width modulation controller of the inverter, where V_{grid} and I_{grid} are the measured voltage and current, respectively, f_{grid} is the network frequency, p_{in} is the injected power, p_{out} is the generated power, ω_{m} is the angular velocity, and V^* is the voltage command. Figure 2.9 depicts its design.

The system formulation is shown in Eq. (2.2). It indicates that the difference between the output and input powers must equal the mechanical power stored in the

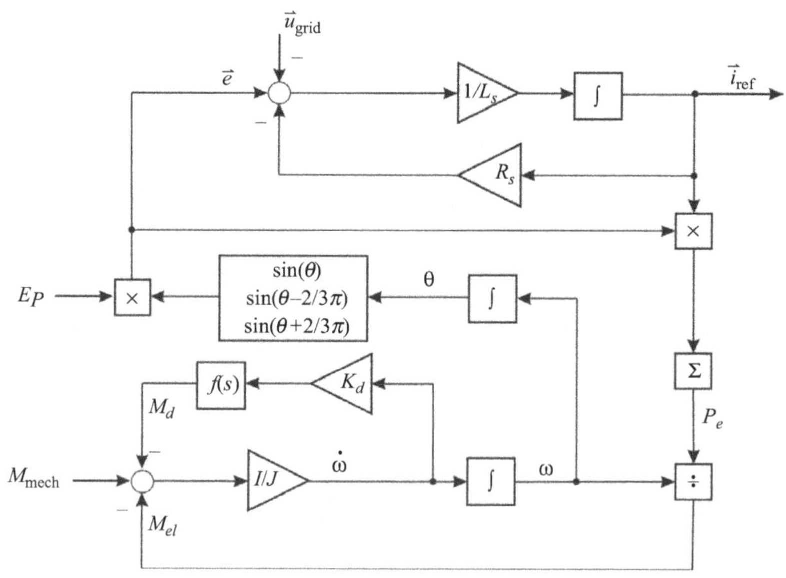

Figure 2.8 IEPE model

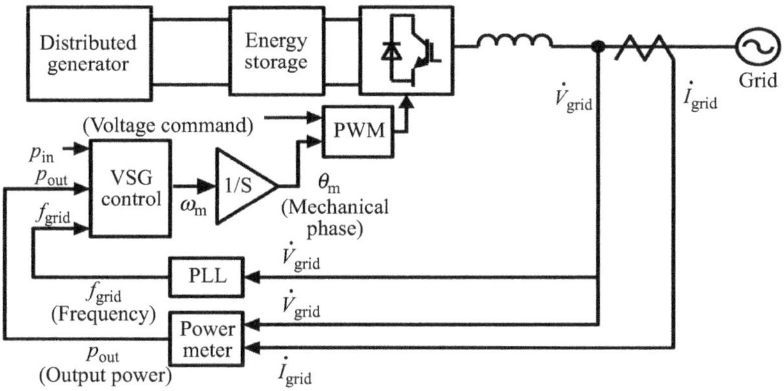

Figure 2.9 Control diagram proposed by the ISE laboratory

rotating mass and the power dissipated by the damping. Thus, D is a constant value and s is the slip with respect to the synchronous frequency.

$$p_{in} - p_{out} = J\omega_m \cdot (d\omega_m/dt) - D \cdot s \tag{2.2}$$

The inertia in power systems comes from the actual behavior of rotating machines. Hence, they have a central participation on the transient response of the network, helping the system to reach a post-fault stable trajectory. As already discussed, the massive incorporation of photovoltaic panels may create some problems, requiring the virtual inertia emulation. Section 2.8 discusses another possibility of virtual inertia, played by plug-in electric vehicles (PEVs).

2.8 Plug-in electric vehicles

PEVs may play a significant role in modern power systems. The concept of PEVs establishes that, beyond working as a load, in emergency conditions they may supply energy to the system. This enables operators to consider PEVs energy while operating islanded microgrids [17], as well as for frequency control and virtual inertia.

Though PEVs present some operational limitations, as described in Table 2.2, that should be taken into account, such as rated battery storage capacity and the maximum depth of discharge (DOD), which characterizes the useable energy from the batteries. Furthermore, these characteristics are particularly important when charging is the issue. If massive penetration of PEVs is featured and none controlled charging methodology is associated, possible undervoltages scenarios will be faced. In this case, the charging of PEVs must be associated with the monitoring of voltage level and system loss, so that the normal operating conditions are respected [17].

Table 2.2 is important because of its technical features. But when it comes to PEVs as a load supply, another aspect should be considered, and that is not addressed in Table 2.2. It is about the minimum charge of its own battery that the owner accepts to take. Even though acting as a virtual inertia equipment may demand a short period

Table 2.2 PEV's key features

Model	Rated energy storage capacity (kWh)	Maximum depth of discharge (DOD)
# 1	27.4	80%
# 2	5.33	60%
# 3	16	65%
# 4	24	65%
# 5	16	65%

of work, depending on the initial state of charge, such a work may take the battery to its lowest acceptable level, disabling the owner to use its own car. This, therefore, is a regulation matter that must be addressed.

2.9 Market aspects

The study of market aspects is mandatory to determine the viability and encourage the development of new and innovative energy storage solutions and projects. For this sake, the market evolution in a structure capable of adequately remunerate storage is critical to support the new requirements arising from the growing role of renewable generation, both at the T&D levels, large-scale penetration of PEVs and the accelerated development of smart initiatives, i.e., smart grids, smart microgrids, and smart cities, which are enabling a complete new level of applications and services.

This new system perspective provides opportunities for agents interested in developing the actual state of technology, as well as business ventures to incorporate more energy storage in the electricity grid. However, a careful analysis considering many aspects should be performed, since both scenarios provide great opportunities and considerable challenges to be overcome.

Among them, direct implications driven by the type of technology and application, regulatory framework, and market structure, must be associated in a holistic perspective capable of addressing the intrinsic effects felt and influenced by ESS on itself and in the market.

2.9.1 Market applications and potential

Energy storages are flexible devices able to perform several types of power system supporting applications that are economically rewardable in almost all power systems' levels [18]. A description of the main applications associated to each level is illustrated in Figure 2.10.

The market potential for energy storages is defined by the existence of an application demand, associated with the technical and economic competitiveness of storage in face of others viable options. The economic viability is particular to each region and is influenced by the resources available, rules, and regulations. In this perspective, storage devices have greater associated value in regions where low

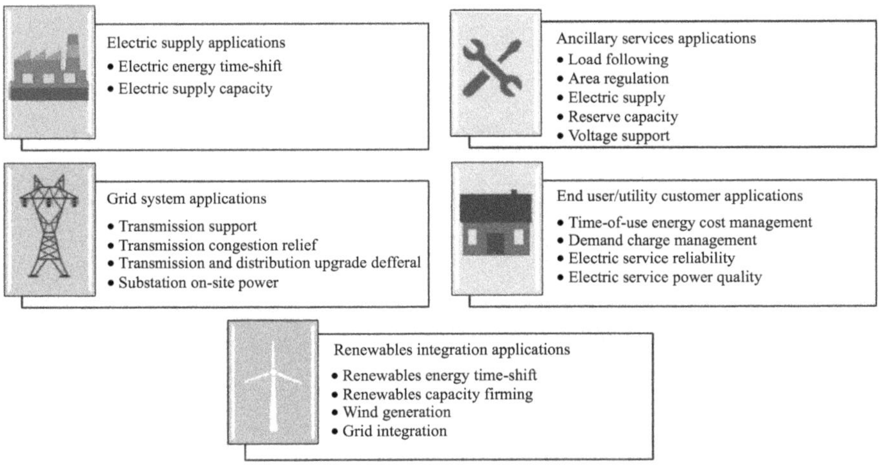

Figure 2.10 Power systems' market applications for energy storages

price off-peak energy is accessible and the system is operating close to technical limits. In addition, additional value to storage is provided in applications taking place in distant and/or remote regions where the cost for fuels' (i.e., coal, gas, etc.) transportation and storage may arise as a significant parcel of generation expenses.

This implicit influence in storages' associated value due to the location of placement and availability of resources is described in Figure 2.11. It presents a comparison between a region with sources providing low off-peak energy costs, as nuclear and hydro power plants, or with generation surplus due to renewable sources without storage capacity (higher value location), compared with another site that requires the operation of power plants fueled by gas for off-peak generation (lower value location).

2.9.2 Energy storage influence in the market equilibrium

In order to allow investors to take economic advantage of the benefits provided, some energy storage applications may require guarantees in price, or risk and reward sharing mechanisms [19]. For example, in a long-term analysis, off-peak energy prices generally increase significantly. In contrast, on-peak rates are volatile and may increase slightly on average, which may demand an associated risk-sharing mechanism to enable this application.

The same perspective applies for power systems' rules and regulations. If they do not provide adequate return to the storage derivatives [20], as deferral of T&D investments, energy time-shifting, regulation, reserve capacity, etc. While there may be technical demand for such an application, there is no market potential.

- Market demand, supply, and equilibrium

The stored energy is generally obtained during off-peak hours, which decreases the effect of fuel prices rising, especially when it is inserted in a higher value location.

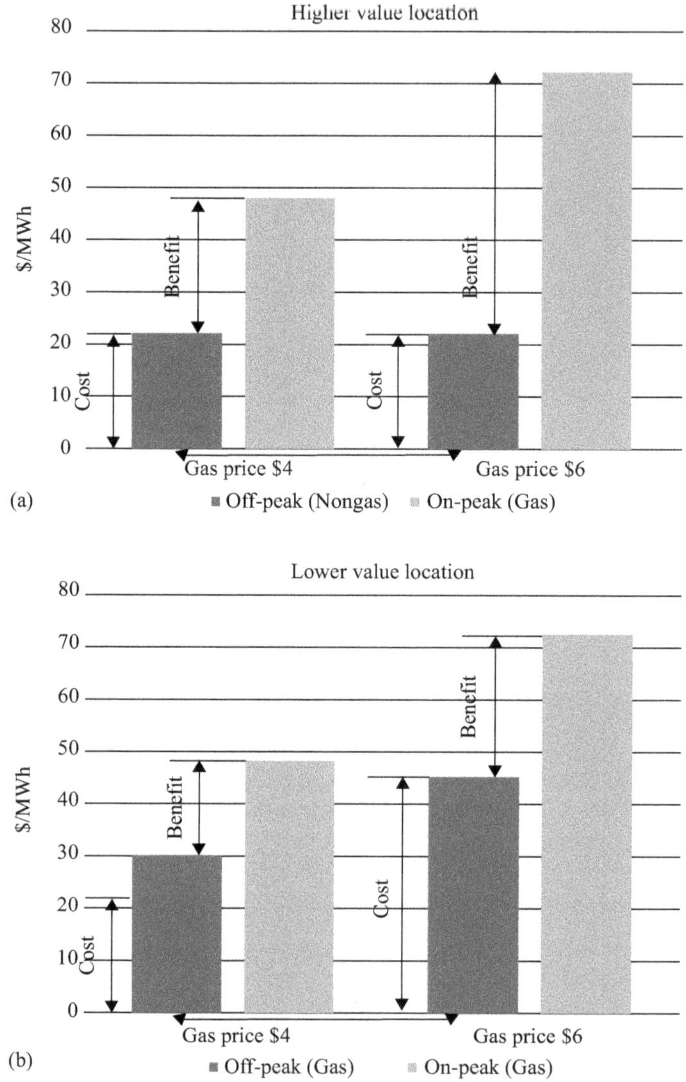

Figure 2.11 Storage associated value for (a) higher value location; (b) lower value location [19]

However, it must be noted that this new demand for low off-peak energy will lead to a marginal rise in the off-peak price. And the increase of power supply during on-peak period provided by storages, will lead to a marginal decrease in the on-peak price.

These variations in the markets' demand for off-peak energy and supply capacity during on-peak period will lead to new market operational equilibriums, which must be assessed to adequate understand the storages' actual market

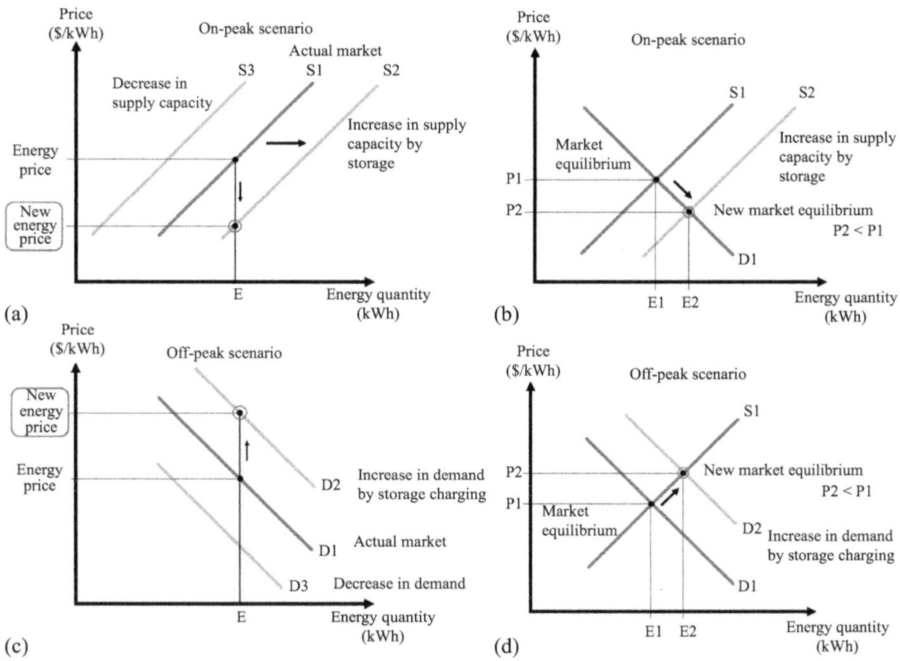

Figure 2.12 Energy storage influence in market demand, supply, and equilibrium for the on-peak (a)-(b) and off-peak (c)-(d) scenarios

potential. To demonstrate these interactions, a linear price/energy relation is adopted to facilitate comprehension, being explained as follows.

Figure 2.12(a) indicates that the storage insertion will lead to an increase in supplying capacity during on-peak period, this feature is represented by the new supply curve (S2) compared to the actual market perspective (S1). It is possible to notice that for the same energy demand requirement (E), a reduced price of energy is presented as there is more availability of supply.

Furthermore, this additional supply will establish a new market equilibrium with lower prices during on-peak operation as described in Figure 2.12(b). Due to the reduction in energy price, consumers are motivated to increase their consumption. This leads to an increase in energy demand (E2 > E1), but still preserving the energy price at a lower energy rate than the previous scenario (P2 < P1).

A similar perspective is faced for the off-peak market depicted in Figure 2.12(c). However, at this time, there is an increase in energy demand due to storages charging (D2) in contrast to the actual market demand (D1). Consequently, there is associated an increase in the energy price for the same energy demand requirement (E).

At last, the new off-peak market equilibrium is shown in Figure 2.12(d). As one can observe, a new marketing equilibrium is reached at a higher energy price (P2 > P1) and demand requirement (E2 > E1). This scenario is obtained by the intersection between the systems' supply (S1) and new demand (D2) curves.

In this sense, investors should be aware and evaluated the feasibility of the economic project for the actual market equilibrium that they will be operating after the insertion of the prospected energy storage units.

2.9.3 Market structures

A market structure able to reward the direct and indirect benefits provided by energy storages is fundamental to increase the volume of storages in the grid [21]. The recognition of technical support through economic compensation leads to a better market outlook for investors interested in developing projects with significant amounts of energy storage. These bonus payments for additional value streams extend the economic profit region making it more attractive for fundraising.

A general representation of the market scenario featuring an energy tariff based on plain energy supplying reward, in comparison with a more comprehensive economic recognition of additional benefits, is depicted in Figure 2.13.

Another factor that greatly influences the market operation regards price regulation imposed by the system operator or government. These practices are usual, and even required, for some markets without open competitiveness due to natural or imposed barriers for new companies' participation.

The definition of a regulated price imposes changes in the market structure, specially impacting the economic profit region and the point of maximum profit. This leads to a decrease in the economic profit region and possible change in the point of maximum profit, which may even become an uneconomical solution, if the price established by the natural point of maximum profit is greater than the regulated price. Furthermore, the real energy price is characterized by the lower value between price and regulated price.

A market perspective featuring price regulation is shown in Figure 2.14.

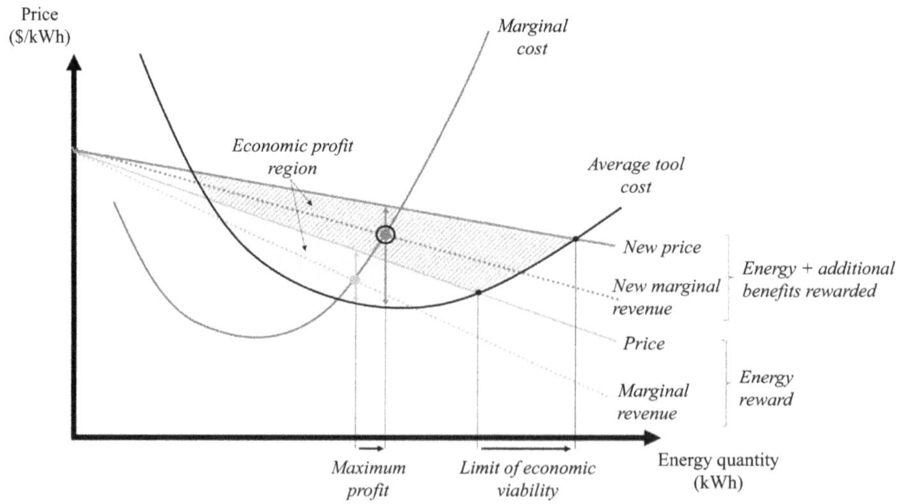

Figure 2.13 Market representation for plain and comprehensive energy tariffs

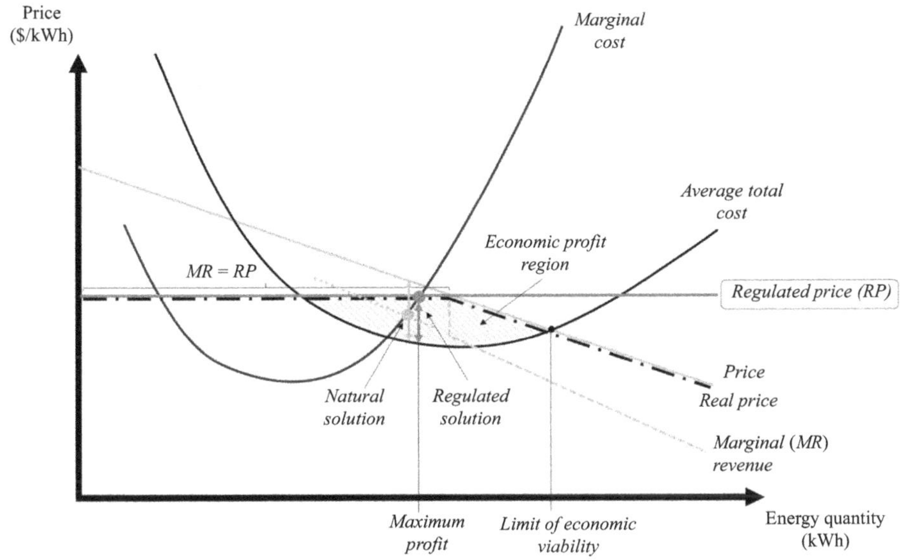

Figure 2.14 Market perspective featuring price regulation

2.9.4 Market scenarios

A typical concern of potential investors in storage projects is the financial recognition of the additional benefits provided by the storage insertion, which can be rewarded or not, and if so rewarded, paid to the wrong agent [19]. In this sense, if these value streams are not remunerated or price guarantees are not offered, projects capable of delivering significant value streams to the system and customers may prove to be a non-economic investment.

Also, different rules and regulations can be faced by different players, leading to distinct economical perspectives for the same investment. A market scenario presenting value streams reward for system operator ventures, but not for private investors is described in Figure 2.15.

One can notice that small quantities of storage are a good investment for both private and system operator stakeholders, since the arbitrage value captured by the energy revenue in the market is sufficient to pay the investor return and provides some short-term benefits for the customers. Yet, small benefits will be felt by the system.

In contrast, a large amount of storage delivers greater social and technical benefits. However, its economic viability requires attention when the investment is performed by the private sector, since value streams may not be rewarded or even create an in-house competition. This scenario is inverted when analyzed by the system operator perspective. In this case, the system operator can pass on part of the costs to the end users making it a win-win scenario in all perspectives. Customers will experience an overall reduction in the energy costs. The system will gain a significant amount of value streams. And the investor will have its financial return paid.

In order to provide a comprehensive market scenario, results indicating the potential US market for energy storage applications are shown in Table 2.3 [22], whereas state-of-the-art projects and discussions can be found in [23].

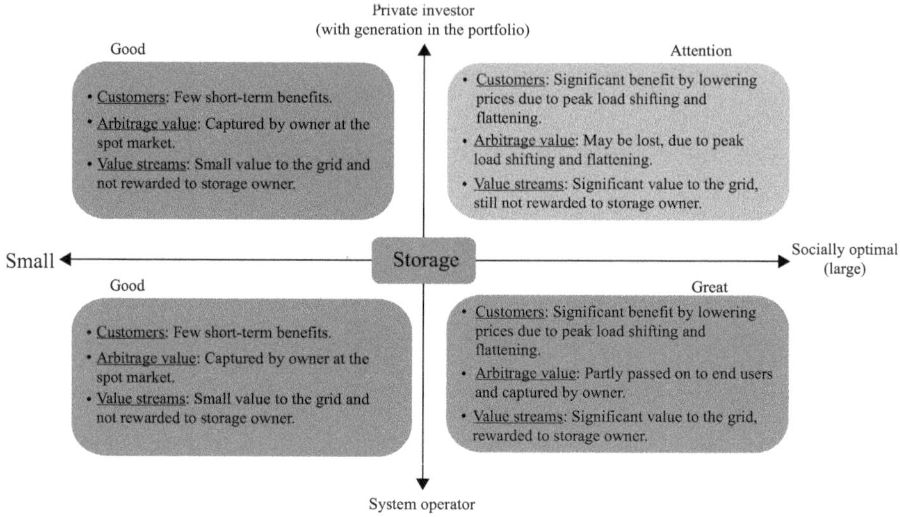

Figure 2.15 Market scenarios value streams

Table 2.3 Maximum market potential estimation [22]

	#	Maximum market potential (US)	(MW, 10 years)	($Million)
Supply	1	Electric energy time-shift	18,417	10,129
	2	Electric supply capacity	18,417	9,398
Ancillary services	3	Load following	36,834	29,467
	4	Area regulation	1,012	1,425
	5	Electric supply reserve capacity	5,986	844
	6	Voltage support	9,209	5,525
Grid applications	7	Transmission support	13,813	2,646
	8	Transmission congestion relief	36,834	3,168
	9.1	T&D upgrade deferral 50th percentile	4,986	2,912
	9.2	T&D upgrade deferral 90th percentile	997	916
	10	Substation on-site power	250	600
Customer applications	11	Time-of-use energy cost management	64,228	78,743
	12	Demand charge management	32,111	18,695
	13	Electric service reliability	9,209	6,154
	14	Electric service power quality	9,209	6,154
Renewables integration	15	Renewables energy time-shift	36,834	11,455
	16	Renewables capacity firming	36,834	29,909
	17.1	Wind generation grid integration, short duration	2,302	1,727
	17.2	Wind generation grid integration, long duration	18,417	8,122

2.10 Conclusions and challenges

A large number of possibilities for the application of energy storage technologies at different voltage levels are available in the market. However, a series of regulatory and market obstacles prevent utilities from capitalizing on these opportunities. The main challenge for the deployment of energy storage devices and systems is the high cost. This reality is evidenced by the existence of a small number of new projects at international level.

The main challenges may be classified in the following categories:

- Regulatory;
- Market;
- Business model;
- Technological;
- Costs.

Emerging technologies, such as response to energy demand and storage, are mixing the functions of each industry with their services and responsibilities. Additional matters such as location, licensing, and security requirements will involve the participation of regulators, which will have to adjust their rules to certify that energy storage solutions can be developed in a technically and economically feasible way.

Among the technical challenges for deploying storage devices associated with renewable and distributed generation, the following items may be stressed:

- Standardization of the system interface with the network;
- Operation and control;
- Planning and design.

Challenges for the use of electrical energy storage devices also include equipment life and environmental issues for batteries and materials for steering wheels. Hybrid metallic nickel batteries have a long life but lower energy density. Sodium-sulfur batteries have been shown to be promising but are currently very expensive. In addition, battery efficiencies are between 70% and 85%, indicating that 15% to 30% of the stored energy is lost.

References

[1] World energy Council, "World energy perspective smart grids: Best practice fundamentals for a modern energy system – Annexes", 2012.

[2] Sandia National Laboratories, "DOE/EPRI 2013 Electricity Storage Handbook in collaboration with NRECA", DOE Energy Storage Systems Program, July 2013.

[3] EPRI Energy Storage Program, "Energy storage valuation tool draft results investigation of cost-effectiveness potential for select CPUC inputs and storage use cases in 2015 and 2020", CPUC Storage OIR Workshop, 2013.

[4] International Energy Agency, "Prospects for large-scale energy storage in decarbonized power grids", Working Paper, December, 2009.
[5] European Commission Directorate-General for Energy, "The future role and challenges of energy storage", 2012.
[6] Boston Consulting Group, "Revisiting energy storage: There is a business case", 2011.
[7] IEC, "Electrical energy storage", White Paper, December, 2011.
[8] K. C. Divya and J. Østergaard, "Battery energy storage technology for power systems – An overview", Electric Power Systems Research, vol. 79, no. 4, pp. 511–520, 2009.
[9] A. Bito, "Overview of the sodium-sulfur battery for the IEEE stationary battery committee", IEEE Power Engineering Society General Meeting, vol. 2, pp. 1232–1235, June, 2005.
[10] M. Beaudin, H. Zareipour, A. Schellenberglabe, and W. Rosehart, "Energy storage for mitigating the variability of renewable electricity sources: An updated review", Energy for Sustainable Development, vol. 14, no. 4, pp. 302–314, 2010.
[11] H. Becker, "Low voltage electrolytic capacitor", US 2800616 A, 1957. Available at: http://www.google.es/patents/US2800616, Accessed: February, 2018.
[12] D. Boos, "Electrolytic capacitor having carbon paste electrodes", US 3536963 A, 1970. Available at: https://www.google.com/patents/US3536963, Accessed: February, 2018.
[13] R. B. Boom and H. A. Peterson, "Superconductive energy storage for power systems", IEEE Transactions on Magnetics, vol. 8, pp. 701–704, September, 1972.
[14] P. F. Ribeiro, B. K. Johnson, M. L. Crow, A. Arsoy, and Y. Liu, "Energy storage systems for advanced power applications", IEEE Proceedings, vol. 89, no. 12, pp. 1744–1756, 2002.
[15] D. H. Doughty, P. C. Butler, A. A. Akhil, N. H. Clark, and J. D. Boyes, "Batteries for large-scale stationary electrical energy storage", Electrochemical Society Interface, vol. 19, no. 3, pp. 48–53, 2010.
[16] Y. Rebours and D. Kirschen, "A survey of definitions and specifications of reserve services", Technical Report, The University of Manchester, 2005.
[17] D. Q. Oliveira, A. C. Zambroni de Souza, and L. F. N. Delboni, "Optimal plug-in hybrid electric vehicles recharge in distribution power systems", Electric Power System Research, vol. 98, pp. 77–85, 2013.
[18] European Commission, "Energy storage – The role of electricity", Commission Staff Working Document, 2017.
[19] T. Jenkin and J. Weiss, "Estimating the value of electricity storage: Some size, location, and market structure issues", Electrical Energy Storage Applications and Technologies Conference (EESAT), October, 2005.
[20] Deloitte, "Electricity storage technologies, impacts, and prospects, Deloitte Center for energy solutions", September, 2015.

[21] Sandia National Laboratories, "Energy storage financing: A roadmap for accelerating market growth", Sandia Report, August, 2016.

[22] Sandia National Laboratories, "Energy storage for the electricity grid: Benefits and market potential assessment guide", Sandia Report, February, 2010.

[23] Sandia National Laboratories, "Electrical energy storage demonstration projects", EESDP13, 2013.

Further Reading

A. Bhuiyan and F. A. Yazdani, "Energy storage technologies for grid connected and off-grid power system applications", IEEE Electrical Power and Energy Conference (EPEC), pp. 303–310, October, 2012. DOI: 10.1109/EPEC.2012.6474970.

A. Khaligh and Z. Li, "Battery, ultracapacitor, fuel cell, and hybrid energy storage systems for electric, hybrid electric, fuel cell, and plug-in hybrid electric vehicles: State of the art", IEEE Transactions on Vehicular Technology, vol. 59, no. 6, pp. 2806–2814, 2010.

A. R. Di Fazio, T. Erseghe, E. Ghiani, M. Murroni, P. Siano, and F. Silvestro, "Integration of renewable energy sources, energy storage systems, and electrical vehicles with smart power distribution networks", Journal of Ambient Intelligence and Humanized Computing, vol. 4, no. 6, pp. 663–671, 2013.

A. P. Wood, "Manchester local section: Some new flywheel storage systems", Journal of the Institution of Electrical Engineers, vol. 39, no. 185, pp. 414–428, 1907.

ABB, "World's Largest Battery Energy Storage System: Case note", 2011.

A. Abele, E. Elkind, and B. Washom, "2020 strategic analysis of energy storage in California". Public Interest Energy Research (PIER) Program, California Energy Commission, November, 2011.

A. Foley, D. Connolly, P. Leahy, *et al.*, "Electrical energy storage: Smart grid technologies to integrate the next generation of renewable power systems", Poceedings of the SEEP2010, Bari, Italy, 2010. http://vbn.aau.dk/en/publications/electrical-energy-storage-smart-grid-technologies-to-integrate-the-next-generation-of-renewable-power-systems(cf3b46e8-0438-4945-ab7a-703eec58d79e).html.

B. Dumaine, "Has the fuel cell car's time finally come?", CNN Money, pp. 457–465, 2013.

B. P. Roberts and C. Sandbergzio, "The role of energy storage in development of smart grids", Proceedings of the IEEE, vol. 99, no. 6, pp. 1139–1144, 2011.

C. A. Hill, M. C. Such, D. Chen, J. Gonzalez, and W. M. Grady. "Battery energy storage for enabling integration of distributed solar power generation", IEEE Transactions on Smart Grid, vol. 3, no. 2, pp. 850–857, 2012.

C. C. Chan and Y. S. Wong, "Electric vehicles charge forward", IEEE Power and Energy Magazine, vol. 2, no. 6, pp. 24–33, 2004.

D. Linzen, S. Buller, E. Karden, and R. W. De Doncker, "Analysis and evaluation of charge-balancing circuits on performance, reliability, and lifetime of supercapacitor systems", IEEE Transactions on Industry Applications, vol. 41, no. 5, pp. 1135–1141, 2005.

D. R. Chvala, "Flywheel energy storage: An alternative to batteries for UPS systems", Energy Engineering: Journal of the association of Energy Engineering, vol. 7, no. 3, pp. 7–26, 2005.

G. Prophet, "Supercaps for supercaches", EDN Europe, pp. 48:16, 2003.

G. D. Rodriguez, "Operating experience with the chino 10 mW/40 mWh battery energy storage facility", Energy Conversion Engineering Conference, vol. 3, pp. 1641–1645, 1989.

H. Qian, J. Zhang, J-S Lai, and W. Yu, "A high efficiency grid-tie battery energy storage system", IEEE Transactions on Power Electronics, vol. 26, no. 3, pp. 886–896, 2011.

J. S. Peck, "Flywheel load equalisers", Journal of the Institution of Electrical Engineers, vol. 43, no. 196, pp. 174–191, 1909.

KIC InnoEnergy, "Thematic Field Smart Grids and Electric Storage: Strategy and Roadmap", 2012.

M. Alizadeh, X. Li, Z. Wang, A. Scaglione, and R. Melton, "Demand-side management in the smart grid: Information processing for the power switch", IEEE Signal Processing Magazine, vol. 29, no. 5, pp. 55–67, 2012.

M. Korpas and C. J. Greiner, "Opportunities for hydrogen production in connection with wind power in weak grids", Renewable Energy, vol. 33, no. 6, pp. 1199–1208, 2008.

N. W. Storer, "A consideration of the inertia of the rotating parts of a train". Transactions of the American Institute of Electrical Engineers, vol. XIX, pp.165–168, 1902.

NGK Insulators, "This is a test entry of type @ONLINE", 2013.

P. Top, M. R. Bell, E. Coyle, and O. Wasynczuk, "Observing the power grid: Working toward a more intelligent, efficient, and reliable smart grid with increasing user visibility", IEEE Signal Processing Magazine, vol. 29, no. 5, pp. 24–32, 2012.

R. Kotz and M. Carlen, "Principles and applications of electrochemical capacitors", Electrochimica Acta, vol. 45, no.15, pp. 2483–2498, 2000.

S. Sheth and M. Shahidehpour, "Geothermal energy in power systems", IEEE Power Engineering Society General Meeting, vol. 2, pp. 1972–1977, 2004.

S. M. R. Tousif and S. M. B. Taslim, "Producing electricity from geothermal energy", Environment and Electrical Engineering (EEEIC), pp. 1–4, 2011. DOI: 10.1109/EEEIC.2011.5874669

Sandia National Laboratories, "ES-Select™ – Documentation and user's manual", Department of Energy, no. 2, December, 2012.

Sandia National Laboratories, "Evaluating utility owned electric energy storage systems: A perspective for state electric utility regulators", DOE/OE Energy Storage Systems Program, November, 2012.

Sandia National Laboratories, "Market and policy barriers to energy storage deployment", A Study for the Energy Storage Systems Program, September, 2013.

Sandia National Laboratories, "Methodology to determine the technical performance and value proposition for grid-scale energy storage systems", A Study for the DOE Energy Storage Systems Program, December, 2012.

W. Saad, Z. Han, H. V. Poor, and T. Basar, "Game-theoretic methods for the smart grid: An overview of microgrid systems, demand-side management, and smart grid communications", IEEE Signal Processing Magazine, vol. 29, no. 5, pp. 86–105, 2012.

Y. Qu, J. Zhu, J. Hu, and B. Holliday, "Overview of supercapacitor cell voltage balancing methods for an electric vehicle". IEEE ECCE Asia, pp. 810–814, 2013. DOI: 10.1109/ECCE-Asia.2013.6579196

Y-F Huang, S. Werner, J. Huang, N. Kashyap, and V. Gupta, "State estimation in electric power grids: Meeting new challenges presented by the requirements of the future grid", IEEE Signal Processing Magazine, vol. 29, no. 5, pp. 33–43, 2012.

Y. Xia, S. C. Douglas, and D. P. Mandic, "Adaptive frequency estimation in smart grid applications: Exploiting non circularity and widely linear adaptive estimators", IEEE Signal Processing Magazine, vol. 29, no. 5, pp. 44–54, 2012.

Y-H Kim, K-H Lee, Y-H Cho, and Y-K Hong, "Comparison of harmonic compensation based on wound/squirrel cage rotor type induction motors with flywheel". Power Electronics and Motion Control Conference, vol. 2, pp. 531–536, 2000.

Chapter 3
Energy storages in microgrids

Hamidreza Nazaripouya[1] and Yubo Wang[2]

3.1 Introduction

Microgrids have recently attracted significant attention as a new structure in electric power systems that can connect to the main grid or operate alone in islanded mode. The motivation to overcome issues such as environmental concerns, depletion of fossil fuels, increase of the population growth rate and energy demand, and air pollution has made researchers and engineers interested in microgrids as an opportunity to utilize renewable energy resources (RES) in the form of distributed generation, and address the energy demand in a reliable and affordable manner.

Although there is no unique and universal definition for microgrid or the size of microgrid regarding energy demand or geographic area, there exists a common understanding based on the definition proposed by the US Department of Energy Microgrid Exchange Group: a microgrid is a single controllable entity composed of interconnected loads and distributed energy resources within the defined electrical boundaries which can operate in both connected and islanded with respect to the main grid [1].

Microgrids can play an imperative role in secure, reliable, and resilient operation of the power grid by providing more flexibilities in operation, supplying uninterruptable power to the critical loads, and proper reconfiguration at the time of severe contingencies such as extreme weather or natural disasters [2, 3].

However, there are still several technical challenges regarding the control, planning, and operation of microgrids due to intermittent nature of RES, uncertainties in loads, and lack of inertia proposed by inverter-connected distributed energy resources. Technical concerns associated with control and operation of microgrid include voltage regulation, system stability and reliability, and power quality [4, 5].

Energy storage systems are considered as a solution to improve the power quality, dynamic stability, reliability, and controllability, of microgrids in the presence of RES [6]. The energy storage can act as an energy buffer to compensate

[1]Electrical and Computer Engineering, Winston Chung Global Energy Center (WCGEC), University of California, Riverside, USA
[2]Mechanical Engineering, University of California, Los Angeles, USA

renewable intermittency and load uncertainties. It enhances the stability of the microgrid by providing the virtual inertia for the system. Energy storage systems can also enhance the microgrids' efficiency by managing the power flow and reducing operational loss. They benefit electricity customers by maximizing renewable energy utilization and minimizing electricity bills through optimal energy management. This chapter discusses microgrid challenges, energy storage technologies, and applications of energy storage within a microgrid.

The outline of this chapter is organized as follows: In Section 3.2, microgrid challenges are reviewed. In Section 3.3, energy storage technologies are presented based on their features and capabilities. Section 3.4 discusses the applications of energy storage in microgrid. Challenges and barriers of widespread applications of energy storage in microgrid are discussed in Section 3.5. Finally, Section 3.6 draws the conclusion.

3.2 Microgrid challenges

Although microgrid is an asset for power system to increase system reliability and flexibility, reduce the investment for power transmission, and promote renewable energy deployment, challenges still widely exist. Some of the main operational challenges of microgrids include system stability, power management, power quality, and system reliability. In this section, the operational challenges of micro grid are discussed in detail.

3.2.1 Microgrid stability

Unlike the traditional power systems where the dynamic behavior of synchronous generator is dominant in power system stability analysis, the microgrids are mainly supplied by inverter-interfaced distributed generation. Therefore, the operating characteristics of microgrid including the time frame, inertia, and overcurrent capability are completely different from traditional grids. Referring to the microgrid definition, a microgrid is able to operate in two operation modes: (1) grid-connected mode and (2) islanded mode. Therefore, the stability analysis in microgrid can be categorized in grid-connected stability analysis and islanded stability analysis [7].

In grid-connected mode, the main grid resolves the mismatch between generation and load. That is, when there is a shortage of energy, microgrids draw energy from the grid, and when there is surplus of energy, microgrid injects it back to the grid. In this mode, the main grid maintains the frequency stability of the entire system and the disturbance in the microgrid barely impacts the main grid frequency. Therefore, the frequency and rotor angle stability are not studied in grid-connected microgrids, although the voltage stability for small and large disturbances (transient analysis) is considered. On the other hand, when microgrids operate in islanded mode, the disconnected microgrid is responsible for maintaining the system stability by managing the resources within the microgrid. In islanded mode, the small disturbance stability and transient stability are worth to investigate for both voltage and frequency. Figure 3.1 depicts the microgrid stability classification.

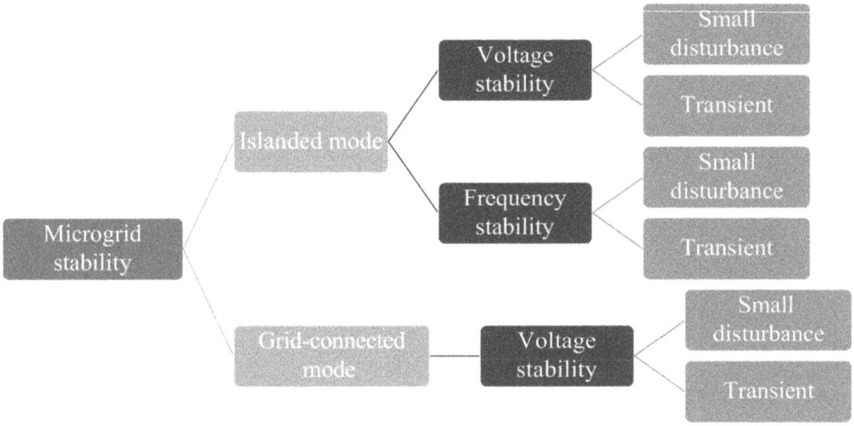

Figure 3.1 Microgrid stability classification

One of the main concerns for microgrid stability is the lack or low inertia of inverter-interfaced distributed generators (DGs) [8]. In traditional power system, the synchronous generators can store a considerable amount of energy on the rotating mass in the form of kinetic energy. The power grids can maintain the frequency and voltage stability by employing the kinetic energy stored in the rotor to compensate the mismatch between supply and demand. However, there is no such an inertia for inverter-based DGs, and voltage and frequency instability due to low inertia is an often time a crucial issue in microgrids.

In addition, the operation of microgrids is different from that of the traditional power system. First, the regulation of frequency and voltage in microgrid is highly dependent and affected by DGs control strategies. Therefore, the dynamic characteristics of DGs affect the dynamic behaviors of microgrid. Second, time response spectrum of microgrids is broader than traditional grid due to the existence of a wide range of DGs, including both inverter-interfaced and traditional DGs. Finally, the small output impedance and small over-current capability of inverter-interfaced DGs lead to faster-acting protection system. These characteristics make the stability analysis and operation process of microgrids different and complicated.

On microgrid stability standardization end, subclasses 8.6.3, 8.6.4, 8.6.5, and 8.6.6 in the Institute of Electrical and Electronics Engineers (IEEE) Standard 399-1997 [9] provide general guidance on accepted industry standards for performing stability studies to determine if a proposed system will meet a criterion of remaining dynamically stable under credible operating conditions. While IEEE 1547.4 [10] covers key consideration for planning and operating microgrids, which includes identifying steady-state and transient conditions.

In [11], the author investigates the transient stability of typical microgrid at the time of islanding. Figure 3.2 depicts the single-line diagram of the studied microgrid. The microgrid is composed of three DGs including one synchronous generator and two voltage source converter-interfaced sources, impedance load, and motor load.

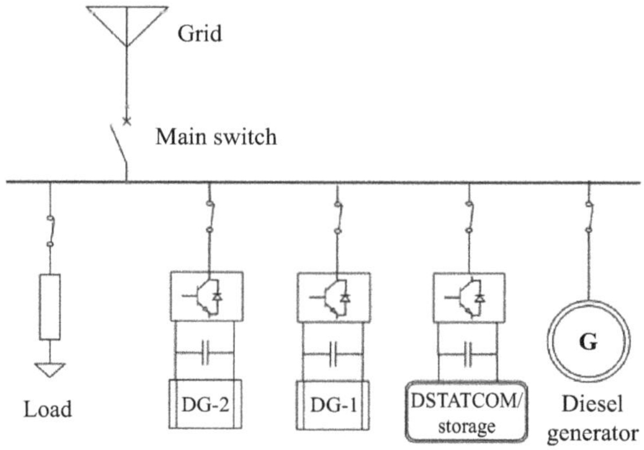

Figure 3.2 Microgrid single-line diagram in islanding transition

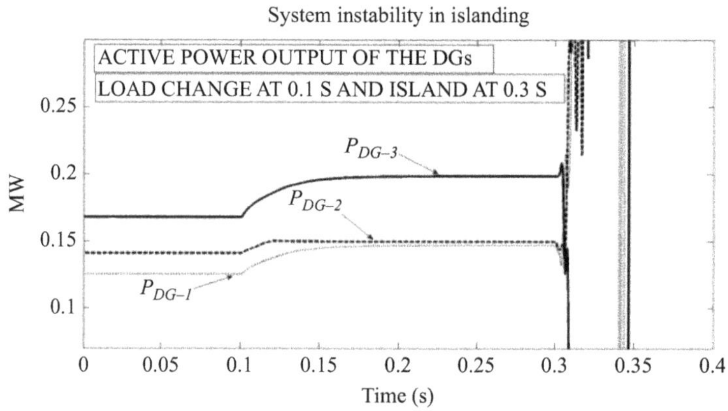

Figure 3.3 System instability during islanding due to power imbalance [11]

Figure 3.3 shows the system instability in islanding. This figure reports the active power of DGs when the load changes at $t = 0.1$ s and microgrid transients to islanded mode at $t = 0.3$ s. The results demonstrate that the microgrid can easily loose stability due to the power imbalance and that might happen before the loads have a chance to be cut off for power imbalance compensation.

3.2.2 Microgrid power management

Microgrid technology requires autonomous and robust power management system to be broadly adopted by utility companies and customers. The main responsibility of power management systems is to guarantee sustainable power delivery to local loads, while optimizing allocated objectives such as energy production, operational cost, energy efficiency, and power loss [12, 13]. However, compared with the conventional power systems, power management in microgrids faces new challenges due to hybrid AC/DC architecture [14], emerging of uncontrollable sources such as wind and solar, emerging of stochastic loads such as electric vehicles [15], deployments of power electronic devices, and multioperational modes including connected mode and islanded mode.

The purview of microgrid power management systems is not limited to supply local demands, but also optimal operation of microgrid reaching minimum cost, loss, and maintenance, and maximum efficiency, reliability, and sustainability in the grid, while there are several constraints imposed by the physics of the system and environment. Therefore, power management in microgrids faces complicated multiobjective optimization problem [16], where the global optimum of any individual objective function might not be in satisfactory range of the others.

The physical configuration of microgrids also imposes additional complexity to the power management systems. In AC/DC hybrid microgrid, more efforts in coordination are required between AC and DC subsystems including power balancing within DC bus, AC bus and load/grid, controlling DC bus voltage, controlling AC bus voltage and frequency, and synchronization.

In traditional power systems, the planning and operation is based on power flow analysis of a limited number of scenarios. However, in a microgrid composed of intermittent RES and uncertain loads, with frequent power flow reversing, power management and planning requires analysis of a considerable number of scenarios to capture all dynamics of the system. To this end, more advanced techniques are needed to model and predict the fast-dynamic behavior of the system [17], which makes the power management more challenging.

Inverter-interfaced DG units, which form a considerable portion of resources in a microgrid, are connected to the network via power electronic converters. However, the control concept, control strategy, and characteristic of power electronic converters are considerably different from those of the conventional rotating DGs [18]. Therefore, the dynamic behavior of a microgrid and power management, especially in islanded mode of operation is noticeably different from that of a conventional power system.

Finally, the power management system of microgrids should consider the transition of the microgrid between grid-connected mode and islanded mode. The transition should be smooth and seamless to minimize voltage and frequency deviations and disturbances, and balance supply and demand to avoid system collapsing. To this end, the power management system needs to adopt an appropriate control strategy for each operation mode and properly switch between control strategies.

As an example, the Consortium for Electrical Reliability Technology Solutions (CERTS) Microgrid with Large-Scale Energy Storage and Renewables at Santa Rita Jail, CA, USA, is a real-world microgrid project demonstration [19]. CERTS microgrid is considered as a self-control entity with capability of autonomous load following, control of voltage and frequency, islanding and resynchronizing to the network, circulating reactive power reduction, and stable operation of microgrids with multiple distributed energy resources units. The single-line diagram of the microgrid is depicted in Figure 3.4. This microgrid integrates 1.2 MW solar

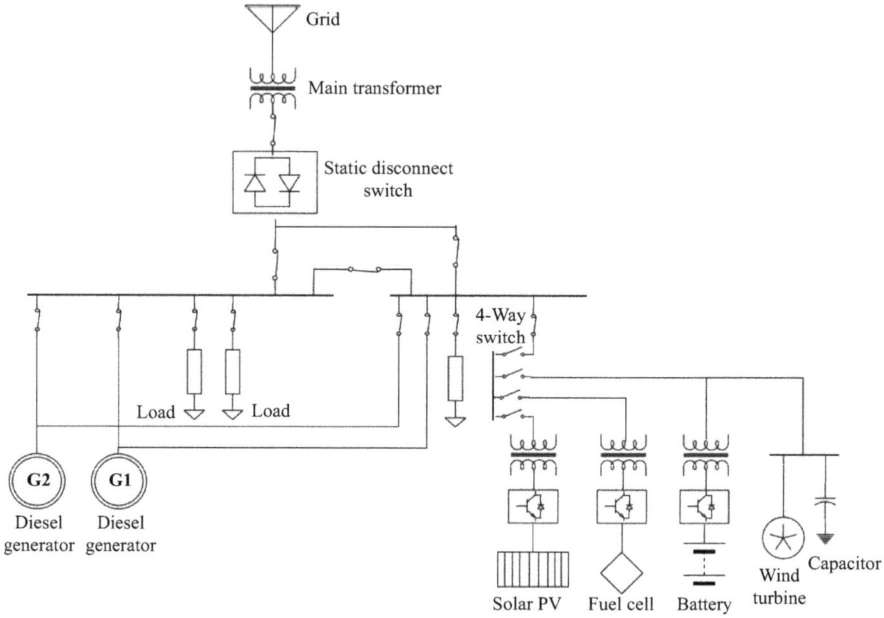

Figure 3.4 Single line diagram of Santa Rita Jail microgrid

photovoltaic, 1 MW fuel cell, two 1.2 MW conventional diesel generators with 2 MW, 4 MWh battery energy storage, a static disconnect switch, a 900 kVAR capacitor bank, and an 11.5 kW wind power system.

In this project, one of the technical challenges is to develop a noncomplex control technique to manage the distributed energy resources with high reliability. The power management system deployed in this project is a comprehensive central control system termed Distributed Energy Resources Management System (DERMS). The DERMS monitors and records information including energy consumption by feeder, electric utility rates, available generation sources, active and reactive power flow, battery condition, and power quality monitoring. In addition, it forecasts the energy usage and the status of the onsite energy resources including battery. The DERMS utilizes this information to determine how to economically dispatch the energy storage system and minimize the monthly utility costs.

The CERTS microgrid is based on plug-and-play functionality with no communication [20]. That is, in both grid-connected and islanded modes, each CERTS source regulates voltage and frequency. To this end, in islanded mode, each CERTS-controlled source seamlessly balances the power using an active-power/frequency droop controller, and storage inverters and the diesel generators also control the voltage using reactive-power/voltage droop controller. When microgrid is in grid-connected mode, the charge and discharge of the battery is managed by the DERMS for economic operation of the system.

3.2.3 Microgrid power quality and reliability

In electric power system, the term "power quality" is defined as maintaining the bus voltage near a sinusoidal waveform at rated magnitude and frequency [21]. When supplying the critical and sensitive loads, one of the basic expectations of the microgrid users is to access better power quality than what power grid offer. Therefore, providing highpower quality is one of imperative concerns in microgrids.

Microgrids can be characterized as a collection of DGs that are usually integrated to the electrical network through power electronic devices such as voltage-sourced inverter (VSI). However, interconnection of these resources using switching VSI interfaces with nonlinear voltage–current characteristics generates harmonics in the system, which adversely affects the power quality in microgrids. In addition, the DG units might be formed based on renewable energy sources such as solar energy and wind energy. The high penetration of DGs degrades the electric power quality due to the intermittent nature of RES and reverse power flow [22]. Moreover, the transition between grid-connected mode and islanded mode, non-linear loads, stochastic loads, and loads with considerable reactive power demand causes more power quality issues in microgrids. Therefore, challenges such as power fluctuation, voltage interruption, voltage flickering, voltage sag, voltage swell, frequency deviation, poor power factor, harmonics, unbalance voltage, and unbalance current are some of the concerns in microgrids [23].

In the IEEE Standard 1250: Guide for Identifying and Improving Voltage Quality in Power Systems [24], power quality issues are classified into two basic categories: (1) steady-state (continuous) voltage quality characteristics and (2) disturbances.

Steady-state voltage quality deals with the voltage magnitude variation from the nominal value, voltage imbalance among three-phase voltages, voltage frequency variation, and voltage distortion. In terms of harmonic distortion levels, the microgrid should comply with the IEEE 519 standard for harmonic control in electric power systems [25] and the IEEE 1547 standard for interconnecting distributed resource with electric power systems [26].

On the other hand, disturbances refer to voltage quality variations that occur at random intervals and are not associated with the continuous characteristics of the voltage. These variations include sustained interruptions (reliability), momentary interruptions, voltage sags (and swells), and transients.

The potential of improving power system reliability is an initial motivation to develop and deploy of microgrids. However, like conventional power systems, microgrids are exposed to a wide range of fault and failures events such as equipment failure, animal/tree contacts, lightning strikes, falling trees, and malicious attaches. In addition, the ability of microgrids to improve the power system reliability depends on the sufficiency and availability of distributed energy resources and dynamic response of the local generation units to quickly recover or ride through a fault. As considerable number of distributed energy resources with microgrids are intermittent and nondispatchable resources such as solar and wind, reliability improvement in microgrid is still challenging as multistochastic parameters including loads, supply, and the failure events need to be managed [27].

Moreover, the microgrids should be equipped with proper reconfiguration mechanics to avoid the cascading failure.

In an experimental attempt to analyze the power quality in a microgrid, a detailed study is performed on a real microgrid in Commonwealth Scientific and Industrial Research organization (CSIRO)-Renewable Energy Integration Facility (REIF), Newcastle-Australia [23]. In this study, the investigated challenges include voltage and frequency variations in both grid-connected and islanded modes at varying load and generation conditions, as well as current and voltage total harmonic distortion (THD) at different PV power levels, and at unbalanced PV generation and load distribution.

The testbed microgrid is composed of three single-phase solar PV panels with 7 kW, 7 kW, and 11 kW capacity, respectively; two 12.5 kW solar PV units integrated through 15 kVA three-phase inverter, 30 kW gas turbine, three single-phase battery storage units with 5kW/24 kWh capacity per each, and 64 kVA resistor-inductor-capacitor (RLC) load bank. The configuration of CSIRO REIF microgrid is depicted in Figure 3.5.

Table 3.1 demonstrates the obtained results from variation range of voltage and frequency and deviation from nominal value due to PV power variation during grid-connected and islanded operation modes of the microgrid.

The results in Table 3.1 show that in the islanded mode due to the lack of inertia the voltage and frequency variations are greater than those of grid-connected mode.

In another investigation, the unbalance voltage and neutral current under unbalanced generation of single-phase solar PV resources and unbalanced loads are analyzed. The obtained results are summarized in Table 3.2. The results depict that

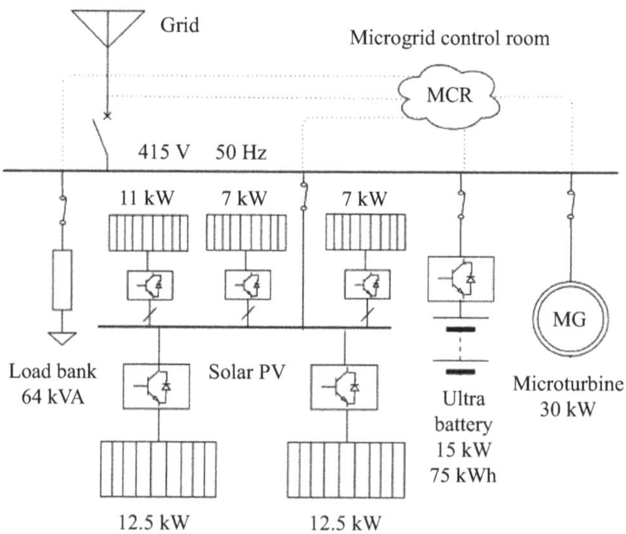

Figure 3.5 Single-line diagram of CSIRO REIF microgrid

Table 3.1 Frequency and voltage range and deviation results [23]

Frequency and voltage variation Under PV power variation	Frequency (Hz)				Voltage (V)			
	Range (Hz)		Deviation (Hz)		Range (V)		Deviation (%)	
	Min.	Max.	Min.	Max.	Min.	Max.	Min.	Max.
Constant load in grid-connected mode	49.98	50.12	−0.02	+0.12	403	409	−2.9	−1.4
Constant load in islanded mode	49.86	50.14	−0.14	+0.14	404	432	−2.7	−4.1
Varying load in grid-connected mode	49.94	50.12	−0.06	+0.12	405	412	−2.4	−0.7

Table 3.2 Unbalance voltage factor and neutral current results [23]

Unbalance voltage analysis Under unbalanced PV generation	PV generation (kW)			Load power (kW)			Unbalance voltage factor (%)	Neutral current (A)
	Phase A	Phase B	Phase C	Phase A	Phase B	Phase C		
No load	10.3	8.9	8.8	0	0	0	0.49	9.2
Unbalanced load	10.3	8.4	4.6	2.1	9.8	9.9	1.1	33

under the condition of unbalanced generation when the loads are uneven, the unbalance voltage is more compared with no-load condition. In this analysis, the unbalance voltage factor (UVF) is calculated as:

$$\mathrm{UVF} = \frac{V_{\text{mean}} - \max(V_{ab}, V_{bc}, V_{ca})}{V_{\text{mean}}} \tag{3.1}$$

where V_{ab}, V_{bc} and V_{ca} are the line voltages, and V_{mean} is the average value of V_{ab}, V_{bc} and V_{ca}. In addition, the neutral current (I_{N}) is calculated as:

$$I_{\text{N}} = \sqrt{I_A^2 + I_B^2 + I_C^2 - I_A I_B - I_B I_C - I_C I_A} \tag{3.2}$$

where I_A, I_B and I_C are the RMS values of phase currents.

Finally, the results of current and voltage THD based on the level of PV output power in each phase are shown in Figures 3.6 and 3.7, respectively. In the figures, the level of solar power is presented according to the ratio of PV power output (P_o) to the full power rating (P_r). The THD limit is also shown with dashed lines.

The odd-order harmonics of current during the maximum PV power generation of 44 kW and minimum PV power generation of 13 kW with reference to the appropriate standard THD limit are depicted in Figure 3.8. The results show that at the minimum level of PV penetration, the odd-order components of current harmonics are high and in some cases exceed the THD tolerance level.

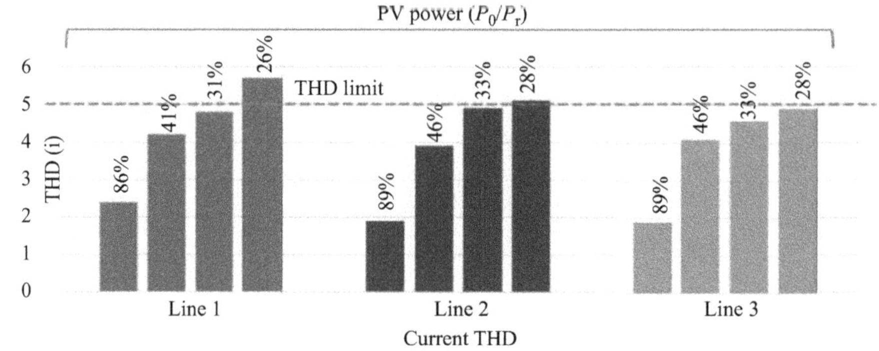

Figure 3.6 Current THD level at different PV power generation level [23]

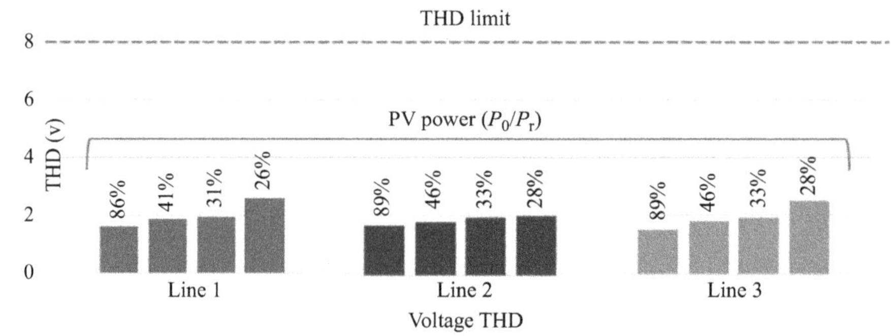

Figure 3.7 Voltage THD level at different PV power generation level [23]

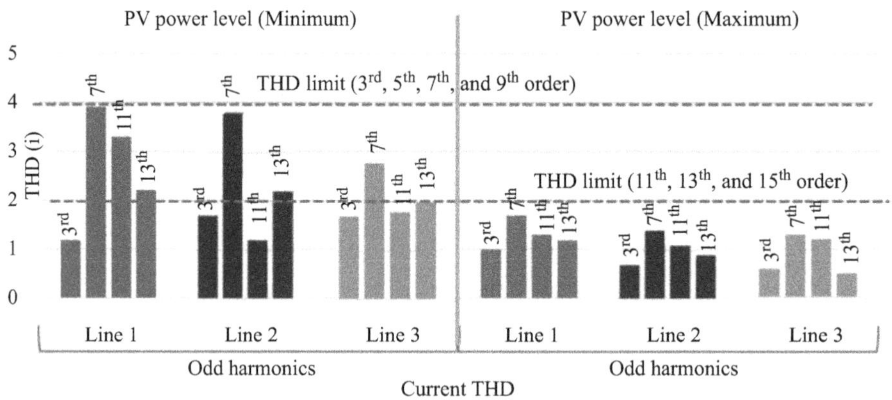

Figure 3.8 Current THD of odd harmonics [23]

3.3 Energy storage technologies

The applications of energy storage systems are highly correlated with the technology of these systems. The characteristics such as response time, power and energy ratings, operating temperature, weight, and volume are the basis for energy storage selection in an application. Therefore, before investigating the energy storage applications in microgrids, the available technologies of energy storage systems are discussed.

Energy storage systems can be defined as a device that can convert the electricity into a storable form, store the energy, and again convert it back to the electricity. To this end, the energy storage technologies based on the form of stored energy can be classified into four main categories and several subcategories as follows.

1. Electrical energy storage (EES): This category can be divided into:
   (a) Magnetic/current energy storage such as superconducting magnetic energy storage (SMES).
   (b) Electrostatic energy storage such as capacitors and supercapacitors.
2. Mechanical energy storage (MES): This category can be also divided into:
   (a) Kinetic energy storage such as flywheel.
   (b) Potential energy storage such as compressed air energy storage (CAES) or pumped hydroelectric storage (PHS).
3. Chemical energy storage (CES): This class of energy storage includes:
   (a) Electrochemical energy storage such as conventional batteries (lead-acid, nickel–cadmium, nickel–metal hydride, lithium ion) and flow-cell batteries (vanadium redox).
   (b) Chemical energy storage such as fuel cells.
   (c) Thermochemical energy storage such as solar hydrogen and solar metal.
4. Thermal energy storage: This type of technology also includes:
   (a) Low-temperature energy storage such as cold aquifer thermal energy storage and cryogenic energy storage.
   (b) High-temperature energy storage such as steam accumulator, molten salt storage, graphite, hot rocks and concrete, and latent heat thermal energy storage.

Based on the presented energy storage technology classification, the characteristics of energy storage systems including power and energy rating, dynamics, and space requirements, and consequently their functionality are different.

The rating characteristics of storage are defined according to two key factors: power rating and energy rating. The power rating represents the maximum output/input power under normal operating conditions and it is referred by discharge/charge rate. While the energy rating represents the amount of energy that can be delivered to loads without recharging, it is expressed by the discharge duration at the rated output power. The dynamics of energy storage can be also evaluated based on the response time and ramp rate. The response time is the time duration that energy storage goes from zero discharge to full. The ramp rate is the rate of changing of the output power. As an example, for power quality applications, the energy storage systems with relatively high power density up to 10 MW, relatively small energy density with discharge duration in the range of seconds to minutes, and fast dynamics with response time of milliseconds are required. However, energy management applications require storage with high energy density and a discharge duration in range of hours, and for reliability application, discharge duration of the storage in the range of minutes to an hour is adequate. Some other key factors such as storage footprint and space requirements are also dependent on energy storage technology and defined according to the energy and power density of storage.

In the following, some of the common energy storage systems deployed in microgrids and their features are elaborated [28–30].

Battery energy storage: Battery is the most well known energy storage technology. More familiar types include lithium-ion (Li-ion), lead-acid, nickel-cadmium (NiCad), sodium/sulfur (Na/S), zinc/bromine (Zn/Br), vanadium-redox, nickel-metal hydride (Ni-MH), and sodium/nickel chloride (ZEBRA). Battery energy storage systems (BESS) offer several operational benefits to the power network such as high round trip energy efficiency, fast response to the system changes, high energy density, high reliability, and cycling capabilities.

Other advantages of batteries include being modular, nonpolluting, and quiet. Battery energy storage can be used in a wide range of applications including load leveling, voltage regulation, frequency regulation, spinning reserve, and power factor correction. However, relatively low cycling time, high maintenance costs, and safety issue in some types are the challenges.

Capacitors (Supercapacitors): Compared with conventional batteries, capacitors can store energy significantly faster and become fully charged. They are performing charging/discharging cycles tens of thousands of times with a high efficiency without material degradation. The main shortcoming of conventional capacitors is low energy density. However, advancement in supercapacitors increases the energy density and capacitance of these devices by thousands of times compared with those of conventional capacitors. The challenges with capacitors, such as short discharge duration, and high energy dissipation due to self-discharging loss, limit their deployment mostly to short timescale applications including voltage and frequency transient stability, and power quality applications such as bridging and ride-through.

Flywheel energy storage: Flywheels are composed of a spinning cylinder on magnetic bearings coupled with an electrical machine, which stores the electricity in the form of kinetic energy, and charged and discharged through switching between motor/generator modes. The flywheels can be characterized by high round trip efficiency, fast response, high power density, and larger number of charge/discharge cycles with longer lifespan than batteries. But similar to capacitors, they have short discharge duration, high self-discharge and frictional loss, and low energy density which prevent them from being utilized in energy management applications. Therefore, the main applications of flywheels can be summarized in power quality improvements, voltage support, uninterruptable power supply (UPS), frequency regulation, and demand reduction. Large-scaled flywheels in the range of MW are also used for spinning reserve and reactive power support.

Fuel cells: Fuel cells are similar to batteries in terms of storing energy in the form of chemical energy. One of the differences between fuel cell and batteries is that although in a battery the energy is stored in a close system, the fuel cells should be replenished by adding reactants. Although the energy efficiency of fuel cells is low, they are able to store large amount of energy up to the terawatt-hour range. Therefore, they are used for backup power.

Compressed air energy storage (CAES): CAES stores the energy in form of compressed air for later use, which can be converted back to electricity. CAES has relatively high efficiency and high energy capacity. The small–medium scale CAES are more suitable for microgrid applications. Among the applications of CAES, the grid support for load leveling is more notable.

Superconducting magnetic energy storage (SMES): In SMES, the energy is stored in the magnetic field of a superconducting coil formed by direct current flowing through the coil. SMES is well known for the high storage efficiency. It can provide both active and reactive power. More advances of SMES include fast response with the response time in the range of milliseconds, high power density, and high cycle life. However, the major challenges of SMES are high system cost, short discharge duration, and the environmental concerns of strong magnetic field. SMES is a good fit for applications such as voltage stability improvement and power quality improvement.

Pumped hydroelectric storage (PHS): PHS is a mature technology of storage with high efficiency and long storage period. The main role of PHS is to store off-peak electricity from the grid in the form of potential energy and deliver it back during the peak demand. It is also utilized in energy management applications. However, nowadays, its application has been expanded to providing ancillary services such as frequency regulation.

The key characteristics and applications of the energy storage technologies discussed in this section are summarized in Table 3.3 [31–33].

3.4 Energy storage applications in microgrids

A microgrid needs to balance power demand and supply continuously at all time. The way that the electricity is generated and consumed has significantly changed nowadays. By merging loads with stochastic behavior such as electric vehicles to the microgrids, the power demand in microgrids is unpredictable. On the other hand, the integration of intermittent RES such as wind and solar has completely changed the power generation trend by switching from controllable and dispatchable recourses to uncontrollable and nondispatchable ones. In this situation, the energy storage is a critical and important ingredient for microgrids to balance the mismatch between the unpredictable power generation and consumption.

Table 3.3 Technical features and application examples of energy storage technologies

Energy storage technology	Rating characteristic		Dynamics		Space requirements		Performance	Example of applications
	Discharge/ charge rate (MW)	Response time	Ramprate	Dischar geduration	Energy density (Wh/kg)	Power density (W/kg)	Efficiency (%)	
Battery energy storage	0–40	Minutes–hours	Milliseconds	MW/sec	10–250	70–300	70–90	• Energy management • Grid stabilization • Power quality • Renewable source integration
Capacitors (Supercapacitors)	0.01–0.05 (0.01–0.3)	Milliseconds –1 hour	Milliseconds	MW/sec	0.05–5 (0.1–15)	100,000 (500 – 10,000)	60–90 (75–95)	• Power quality • Frequency regulation
Flywheel energy storage (FES)	0.002–0.25	Milliseconds–15 min	Instantaneous	MW/min	5–130	400–1500	90–95	• Load leveling • Frequency regulation • Peak shaving and off-peak storage • Transient stability
Fuel cells	0.001–50	Sec–24+ hour	Milliseconds	MW/min	800–10,000	500+	20–90	
Compressed air energy storage (CAES)	0.1–300	1–24+ hour	Minutes	MW/min	30–60	–	42–89	• Energy management • Backup and seasonal reserves • Renewable integration
Superconducting magnetic energy storage (SMES)	0.1–10	Milliseconds–10 sec	Instantaneous	MW/ms	0.5–5	500–2000	> 97	• Power quality • Frequency regulation
Pumped hydroelectric storage (PHS)	0.1–5,000	1–24+ hour	Seconds – Minutes	MW/sec	0.5–1.5	–	71–85	• Energy management • Backup and seasonal reserves • Regulation service

Energy storage systems facilitate the integration of RES and unpredictable loads into microgrids, ensure high reliability of microgrid, improve the stability and power quality in microgrids, lower the operational cost and enhance the efficiency, and provide robust and resilient electricity delivery within a microgrid.

So far, significant studies have been conducted on the different applications of energy storage systems in the power systems including main grid, areas that experience reliability and power quality issues, isolated areas, and remote areas [34, 28, 35]. In this section, based on the main challenges in microgrid, and the capability of different energy storage technologies, discussed in the previous sections, we investigate different values and services that energy storage system can provide for microgrids to address the existing challenges.

3.4.1 Stability enhancement

With adoption of power electronic devices in power systems, the energy storage systems can be controlled properly to operate in four quadrants and effectively exchange active and reactive power with electrical grid. To this end, energy storage systems are utilized in microgrids to maintain the voltage stability during both connected and islanded modes of operation, and frequency stability during the operation in islanded mode.

3.4.1.1 Voltage stability

Maintaining the microgrid voltage within an allowable range or returning it back immediately after faults or disturbances occur in the system is an imperative requirement of stable operation of microgrids. Therefore, voltage stability fulfillment is essential for acceptable operation of microgrid in both connected and islanded modes. In a microgrid, issues such as reactive power limits, dynamics of load and generation, and operation of load tap changers and voltage regulators are the main sources of voltage instability.

Unlike the transmission systems, in a microgrid, the line resistance to line reactance ratio (*R/X*) is considerable. As a result, the impact of active and reactive power, respectively, on frequency and voltage is not decoupled [36]. Therefore, the BESS as a source of active and reactive power can be controlled in four quadrants to compensate the active and reactive imbalance with a fast dynamics.

3.4.1.2 Frequency stability

Unlike synchronous generators in the main grid, RES connected to the microgrids such as photovoltaic arrays and modern variable-speed wind turbines do not contribute to maintain the stability of frequency due to low or lack of inertia. It provides an opportunity for energy storage system to act as virtual inertia in the microgrids [37, 38]. That is, the energy stored in the storage systems emulates the kinetic energy stored in the rotor of synchronous generators which can be released in the events of disturbance or drastic demand–supply imbalance.

BESS, supercapacitors, SMES, and FES are suitable candidates for dampening the frequency oscillations in microgrids due to their fast dynamics.

3.4.1.3 Case studies

There exist several microgrid projects around the world that deploy energy storage to provide system stability. As an example, the Kodiak Island microgrid in Alaska utilizes two 1 GW flywheels and two 1.5 MW lead-acid battery banks to address the stability challenges in the microgrid [39]. The total capacity of this microgrid is 80 MW supplied by solar PV, diesel, and 42 MW exclusively from hydropower and wind. The fast-acting flywheel is able to store short-term energy, and exchange both real and reactive power with the microgrid through the inverter. This technology enables the stable operation of the microgrid under renewable penetration up to 100 percent.

In another example in Indonesia, 500 kWh vanadium redox flow battery is integrated through two 240 kW inverters into the Sumba Island microgrid to improve the stability of the system and increase the renewable energy penetration capacity [40]. The microgrid is composed of 50 kW solar photovoltaic, and 2×135 kVA diesel generator power plant. In this microgrid, the primary services of the storage is to enhance the system stability through frequency and voltage regulation. The energy storage systems in this microgrid can compensate the renewable intermittency by providing ramp rating and smoothing the renewable output power and help with the microgrid stability.

3.4.2 Energy management

Energy management is one of the challenging issues in a microgrid. Within a microgrid, the energy management can be described as a strategy which determines how the distributed energy resources should be dispatched in the most economic manner, and how the loads need to be managed maintaining the most comfort, such that the operational cost is minimized, load demand is addressed, and operational constraints are met. Due to the intermittent nature of distributed RES and stochastic behavior of dynamic loads in microgrids, the role of energy storage system in balancing the supply and demand and cost-effective operation of the system is indispensable [41]. The energy storage can be considered as a dispatchable source/ controllable load in energy management planning and provides more flexibilities for relaxing the constraints and optimizing the objective function in energy management problems. In addition, energy storage systems can offer different services to facilitate the microgrid energy management. In the below sections, we discuss about some of these services.

3.4.2.1 Reducing the complexity of energy management problem

Microgrids noticeably rely on distributed RES, while these resources are non-dispatchable and intermittent. On the other hand, merging new dynamic loads such as electric vehicle into microgrids leads to uncertainties on the demand side. Unpredictable nature of renewable resources and load uncertainties change the microgrid energy management problem from a deterministic problem to a more challenging stochastic problem. The energy storage systems by smoothing the renewable energy power generation and managing the demand side can facilitate the planning and scheduling task in microgrids. They help with balancing demand and supply at the presence of seasonal and weekly load and generation fluctuations, intermittency of renewables, dynamic load variation, and load imbalance [42].

3.4.2.2 Time-shifting of generation and demand

Energy storage systems can shift the power demand over time. That is, they can serve as a source of energy and provide the load through discharging, and subsequently act as a load and consume the energy through charging. Energy storage systems can also shift the power generation over time. Many RES might generate considerable amount of energy when there is no demand for it, or when the financial value of it is low [43]. Thus, the energy storage can be employed to store the low-value/surplus energy generated by RES and use it when there is energy shortage or the value is high. These capabilities allow energy storage systems to offer microgrids services such as load leveling, peak shaving, energy arbitrage, and also prevent the renewable energy from being curtailed [44].

3.4.2.3 Reducing power loss and improving efficiency

The energy storage systems, if they are optimally placed within a microgrid, can maximize the local energy utilization and reduce the transferred power from the main grid, and consequently reduce the power loss over lines, and the power loss due to the line congestion. Moreover, the energy loss through conductors and electrical equipment in microgrids is proportional to the square of the current (RI^2). Therefore, the energy storage systems by distributing the total load evenly over time and preventing the microgrid from being heavily loaded can reduce the energy loss due to the peak current reduction. Reactive power compensation is another option for energy storage systems to reduce the power loss in a microgrid. Supplying reactive power locally through energy storage systems reduces the current drawn by loads from resources and decreases the power loss over lines.

3.4.2.4 Case studies

Annobon Island Microgrid in Africa is one of the largest self-sufficient, solar-based microgrids [45]. Annobon province is an island that belongs to Equatorial Guinea in west central Africa with approximately 5,000 residents. The microgrid is a solution for the residents to provide them with uninterruptable and reliable electricity. The microgrid is composed of 5 MW solar PV, 3 MW/6 MWh sodium-nickel-chloride BESS, and 5 MW diesel generators [46]. The microgrid is also equipped with advanced energy management system. The energy management system uses battery storage to prioritize the solar energy and minimize the generators operating hours by shifting the demand and generation over time. It also manages the energy storage system for optimal and cost-effective operation of the microgrid.

3.4.3 Power quality improvement

Energy storage systems can be employed within a microgrid to assist in power quality improvement. Power quality in a microgrid is subject to degradation due to load characteristics, equipment malfunctioning, switching devices operation, and generation variability. Low power quality adversely impacts on the operation, efficiency, and maintenance cost of microgrid components and loads. By providing services such as voltage and frequency support, power factor correction, phase balancing, harmonic compensation, and renewable intermittency compensation, the energy storage systems can protect on-site load against the events that impact power quality. Typically this service requires fast response energy storage systems which are able to track and compensate the high dynamic events.

3.4.3.1 Renewable intermittency compensation

A fast response energy storage system integrated with the RES, can effectively act as a buffer to compensate the intermittency of these resources [47]. When there is a sudden change in power generated by renewables, the energy storage unit can automatically smooth the variations by instantaneously exchanging energy to control the ramp rate [48]. The energy storage by limiting up and down ramps of the power production enables the injection of a smoother power profile to the microgrid. This capability increases the power quality of RES and help microgrid to manage sudden changes in renewable power. Technologies such as advanced battery, super capacitor, and flywheel are suitable for this application.

3.4.3.2 Voltage support

One of the power quality issues in microgrid is voltage sag/rise. Voltage sag happens when demand exceeds the capacity of the system to deliver the required energy. Voltage rise happens when the local generation exceeds the consumption and reverse power flow occurs. The voltage sag/rise causes malfunctioning or shutting down the electric equipment if it is sufficiently long and large. In microgrids, the line impedances might mostly be resistive and the ratio of the resistance to reactance (R/X) be considerably large. Therefore, the impact of active power on voltage is not negligible. Energy storage systems integrated into microgrid through power electronic devices can operate in four quadrants (charging, discharging, leading, or lagging). This allows them to control both active and reactive power in the microgrid to effectively regulate the voltage [36].

3.4.3.3 Power factor correction

Microgrids often require to compensate for reactance and improve power factor. Energy storage systems with power converter are well suited to provide power factor correction and offset effects from localized reactance. Distributed BESS are one example which can be deployed for enhancing the power factor in a microgrid besides the other services they might provide [49].

3.4.3.4 Phase balancing

A three-phase microgrid might contain single-phase loads, generation, and storage. Balancing the loads among phases is a requirement for safe and efficient operation of a microgrid. To this end, single-phase energy storage systems can be integrated separately to each phase and mitigate the phase imbalance by independently exchange active and reactive power in each phase [50].

3.4.3.5 Harmonic compensation

Integration of RES such as wind and solar as well as nonlinear loads such as motor drivers, fluorescent lamp, electric welding machine produces voltage, and current harmonics in the system. Energy storage systems connected to the microgrid through PWM power converters, besides their power compensation functionality, can act as active filters to compensate the harmonics [51]. That is, by proper switching of power converter the integrated storage system with power electronics produces the inverse of the existing harmonics in the system and cancel them out.

In addition, energy storage systems such as battery energy storage are utilized as UPS which provides pure sinusoidal waveform for sensitive loads.

3.4.3.6 Case studies

Borrego Springs microgrid located at a remote area about 75 miles northeast of the city of San Diego within the San Diego Gas & Electric service territory, serves 615 customers with a peak load of around 4.6 MW [52]. The microgrid is composed of two 1.8 MW diesel generators, one 500 kW–1,500 kWh substation BESS, three 25 kW–50 kWh community energy storage systems, and six 4 kW–8 kWh energy storage units at home level. The substation storage is considered for peak shaving, load following, renewable smoothing, and supporting islanding operation. The community energy storage is used for peak shaving, renewable smoothing, and voltage support.

In one application, the community energy storage smooths the intermittency of solar PV power through ramp rate control. Figure 3.9 shows the result of solar power smoothing by energy storage in Borrego Springs microgrid.

Power factor correction is another example of power quality improvement performed by substation energy storage in this microgrid [53]. According to Figure 3.10, at 11:55 PM, the diesel generators operate at power factor around 0.9. After the reactive power of the energy storage is set to 100 kVAr at 12:00 PM, the

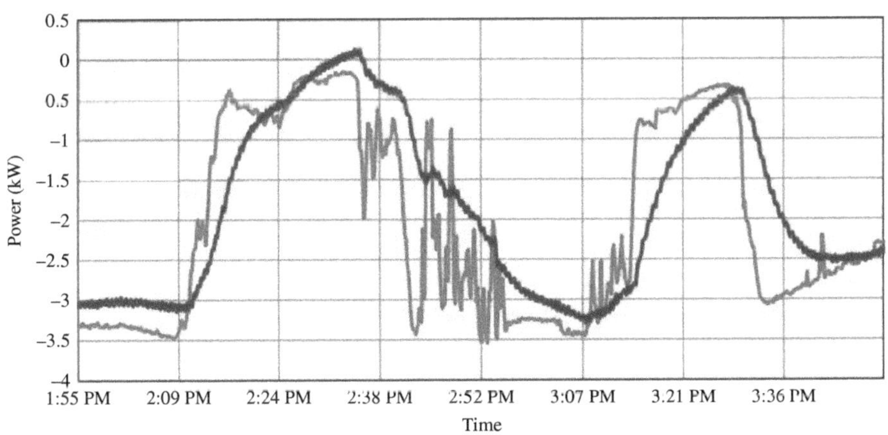

Figure 3.9 Solar power smoothing by energy storage in Borrego Springs microgrid [51]

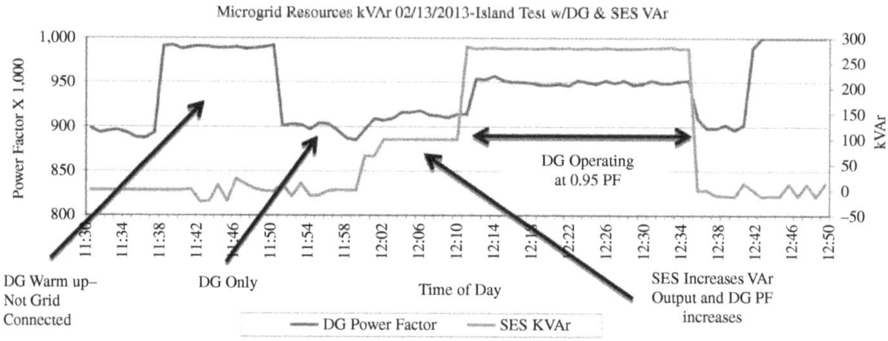

Figure 3.10 Reactive power control for power factor correction [52]

power factor increases to 0.92. Again by injecting more reactive power up to 275 kVAr into the microgrid, the generator power factor reaches 0.95 at 12:10 PM. This result shows how energy storage systems can contribute in power factor correction by exchanging the reactive power with microgrid.

3.4.4 Reliability improvement

One of the main motivations of forming microgrids is to improve the reliability and availability of electricity services. Microgrid reliability can be defined as the ability of the microgrid to meet the consumers' electricity demand in the quantity and with the quality requested by them. According to reliability indices defined by the IEEE [54] power interruption is one of the key factors in reliability degradation. The stored energy in energy storage systems offers backup power to customers, reduces their power interruption, and improves their reliability.

3.4.4.1 Ride-through and bridging

In the event of power outage, it normally takes time (seconds/minutes) for the system to transfer from the main grid to the on-site generation resources, adjust the load, and continue the normal operation. The energy storage can provide the required backup power for electricity consumers to ride-through the outage until the backup generation start up or/and the orderly shutdown of equipment is completed [55]. The energy storage systems by providing ride-through and bridging services during an islanded operation of microgrid can improve the microgrid reliability by avoiding the power interruption. Depending on whether bridging the gap from one source to another is intended or orderly shutdown of equipment, the required discharge duration of energy storage varies from a few seconds to couple of hours.

3.4.4.2 Resource adequacy

Available capacity and resource adequacy is an important factor in a microgrid that impacts on the system reliability. Microgrids normally need to maintain a generation capacity of about 115 percent of their forecasted peak demand. The common practice of providing generation capacity is to integrate more dispatchable generation. However, energy storage system is a competitive alternative to integrating new generation resources as its value stream can be expanded to other services especially at light load.

3.4.4.3 Case studies

In a microgrid project at Presidio, Texas, the local utility Electric Transmission Texas installed 4 MW sodium-sulfur BESS mainly as a backup power to improve the system reliability by mitigating short-term power outage [56]. Presidio is located at the end of 60 miles 69 kV transmission line. This line is subject to frequent storms and lightning causing power failure and reliability degradation. Since it is hard to access and repair the line, residents might experience extended power outages. To this end, the integrated energy storage provides the power up to 8 hours for the entire town at the time of power outage and provides the required time to repair the line. The main advantage of the energy storage technology used in this application is the fast response time to address the sudden power outage. The energy storage is also equipped with a fast response automatic power converter which controls the power flow of storage.

In another example, the small town called Glacier, located at the north central part of Washington State, faced frequent power outage due to being supplied by a long and exposed 55 kV transmission line [57]. This town including around 1,000 customers experienced about three outages per year, averaging 7 hours long. In one incident, the outage lasted 33.5 hours. Puget Sound Energy decided to improve the electric service reliability of this area by integration of BESS. This project involves installation of 2 MW, 4.4 MWh lithium-ion BESS to serve as backup power during outages. Figure 3.11 depicts the single-line diagram of the Glacier's power network. When the

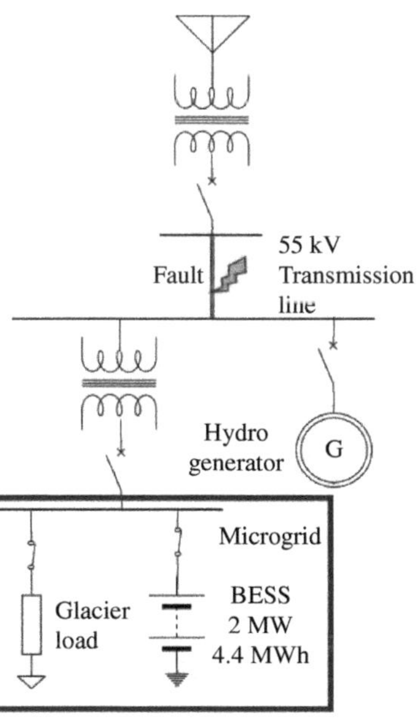

Figure 3.11 Single-line diagram of the Glacier's power circuit

fault happens at the 55 kV transmission line, the microgrid switches to the islanded mode and provides the load through BESS. The battery is able to support 2 MW of the town load for about 2.2 hours. In addition to the backup power application, the energy storage is used to reduce the peak demand, and balance energy demand and supply, helping to support greater accommodation of renewable generation.

3.4.5 Resiliency improvement

Microgrid resiliency refers to the system capability to withstand and recover from disruptive events and to minimize the duration, intensity, and the negative impacts of unfavorable events [58]. At the time of contingencies in microgrids, energy storage systems can be properly managed to control network flow, maintain supply–demand balancing, and avoid instability in the microgrid [59]. Energy storage systems owing to the amount of energy they store offer virtual inertia to microgrids, which increases the tolerance and robustness of the entire system to sudden changes. The integration of distributed energy storage systems in a microgrid enhances the flexibility of microgrid for immediate reconfiguration in response to unforeseen events. In addition, incorporation of RES like solar with energy storage systems, turns renewable generation to reliable source to provide power during grid outages.

3.4.5.1 Case studies

In response to the hurricane Sandy which caused power outage for around 8.5 million customers in the north-eastern United States in 2012, and frequent violent weather events, the Connecticut Department of Energy and Environmental Protection initiated a $48 million/three years pilot program to support development of resilient power technology. Town of Windham as part of this program is developing a microgrid that consists of 130 kW natural gas generator, 250 kW solar PV, and 200 kWh battery [60]. This microgrid supports two critical facilities: Windham Middle School and Sweeney School at the time of emergency. The microgrid is capable of providing electricity 24/7 for up to 4 weeks if the contingent event occurs.

Providing resilient power, "solar + storage" microgrid project in Rutland, Vermont, is another attempt [61]. This project is one of the first microgrids in the United States which is exclusively powered by solar. This area also suffers from frequent loss of power due to storms. The main goal of this project is to help community to cope with power outage in an unforeseen disaster. This microgrid combines 2.5 MW solar PV with 4 MW/3.4 MWh lithium-ion and lead-acid battery storage system to guarantee resilient power in case of a grid outage. It serves the resilient power to a Rutland school as a public emergency shelter.

3.5 Challenges and barriers

In this section, some of the barriers to broad deployment of energy storage system in microgrids are discussed. Addressing these challenges helps with widespread utilization of energy storage system within microgrids. These barriers can be summarized as (1) cost, (2) type, size, and placement (3) rules, regulations, and standards.

Cost: One of the main factors to promote the deployment of energy storage systems is to be cost competitive with other alternatives available to the microgrid. To this end, the value stream and extra revenue that energy storage systems can create within a microgrid should be explored and clearly quantified. Moreover, the cost of energy storage deployment should be reduced. It is being understood that the total cost of energy storage systems is limited not only to the cost of technology but also to the cost of supplementary equipment, installation, integration, and commissioning. In some cases, the secondary cost reaches up to 60–70 percent of the total cost [34], which shows the importance of targeting all aspects of energy storage deployment cost.

Type, size, and placement: The existence of a wide range of energy storage technologies in the market, each with its own characteristics, makes the energy storage users confused to select the best-suited technology for their applications. Even if they select one technology, it would be hard to properly size the required energy storage. Therefore, in order to facilitate the energy storage utilization, creating comprehensive standards to evaluate and compare the quality and performance of energy storage systems is the first step. The second step is to develop a transparent guideline to help energy storage users with optimal selection of energy storage type and size and location in order to gain the maximum benefit.

Rules, regulations, and standards: Proper rules and regulation can encourage stakeholders to invest more in the energy storage industry. This regulatory environment should provide investors with clear market models as well as incentives, while removing the regulatory restrictions which prevent stakeholders to collect revenue. Moreover, the stakeholders require clear and accurate vision about the performance of available energy storage systems before they could perform investment calculation. Therefore, in order to alleviate the existing uncertainties over the energy storage capabilities and make the stakeholders confident for investment, a unified framework for reporting and evaluation of energy storage performance, and also codes and standards which define the desired performance criteria are required.

3.6 Conclusion

This chapter presents an overview of the existing challenges in control and operation of microgrids according to the accumulated experience of pilot projects around the world. Stability, power management, power quality, and reliability are some challenges that microgrids are facing. This chapter also presents the existing energy storage technologies according to their characteristics and services that they can offer to microgrids. Different storage technologies based on rating characteristics, dynamics, and space requirements are compared. The roles of energy storage systems to improve the stability, reliability, power quality, resiliency, and also facilitating the energy management within microgrids are investigated and real case studies are presented. Finally, in this chapter, the barriers to increasing deployments of energy storage in microgrids are discussed. High cost, and lack of rules, regulation, and standards are the challenges which should be overcome to enable the widespread deployment of energy storage. Nevertheless, there is no doubt that energy storage will be an inseparable component of future microgrids, and with rapid reduction of cost and improvement of performance, new market, and applications will open for energy storage.

References

[1] D. T. Ton and M. A. Smith, "The U.S. Department of Energy's microgrid initiative," *The Electricity Journal*, vol. 25, no. 8, pp. 84–94, 2012.
[2] "Microgrids—Benefits, models, barriers and suggested policy initiatives for the Commonwealth of Massachusetts," *KEMA, Inc.*, Burlington, MA, 2014.
[3] N. Hatziargyriou, H. Asano, R. Iravani, and C. Marnay, "Microgrids," *IEEE Power and Energy Magazine*, vol. 5, no. 4, pp. 78–94, 2007.
[4] S. Parhizi, H. Lotfi, A. Khodaei, and S. Bahramirad, "State of the art in research on microgrids: A review," *IEEE Access*, vol. 3, pp. 890–925, 2015.
[5] M. Soshinskaya, W. H. Crijns-Graus, J. M. Guerrero, and J. C. Vasquez, "Microgrids: experiences, barriers and success factors," *Renewable and Sustainable Energy Reviews*, vol. 40, pp. 659–672, 2014.
[6] Q. Fu, A. Hamidi, A. Nasiri, V. Bhavaraju, S. B. Krstic, and P. Theisen, "The role of energy storage in a microgrid concept: Examining the opportunities and promise of microgrids," *IEEE Electrification Magazine*, vol. 1, no. 2, pp. 21–29, 2013.
[7] Z. Shuai, Y. Sun, Z. Shen, W. Tian, C. Tu, Y. Li, and X. Yin, "Microgrid stability: Classification and a review," *Renewable and Sustainable Energy Reviews*, vol. 58, pp. 167–179, 2016.
[8] R. Zamora and A. K. Srivastava., "Controls for microgrids with storage: Review, challenges, and research needs," *Renewable and Sustainable Energy Reviews*, vol. 14, no. 7, pp. 2009–2018, 2010.
[9] IEEE Std 399–1997, "IEEE recommended practice for industrial and commercial power systems analysis (brown book)," *IEEE Std 399-1997*, pp. 1–488, 1998 [Online]. Available: https://www.techstreet.com/ieee/standards/ieee-399-1997?gateway_code=ieee&vendor_id=613&product_id=18355#jumps
[10] IEEE Std 1547.4, "IEEE guide for design, operation, and integration of distributed resource island systems with electric power systems," *IEEE Std 1547.4-2011*, pp. 1–54, July 20, 2011 [Online]. Available: https://ieeexplore.ieee.org/document/5960751/
[11] R. Majumder, "Some aspects of stability in microgrids," *IEEE Transactions on Power Systems*, vol. 28, no. 3, pp. 3243–3252, 2013.
[12] Y. Wang, B. Wang, C.-C. Chu, H. Pota, and R. Gadh., "Energy management for a commercial building microgrid with stationary and mobile battery storage," *Energy and Buildings*, vol. 116, pp. 141–150, 2016.

[13] Y. Wang, W. Shi, B. Wang, C.-C. Chu, and R. Gadh, "Optimal operation of stationary and mobile batteries in distribution grids," *Applied Energy*, vol. 190, pp. 1289–1301, 2017.

[14] F. Nejabatkhah and Y. W. Li, "Overview of power management strategies of hybrid AC/DC microgrid," *IEEE Transactions on Power Electronics*, vol. 30, no. 12, pp. 7072–7089, 2015.

[15] B. Wang, Y. Wang, H. Nazaripouya, C. Qiu, C.-C. Chu, and R. Gadh, "Predictive scheduling framework for electric vehicles with uncertainties of user behaviors," *IEEE Internet of Things Journal*, vol. 4, no. 1, pp. 52–63, 2017.

[16] C. M. Colson and M. H. Nehrir, "A review of challenges to real-time power management of microgrids," in *2009 IEEE Power & Energy Society General Meeting*, Calgary, AB, Canada, 2009.

[17] H. Nazaripouya, B. Wang, Y. Wang, P. Chu, H. R. Pota, and R. Gadh, "Univariate time series prediction of solar power using a hybrid wavelet-ARMA-NARX prediction method," in *2016 IEEE/PES Transmission and Distribution Conference and Exposition (T&D)*, Dallas, TX, USA, 2016.

[18] F. Katiraei, R. Iravani, N. Hatziargyriou, and A. Dimeas, "Microgrids management," *IEEE Power and Energy Magazine*, vol. 6, no. 3, pp. 54–65, 2008.

[19] E. Alegria, T. Brown, E. Minear, and R. H. Lasseter, "CERTS microgrid demonstration with large-scale energy storage and renewable generation," *IEEE Transactions on Smart Grid*, vol. 5, no. 2, pp. 937–943, 2014.

[20] E. Alegria, A. Ma, and O. Idrees, "CERTS microgrid demonstration with large-scale energy storage and renewables at Santa Rita Jail," *Chevron Energy Solutions Company*, A Division of Chevron U.S.A. Inc., San Francisco, CA, 2014.

[21] E. Fuchs and M. Masoum, "*Power Quality in Power Systems and Electrical Machines*," Academic Press, London, 2011 [Online]. Available: https://www.elsevier.com/books/power-quality-in-power-systems-and-electrical-machines/fuchs/978-0-12-800782-2

[22] K. Rahimi, S. Mohajeryami, and A. Majzoobi, "Effects of photovoltaic systems on power quality," in *2016 North American Power Symposium (NAPS)*, Denver, CO, USA, 2016.

[23] A. Vinayagam, K. Swarna, S. Y. Khoo, and A. Stojcevski, "Power quality analysis in microgrid: An experimental approach," *Journal of Power and Energy Engineering*, vol. 4, no. 4, pp. 17–34, 2016.

[24] "IEEE guide for identifying and improving voltage quality in power systems," *IEEE Std 1250–2011 (Revision of IEEE Std 1250–1995)*, pp. 1–70, 2011.

[25] "IEEE recommended practice and requirements for harmonic control in electric power systems," *IEEE Std 519–2014 (Revision of IEEE Std)*, pp. 1–29, 2014.

[26] "IEEE standard for interconnecting distributed resources with electric power systems," *IEEE Std 1547–2003*, pp. 1–28, 2003.

[27] S. Kennedy and M. M. Marden, "Reliability of islanded microgrids with stochastic generation and prioritized load," in *2009 IEEE Bucharest PowerTech*, Bucharest, Romania, 2009.

[28] D. Rastler, *"Electricity energy storage technology options: A white paper primer on applications, costs, and benefits,"* Electric Power Research Institute (EPRI), Palo Alto, California, 2010.

[29] S. Vazquez, S. M. Lukic, E. Galvan, L. G. Franquelo, and J. M. Carrasco, "Energy storage systems for transport and grid applications," *IEEE Transactions on Industrial Electronics*, vol. 57, no. 12, pp. 3881–3895, 2010.

[30] H. Chen, T. Cong, W. Yang, C. Tan, Y. Li, and Y. Ding, "Progress in electrical energy storage system: A critical review," *Progress in Natural Science*, vol. 19, no. 3, pp. 291–312, 2009.

[31] J. Vinois. "DG ENER Working Paper: The future role and challenges of energy storage," European Commission, Director-General for Energy, 2012 [Online]. Available: https://ec.europa.eu/energy/sites/ener/files/energy_storage.pdf

[32] A. Castillo and D. F. Gayme, "Grid-scale energy storage applications in renewable energy integration: A survey," *Energy Conversion and Management*, vol. 87, p. 885–894, 2014.

[33] D. Akinyele and R. Rayudu, "Review of energy storage technologies for sustainable power networks," *Sustainable Energy Technologies and Assessments*, vol. 8, pp. 74–91, 2014.

[34] I. Gyuk, M. Johnson, J. Vetrano, K. Lynn, W. Parks, R. Handa, L. Kannberg, S. Hearne, K. Waldrip, and R. Braccio. "Grid energy storage," in *US Department of Energy*, 2013 [Online]. Available: https://www.energy.gov/sites/prod/files/2014/09/f18/Grid%20Energy%20Storage%20December%202013.pdf

[35] J. Eyer and G. Corey, *"Energy storage for the electricity grid: Benefits and market potential assessment guide,"* Sandia National Laboratories, Albuquerque, New Mexico and Livermore, California, 2010.

[36] H. Nazaripouya, Y. Wang, P. Chu, H. R. Pota, and R. Gadh, "Optimal sizing and placement of battery energy storage in distribution system based on solar size for voltage regulation," in *2015 IEEE Power & Energy Society General Meeting*, Denver, CO, USA, 2015.

[37] S. Lamichhane, H. Nazaripouya, and S. Mehraeen, "Micro grid stability improvements by employing storage," in *2013 IEEE Green Technologies Conference (GreenTech)*, Denver, CO, USA, 2013.

[38] T. Shintai, Y. Miura, and T. Ise, "Oscillation damping of a distributed generator using a virtual synchronous generator," *IEEE Transactions on Power Delivery*, vol. 29, no. 2, pp. 668–676, 2014.

[39] G. Holdmann, "Energy storage in Islanded Microgrids," Alaska Center for Energy and Power; University of Alaska Fairbanks, 27 September 2016.

[Online]. Available: http://www.sandia.gov/ess/docs/pr_conferences/2016/Monday_Presentations/13_Gwen_Holdmann.pdf.

[40] A. Eller and D. Gauntlett, "Energy storage trends and opportunities in emerging markets," *IFC and ESMAP*, 2017.

[41] Y. Wang, B. Wang, T. Zhang, H. Nazaripouya, C.-C. Chu, and R. Gadh, "Optimal energy management for microgrid with stationary and mobile storages," in *2016 IEEE/PES Transmission and Distribution Conference and Exposition (T&D)*, Dallas, TX, USA, 2016.

[42] H. Geramifar, M. Shahabi, and T. Barforoshi, "Coordination of energy storage systems and DR resources for optimal scheduling of microgrids under uncertainties," *IET Renewable Power Generation*, vol. 11, no. 2, pp. 378–388, 2017.

[43] A. Majzoobi and A. Khodaei, "Application of microgrids in supporting distribution grid flexibility," *IEEE Transactions on Power Systems*, vol. 32, no. 5, pp. 3660–3669, 2017.

[44] P. Denholm, M. O'Connell, G. Brinkman, and J. Jorgenson, "*Over-generation from solar energy in California: A field guide to the duck chart,*" National Renewable Energy Laboratory, Golden, CO, 2015.

[45] P. K. Ainah and K. A. Folly, "Development of micro-grid in Sub-Saharan Africa: An overview," *International Review of Electrical Engineering (I.R. E.E.)*, vol. 10, no. 5, pp. 633–645, 2015.

[46] "DOE Global Energy Storage Database," *Sandia Corporation* [Online]. Available: http://www.energystorageexchange.org/projects/1354.

[47] M. J. E. Alam, K. M. Muttaqi, and D. Sutanto, "A novel approach for ramp-rate control of solar PV using energy storage to mitigate output fluctuations caused by cloud passing," *IEEE Transactions on Energy Conversion*, vol. 29, no. 2, pp. 507–518, 2014.

[48] H. Nazaripouya, C. C. Chu, H. R. Pota, and R. Gadh, "Battery energy storage system control for intermittency smoothing using an optimized two-stage filter," *IEEE Transactions on Sustainable Energy*, vol. 9, no. 2, pp. 664–675, 2018.

[49] A. Moeini, I. Kamwa, and M. D. Montigny, "Power factor-based scheduling of distributed battery energy storage units optimally allocated in bulk power systems for mitigating marginal losses," *IET Generation, Transmission & Distribution*, vol. 10, no. 5, pp. 1304–1311, 2016.

[50] S. Sun, B. Liang, M. Dong, and J. A. Taylor, "Phase balancing using energy storage in power grids under uncertainty," *IEEE Transactions on Power Systems*, vol. 31, no. 5, pp. 3891–3903, 2016.

[51] S. W. Mohod and M. V. Aware, "A STATCOM-control scheme for grid connected wind energy system for power quality improvement," *IEEE Systems Journal*, vol. 4, no. 3, pp. 346–352, 2010.

[52] N. Bartek, *"Borrego Springs microgrid demonstration overview,"* Society of American Military Engineers, San Diego, CA, 2015.

[53] T. Bialek, *"Borrego Springs microgrid demonstration project,"* San Diego Gas & Electric, San Diego, CA, 2013.

[54] T. a. D. Committee, "IEEE guide for electric power distribution reliability indices," IEEE Std 1366, 2003.

[55] R. Arghandeh, M. Pipattanasomporn, and S. Rahman, "Flywheel energy storage systems for ride-through applications in a facility microgrid," *IEEE Transactions on Smart Grid*, vol. 3, no. 4, pp. 1955–1962, 2012.

[56] "Electric Transmission Texas: Projects & Project Plans: Presidio NaS Battery Project," *Electric Transmission Texas*, [Online]. Available: http://www.ettexas.com/projects/presnas.asp.

[57] "Glacier Battery Storage Project," *Puget Sound Energy*, 2014. [Online]. Available: https://pse.com/inyourcommunity/pse-projects/system-improvements/Documents/Glacier_Battery_Storage_factsheet.pdf.

[58] S. Chanda and A. K. Srivastava, "Defining and enabling resiliency of electric distribution systems with multiple microgrids," *IEEE Transactions on Smart Grid*, vol. 7, no. 6, pp. 2859–2868, 2016.

[59] C. Gouveia, J. Moreira, C. L. Moreira, and J. A. P. Lopes, "Coordinating storage and demand response for microgrid emergency operation," *IEEE Transactions on Smart Grid*, vol. 4, no. 4, pp. 1898–1908, 2013.

[60] "Safe Haven Microgrid™ for Windham Schools," *Town of Windham*, [Online]. Available: https://www.windhamct.com/resources/safe_haven_fact_sheet.pdf.

[61] "Stafford Hill Solar Farm and Microgrid," *Clean Energy Group*, [Online]. Available: https://www.epa.gov/sites/production/files/2016-07/documents/re_on_cl_rutland_case_study.pdf.

Chapter 4
Energy storage in electricity markets

Hrvoje Pandžić[1] and Yury Dvorkin[2]

4.1 Introduction

Unlike in vertically integrated power systems, where energy storage is coordinated with the rest of the system to minimize the overall generation cost [1], operations of energy storage in systems with electricity markets are driven by preferences of their owners [2]. The owners are typically profit-seeking entities that aim to collect the maximum profit possible for given system conditions. These preferences of the storage owners do not necessarily lead to the minimization of the overall generation cost in the system. Generally, merchant storage devices will provide services that can be monetized and that are most profitable. These include energy arbitrage, frequency regulation, reserve provision, voltage support, etc. [3, 4]. On the other hand, energy storage may be beneficial to the system in terms of deferred investment in transmission lines or generating units, reduced cycling of thermal units, reduced curtailment of renewable generation, etc. [5]. However, since energy storage does not always receive a payment for providing these services, they are merely consequences of an energy storage providing services for which it receives remuneration.

4.2 Energy storage in energy market

Energy storage can be seen as a resource that sequentially exhibits both producer and consumer behaviors at different periods of time. However, this representation is oversimplified and does not represent the effect of this resource on the system accurately [6]. As a matter of fact, using energy storage increases the overall energy consumption in the system because its roundtrip efficiency is below 100% and storing energy causes self-discharge losses. Therefore, in order to benefit from energy storage, there is a need to couple it with physical energy producers that produce sufficient electricity to cover the losses of energy storage in addition to

[1]Department of Energy and Power Systems, Faculty of Electrical Engineering and Computing University of Zagreb, Croatia
[2]Department of Electrical and Computer Engineering, Tandon School of Engineering, New York University, USA

providing electricity to end consumers. The physical energy producers can be both conventional and renewable generation resources.

During some time periods, energy is more valuable than during others, which may result in significant price differences throughout the day. If the price difference is sufficient to cover for roundtrip inefficiency and losses, energy storage will take advantage of that and charge (purchase electricity in the energy market) when the price is low, and discharge (sell electricity in the energy market) when the price is high. Depending on its capacity and network placement, energy storage operation may or may not affect market prices. If its capacity is insignificant and storage operations do not alter market prices, energy storage is considered a price taker. Modeling energy storage operations as a price taker is much easier, since the only external input parameter are forecast market prices. On the other hand, if its charging and discharging decisions affect market prices, energy storage is a price maker. This makes modeling of energy storage operation far more complex. In order to capture the impact of energy storage on market prices, one needs to consider the entire transmission network, all generation resources and demand in the system.

4.2.1 Energy storage as price taker

4.2.1.1 Deterministic model

Maximizing the profit of a price-taker energy storage reduces to identifying price differences between time periods and deriving optimal charging and discharging schedule. The deterministic profit-maximizing model is formulated after the notation:

η^{ch} efficiency of the charging process
η^{dis} efficiency of the discharging process
λ_t price in the day-ahead electricity market at time period t (€/MWh)
p_t^{ch} power charged at time period t (MWh)
p_t^{cmax} charging capacity limit (MW)
p_t^{dis} power discharged at time period t (MWh)
p_t^{dmax} discharging capacity limit (MW)
soc_t state of charge at time period t (MWh)
soc^{min} minimum storage state of charge (MWh)
soc^{max} maximum storage state of charge (MWh)
Δt length of time period, usually 1 hour (h)
x_t binary variable to prevent simultaneous charging and discharging

$$\underset{\Xi}{\text{Maximize}} \sum_{t=1}^{T} \lambda_t \cdot \left(p_t^{dis} - p_t^{ch}\right) \tag{4.1}$$

subject to:

$$p_t^{dis} \le p_t^{dmax} \cdot x_t, \qquad \forall t \le T \tag{4.2}$$

$$p_t^{ch} \le p_t^{cmax} \cdot (1 - x_t), \qquad \forall t \le T \tag{4.3}$$

$$soc_t = soc_{t-1} + \Delta t \cdot p_t^{ch} \cdot \eta^{ch} - \Delta t \cdot \frac{p_t^{dis}}{\eta^{dis}}, \qquad \forall 2 \le t \le T \tag{4.4}$$

$$soc_t = soc_0 + \Delta t \cdot p_t^{\mathrm{ch}} \cdot \eta^{\mathrm{ch}} - \Delta t \cdot \frac{p_t^{\mathrm{dis}}}{\eta^{\mathrm{dis}}}, \qquad t = 1 \tag{4.5}$$

$$soc_t \geq soc_0, \qquad t = T \tag{4.6}$$

$$soc^{\min} \leq soc_t \leq soc^{\max}, \qquad \forall t \leq T \tag{4.7}$$

where $\Xi = \{p_t^{\mathrm{ch}}, p_t^{\mathrm{dis}}, soc_t, x_t\}$.

Objective function (4.1) maximizes the overall profit based on market prices and discharging and charging quantities at each time period t. Constraint (4.2) limits the discharged energy with respect to the discharging capacity of energy storage, while constraint (4.3) limits the charged energy with respect to the charging capacity. Equation (4.4) calculates storage state of charge based on the previous state at $t-1$ and charged and discharged quantities. Using binary variable x_t in (4.2) and (4.3) makes it possible to enforce the complementarity constraint $p_t^{\mathrm{ch}} \cdot p_t^{\mathrm{dis}} = 0$, which prevents simultaneous charging and discharging of energy storage. Thus, at any point in time energy storage can be either charging ($p_t^{\mathrm{ch}} \neq 0, p_t^{\mathrm{dis}} = 0$), discharging ($p_t^{\mathrm{ch}} = 0, p_t^{\mathrm{dis}} \neq 0$), or idling ($p_t^{\mathrm{ch}} = 0, p_t^{\mathrm{dis}} = 0$). Paramters η^{ch} and η^{dis} are charging and discharging efficiencies lower than 1. This means that when charging, storage needs to purchase more electricity than charged in the storage due to charging losses. On the other hand, when discharging, more electricity is discharged from the storage than sold in the market due to discharging losses. Discharging and charging losses depend on charging and discharging currents [7] and include power converter losses and losses in the transformer connecting the storage facility to the network. Equation (4.5) refers only to the first time period, when the initial state of charge needs to be considered. The initial state of charge is normally obtained from the state of charge at the last time period of the preceding optimization horizon. In order to preserve the desired state of charge at the end of the optimization horizon, constraint (4.6) relates state of charge at the beginning and at the end of the optimization horizon. Constraint (4.7) limits the amount of energy stored with the minimum and maximum quantities.

The presented model captures the fact that energy storage bears financial losses due to charging and discharging inefficiencies. As a result, the storage owner will be paid only $\lambda_t \cdot p^{\mathrm{dis}}$ when $p^{\mathrm{dis}}/\eta^{\mathrm{dis}}$ will be physically discharged. On the other hand, energy storage will need to pay $\lambda_t \cdot p^{\mathrm{ch}}$, while only $p^{\mathrm{ch}} \cdot \eta^{\mathrm{ch}}$ will be physically charged.

From the optimization perspective, the problem in (4.1)–(4.7) is a mixed-integer linear program (MILP) due to variables x_t. Although excluding variable x_t from (4.2) and (4.3) would turn this problem into convex and thus make it more computationally tractable, there is no convex formulations preserving the complementarity constraint $p_t^{\mathrm{ch}} \cdot p_t^{\mathrm{dis}} = 0$ that guarantees physical accuracy of the optimal solution. Since variable x_t is limited to binary values, i.e. $x_t \in \{0,1\}$, the problem in (4.1)–(4.7) falls within the NP-complete class. To date, there are no efficient algorithms to solve NP-complete problems with rigor performance guarantees. However, a combination of relaxation and cutting-plane-based solution techniques can be used to accelerate computations.

4.2.1.2 Stochastic model

Model (4.1)–(4.7) is useful in case the market prices are perfectly known ahead of the market-clearing process. However, this is not the case in practice and, as a result, the deterministic model can result in much lower profit than anticipated or even losses. Therefore, one needs to appropriately consider uncertainty of energy prices. The most common technique for considering uncertainty of input parameters is stochastic optimization. The stochastic formulation of objective function (4.1) is:

$$\underset{\Xi}{\text{Maximize}} \sum_{p=1}^{P} \pi_p \cdot \sum_{t=1}^{T} \lambda_{t,p} \cdot \left(p_t^{\text{dis}} - p_t^{\text{ch}}\right) \tag{4.8}$$

The stochastic model objective function (4.8), unlike the deterministic model objective function (4.1), considers multiple price scenarios s and their probabilities π_p (sum of all probabilities has to satisfy the integrality condition, i.e. $\sum_{p=1}^{P} \pi_p = 1$). The value of the objective function obtained by stochastic formulation is the expected profit of the energy storage owner. In other words, if the results of the stochastic optimization are tested on a sufficiently large number of randomly generated samples of uncertain parameters, in average the profit will be equal to the optimized expected profit. Since energy storage submits charging/discharging bids before it knows the actual prices, variables p_t^{ch} and p_t^{dis} do not depend on the scenario. Therefore, the obtained market bidding schedule is deemed optimal for a set of scenarios, considering their probabilities, and not any of the scenarios individually. Constraints (4.2)–(4.7) remain the same in the stochastic formulation since they do not include any uncertain parameters, i.e. market prices.

4.2.1.3 Robust model

In addition to stochastic optimization, uncertainty can be modeled using the so-called robust optimization. As the name indicates, the robust approach is immune to any realization of uncertainty within a given uncertain set. The process starts in the same way as in the stochastic optimization—by deriving stochastic scenarios. However, instead of using scenarios in the model itself, they are used to determine the upper and lower bounds of the uncertainty set. This is achieved by finding the maximum and the minimum values of the uncertain parameter, i.e. market prices, at each time period, as shown in Figure 4.1. In case many scenarios are available and used, the upper and lower bounds may be very far and the uncertain interval could be very wide. In this case, it makes sense to leave some of the very outlying scenarios outside the uncertain interval in order to reduce it. For example, setting the bounds at 5th and 95th percentile would leave 5% of scenarios below the lower bound and 95% of scenarios below the upper bound. These bounds of uncertainty are then used in the optimization model instead of stochastic scenarios.

As opposed to stochastic optimization techniques, where boundary scenarios with low probability have low effect on the results, robust optimization does not

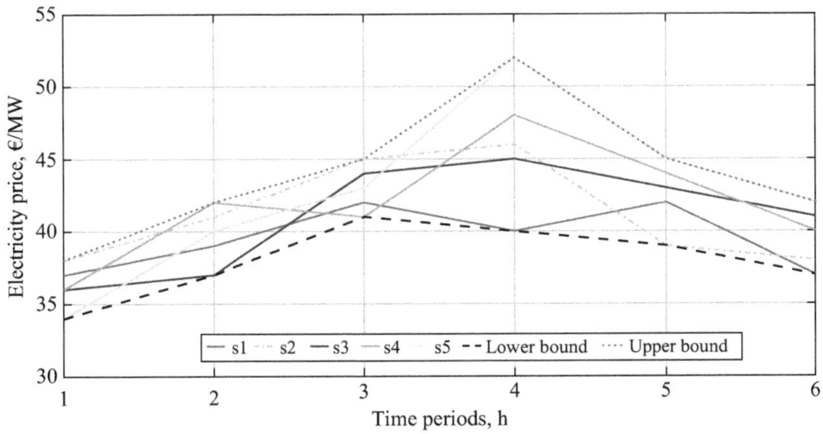

Figure 4.1 Definition of the upper and lower bounds in robust optimization

consider uncertainty distribution and hence considers equally all points within the range of uncertainty. Furthermore, robust optimization is focused on the worst-case realization of uncertainty as the objective function is to maximize the profit under the worst possible outcome of uncertainty. In other words, the objective function is formulated as in the deterministic model, objective function (4.1), but instead of the deterministic *scenario*, the model maximizes the objective function over the worst-case *scenario*.

In order to control conservativeness of the solution, the robust model contains a control parameter, usually denoted as Γ, which takes values in the range $[0,|J||$, where $J = \{j|d_j > 0\}$ Parameter d_j represents deviation from the preferred value. For example, let us consider market price estimated to be between 30 and 35 €/MWh at a specific hour. If energy storage is discharging, the preferred value is 35 €/MWh since it generates higher income. On the other hand, if energy storage is charging, it prefers to charge at a lower cost, so the preferred price is 30 €/MWh. In both cases, $d_j = 5$. In case of charging, $\Gamma = 0$ represents the most optimistic case, 30 €/MWh, and if $\Gamma = 1$, the most pessimistic case is considered, i.e. 35 €/MWh. Analogously, in case of discharging, $\Gamma = 0$ indicates the most optimistic case, i.e. discharging at 35 €/MWh, while $\Gamma = 1$ considers discharging at 30 €/MWh, which is the most conservative case. In case 24-hour time horizon is considered, Γ can take values between 0 and 24, denoting the number of hours with conservative market price deviations.

The robust energy storage offering model is formulated as:

$$\underset{\Xi}{\text{Maximize}} \sum_{t=1}^{T} \lambda_t \cdot (p_t^{\text{dis}} - p_t^{\text{ch}}) - z^{\text{dis}} \cdot \Gamma^{\text{dis}} - \sum_{t=1}^{T} q_t^{\text{dis}} - z^{\text{ch}} \cdot \Gamma^{\text{ch}} - \sum_{t=1}^{T} q_t^{\text{ch}} \quad (4.9)$$

subject to

$$(2) - (7) \tag{4.10}$$

$$z^{\text{dis}} + q_t^{\text{dis}} \geq d_t \cdot y_t^{\text{dis}}, \qquad \forall t \leq T \tag{4.11}$$

$$p_t^{\text{dis}} \leq y_t^{\text{dis}}, \qquad \forall t \leq T \tag{4.12}$$

$$z^{\text{ch}} + q_t^{\text{ch}} \geq d_t \cdot y_t^{\text{ch}}, \qquad \forall t \leq T \tag{4.13}$$

$$p_t^{\text{ch}} \leq y_t^{\text{ch}}, \qquad \forall t \leq T \tag{4.14}$$

$$z^{\text{dis}}, q_t^{\text{dis}}, y_t^{\text{dis}}, z^{\text{ch}}, q_t^{\text{ch}}, y_t^{\text{ch}} \geq 0, \qquad \forall t \leq T \tag{4.15}$$

The above formulation is obtained using methodology from [8], which exploits dual properties of the cost coefficient deviations and uncertainty control level. Similar formulation is presented in [9], which considers optimal bidding of thermal-generating units. Constraints (4.11) and (4.12) are used to model the uncertainty of market prices at discharging. Variables z^{dis} and q_t^{dis} are dual variables of the original deterministic formulation (4.1)–(4.7). Variable y_t^{dis} is used for linearization purposes. Equivalently, constraints (4.13) and (4.14) are used to model market price uncertainty while charging, with corresponding charging dual and auxiliary variables. Both discharging and charging dual and auxiliary variables are nonnegative, as imposed in (4.15).

4.2.2 *Numerical example of energy storage as price taker*

As an illustrative example, we model energy storage with 20 MWh and 10 MW capacity acting in the day-ahead energy market as a price taker. Both charging and discharging efficiencies are 0.95, minimum state of charge is 0.2, and maximum state of charge is 0.95. Initial state of charge is 0.5 and we enforce the same state of charge at the end of the optimization horizon. As in most energy markets, we observe 1-hour time periods and 24-hour market clearing horizon. Forecasted market prices are provided in Table 4.1. In the following subsections, we use deterministic, stochastic, and robust models to derive optimal bidding schedule. General Algebraic Modeling System (GAMS) codes for all three formulations of this example are provided in the Appendix section.

4.2.2.1 Deterministic optimization

In deterministic optimization, we only consider a single scenario (Scen. 1 from Table 4.1). The overall storage profit throughout the day is €470. State of charge and charging/discharging actions of energy storage are shown in Figure 4.2. Since the prices are considered to be perfectly known, storage is taking advantage of all changes in prices. At hour 2, storage is slightly discharged at €25 and then is fully charged at €22, which is the lowest price throughout the day. Storage is again discharged at full power at hour 8, which has a small price spike, and then is charged at the maximum rate in the following hour at €34, which is a local price minimum. Storage is then discharged at high-price hours 12 and 14, charged at low-price hours

Table 4.1 Five market price scenarios (used for stochastic optimization) and their minimum and maximum values (used for robust optimization).

Hour	Scen. 1	Scen. 2	Scen. 3	Scen. 4	Scen. 5	Min.	Max.
1	25	26	24	29	23	23	29
2	22	23	21	28	22	21	28
3	24	25	22	22	25	22	25
4	27	29	28	26	30	26	30
5	29	33	31	35	34	29	35
6	27	26	30	31	26	26	31
7	33	34	35	29	33	29	35
8	39	37	38	35	37	35	39
9	34	36	33	37	37	33	37
10	41	38	39	40	42	38	42
11	44	45	42	43	44	42	45
12	48	47	49	45	47	45	49
13	45	48	49	52	51	45	52
14	47	49	46	52	47	46	52
15	40	44	41	49	39	39	49
16	37	40	38	39	40	37	40
17	38	39	40	42	44	38	44
18	44	41	45	42	49	41	49
19	51	48	49	50	52	48	52
20	57	55	59	55	58	55	59
21	55	56	58	60	57	55	60
22	50	52	48	47	51	47	52
23	39	40	39	42	42	39	42
24	36	37	40	35	37	35	40

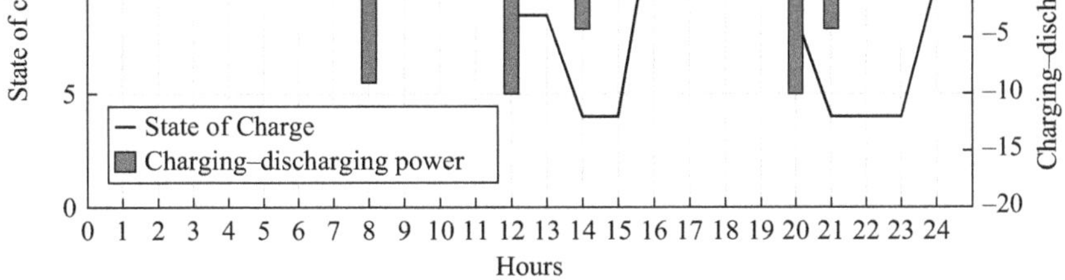

Figure 4.2 State of charge and charging/discharging schedule for deterministic formulation

16 and 17, and discharged again at spike-price hours 20 and 21. In order to reach the initial value, the storage is charged to 0.5 state of charge in the last hour. In the first half of the day, the storage undergoes two half-cycles of charging/discharging because the power capacity is insufficient to fully charge/discharge the storage during a single hour. In the afternoon, the storage charges and discharges two hours in a row, which enables a full charging/discharging cycle. Since deterministic optimization implies perfect information on electricity prices at the day ahead, this result is the upper bound on the profit that energy storage can collect.

4.2.2.2 Stochastic optimization

Stochastic optimization acknowledges imperfect information and addresses this issue by using multiple scenarios with corresponding probabilities. In this example, we use five scenarios from Table 4.1, each with 0.2 probability. The resulting expected profit of energy storage is €444, which is 5.5% lower than in the deterministic case. State of charge and charging/discharging actions of energy storage are shown in Figure 4.3. The obtained schedule is slightly different than in the deterministic case. For instance, in deterministic case, the discharging at hour 8 is followed by charging at hour 9 due to €5 difference in electricity price. However, stochastic optimization results in a schedule where discharging occurs earlier, at hour 5, also followed by charging in the following hour. This occurs because scenarios 2–5 have significantly higher prices at hour 5 and lower prices at hour 8 as compared to scenario 1. For the same reason, the discharging in the stochastic model is moved from hour 12 to hour 13.

4.2.2.3 Robust optimization

Results of the robust optimization model depend on the choice of uncertainty parameters Γ^{ch} and Γ^{dis}, as shown in Figure 4.4. For $\Gamma^{ch} = 0$ and $\Gamma^{dis} = 0$, the obtained schedule is the most optimistic. This means that the storage expects to be charged at the lowest cost possible (penultimate column in Table 4.1) and discharged at the

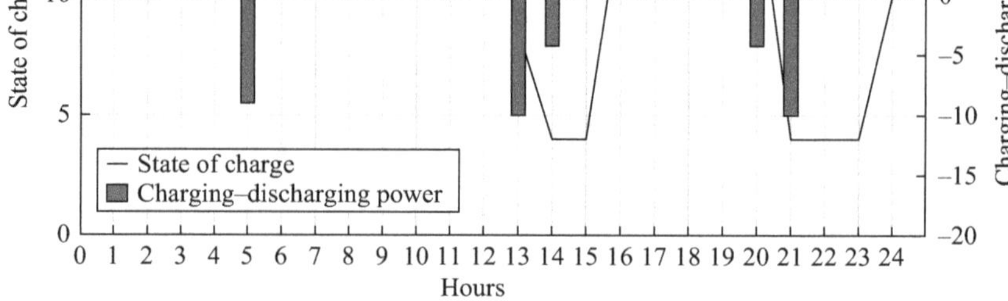

Figure 4.3 State of charge and charging/discharging schedule for stochastic formulation

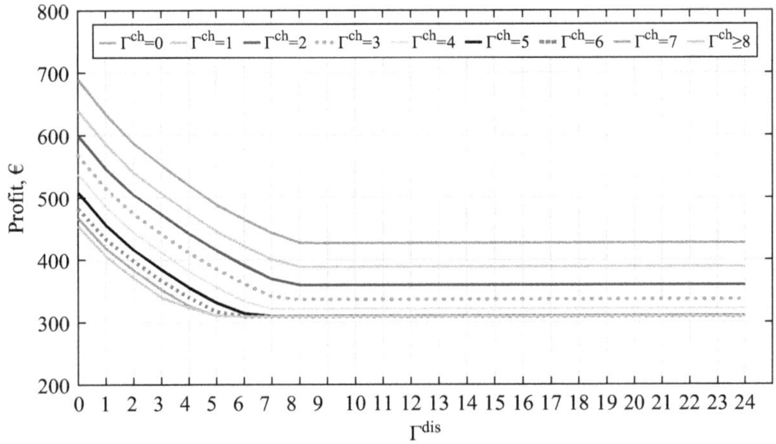

Figure 4.4 Profit when using the robust formulation for different Γ^{ch} and Γ^{dis}

highest cost possible (last column in Table 4.1). The solution is even more optimistic than the deterministic solution and results in €689 profit. If Γ^{ch} is kept at zero, and Γ^{dis} is increased (see Figure 4.4), profit decreases because the price for more and more hours is considered to be at lower bound instead of the upper bound when discharging. The resulting curve is the same for all $\Gamma^{dis} \geq 9$, because the model would not consider discharging during more than 9 h due to the price differences among hours. Therefore, setting the discharge price to the lower bound at more than 9 h does not change the charging/discharging schedule. This saturation point moves to the left as Γ^{ch} increases, and for $\Gamma^{ch} \geq 7$ the saturation point appears at hour six.

A similar behavior is observed when Γ^{dis} is kept at zero and Γ^{ch} increases. For $\Gamma^{ch} \geq 8$ the charging/discharging schedule is the same as the model does not benefit from extra hours of charging.

Here, an important question arises: which Γ^{dis} and Γ^{ch} should be considered at the day-ahead stage? There is no optimal way to determine this. In order to properly assess the solutions, one needs to generate a large number, e.g. 1,000, of out-of-sample scenarios that represent possible realizations of uncertainty. Then, all of the solutions obtained for different Γ^{dis} and Γ^{ch} are applied to each out-of-sample scenario and actual profit in each of those cases is calculated. Then, for each combination of Γ^{dis} and Γ^{ch}, a cumulative distribution function can be derived and the best schedule can be identified. This procedure, just applied to the unit commitment under uncertainty problem, is well elaborated in [10].

4.2.3 Energy storage as price maker

If capacity of energy storage is large, i.e. its quantities can affect market prices, the entire market clearing procedure needs to be modeled. This results in a bilevel problem shown in Figure 4.5. The upper-level problem aims to maximize the storage profit, while the lower-level problem simulates market clearing, i.e. maximizes social welfare, considering all market participants. This problem is formulated after a notation, where all the parameters and variables that appear in the model are defined.

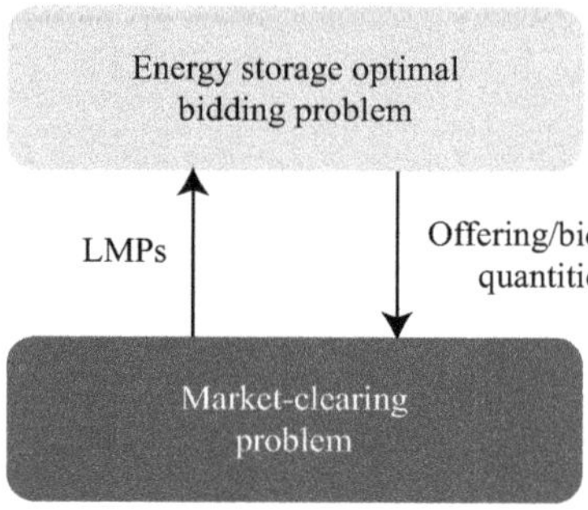

Figure 4.5 Interaction between energy storage and electricity market

4.2.3.1 Notations

Parameters:

$B_{n,\ m}$	susceptance of line l connecting nodes n and m (S)
$D_{t,n,c}^{\max}$	demand block c at bus n during time period t (MW)
$G_{i,b}^{\max}$	capacity of generation block b of generator i (MW)
p^{cmax}	charging capacity limit (MW)
p^{dmax}	discharging capacity limit (MW)
$PF_{l}^{\max}$	maximum power flow through line l (MW)
$SoC^{\min}$	minimum storage state of charge (MWh)
$SoC^{\max}$	maximum storage state of charge (MWh)
Δt	length of time period, usually 1 hour (h)
$WF_{t,w}$	available wind generation at wind farm w during time period t (MW)
η^{ch}	efficiency of storage charging process
η^{dis}	efficiency of storage discharging process
λ_{s}^{CH}	offering price of energy storage s (€/MWh)
$\lambda_{n,c}^{D}$	bidding price of demand block c at bus n (€/MWh)
$\lambda_{s}^{\mathrm{DIS}}$	bidding price of energy storage s (€/MWh)
$\lambda_{i,b}^{\mathrm{G}}$	offering price of generation block b of generator i (€/MWh)

Variables:

$d_{t,n,c}$	cleared demand block c at bus n during time period t (MW)
$g_{t,\ i,\ b}$	output of generator block b of generator i during time period t (MW)
$p_{t,s}^{\mathrm{ch}}$	power charged to storage s at time period t (MW)
$p_{t,s}^{\mathrm{dis}}$	power discharged from storage s at time period t (MW)
$pf_{t,l}$	power flow through line l during time period t (MW)
$SoC_{t,s}$	state of charge of storage s at time period t (MWh)
$wg_{t,w}$	output of wind farm w during time period t (MW)
$\theta_{t,n}$	voltage angle at bus n during time period t (rad)

4.2.3.2 Mathematical formulation

$$\underset{\Xi^{\mathrm{UL}}}{\text{Maximize}} \sum_{t=1}^{T}\sum_{s=1}^{S} \lambda_{t,n|s} \cdot \left(p_{t,s}^{\mathrm{dis}} - p_{t,s}^{\mathrm{ch}}\right) \tag{4.16}$$

subject to:

$$SoC_{t,s} = SoC_{t-1,s} + \Delta t \cdot p_{t,s}^{\mathrm{ch}} \cdot \eta_s^{\mathrm{ch}} - \Delta t.\frac{p_{t,s}^{\mathrm{dis}}}{\eta_s^{\mathrm{dis}}} \qquad \forall t \in T, s \in S \tag{4.17}$$

$$SoC_s^{\min} \le SoC_{t,s} \le SoC_s^{\max} \qquad \forall t \in T, s \in S \tag{4.18}$$

$$\begin{aligned}\underset{\Xi^{\mathrm{LL}}}{\text{Maximize}} \quad & \sum_{t=1}^{T}\sum_{n=1}^{N}\sum_{c=1}^{C}\lambda_{n,c}^{\mathrm{D}} \cdot d_{t,n,c} + \sum_{t=1}^{T}\sum_{s=1}^{S}\lambda_s^{\mathrm{CH}} \cdot p_{t,s}^{\mathrm{ch}} \\ & -\sum_{t=1}^{T}\sum_{i=1}^{I}\sum_{b=1}^{B}\lambda_{i,b}^{\mathrm{G}} \cdot g_{t,i,b} - \sum_{t=1}^{T}\sum_{s=1}^{S}\lambda_s^{\mathrm{DIS}} \cdot p_{t,s}^{\mathrm{dis}}\end{aligned} \tag{4.19}$$

subject to:

$$\begin{aligned}&-\sum_{w=1}^{W|N} wg_{t,w} - \sum_{s=1}^{S|N} p_{t,s}^{\mathrm{dis}} - \sum_{i=1}^{I|N}\sum_{b=1}^{B} g_{t,i,b} + \sum_{l=1}^{L|N} pf_{t,l}^{-} - \sum_{l=1}^{L|N} pf_{t,l}^{+} \\ &+\sum_{n=1}^{N}\sum_{c=1}^{C} d_{t,n,c} + \sum_{s=1}^{S|N} p_{t,s}^{\mathrm{ch}} = 0 \quad : \lambda_{t,n} \quad \forall t \in T, n \in N\end{aligned} \tag{4.20}$$

$$pf_{t,l} = B_{n,m} \cdot \left(\theta_{t,n} - \theta_{t,m}\right) \qquad : \beta_{t,l} \qquad \forall t \in T, l \in L, l \in \{n,m\} \tag{4.21}$$

$$-PF_l^{\max} \le pf_{t,l} \le PF_l^{\max} \qquad : \beta_{t,l}^{\min}, \beta_{t,l}^{\max} \qquad \forall t \subset T, l \subset L \tag{4.22}$$

$$0 \le g_{t,i,b} \le G_{i,b}^{\max} \qquad : \gamma_{t,i,b}^{\min}, \gamma_{t,i,b}^{\max} \qquad \forall t \in T, i \in I, b \in B \tag{4.23}$$

$$0 \le d_{t,n,c} \le D_{t,n,c}^{\max} \qquad : \sigma_{t,n,c}^{\min}, \sigma_{t,n,c}^{\max} \qquad \forall t \in T, n \in N, c \in C \tag{4.24}$$

$$0 \le p_{t,s}^{\mathrm{dis}} \le P_s^{\mathrm{dmax}} \qquad : \phi_{t,s}^{\mathrm{dmin}}, \phi_{t,s}^{\mathrm{dmax}} \qquad \forall t \in T, s \in S \tag{4.25}$$

$$0 \le p_{t,s}^{\mathrm{ch}} \le P_s^{\mathrm{cmax}} \qquad : \phi_{t,s}^{\mathrm{cmin}}, \phi_{t,s}^{\mathrm{cmax}} \qquad \forall t \in T, s \in S \tag{4.26}$$

$$0 \le wg_{t,w} \le WF_{t,w} \qquad : \eta_{t,w}^{\min}, \eta_{t,w}^{\max} \qquad \forall t \in T, w \in W \tag{4.27}$$

$$-\pi \le \theta_{t,n} \le \pi \qquad : \mu_{t,n}^{\max}, \mu_{t,n}^{\max} \qquad \forall t \in T, n \in N \tag{4.28}$$

$$\theta_{t,n_1} = 0 \qquad : \mu_t^{\mathrm{ref}} \qquad \forall t \in T \tag{4.29}$$

Objective function (4.16) maximizes the storage profit. Locational marginal price (LMP) at bus n to which storage s is connected is equal to $\lambda_{t,s}$, the dual variable of the power balance equation (4.20). Profit comes from the difference of the LMPs when storage purchases and sells electricity in the market. Therefore, the revenue is highly dependent on the LMP profile at the connecting bus.

Objective function (4.16) is subject to the upper-level constraints representing storage operation (4.17)–(4.18) and the lower-level problem simulating market-clearing procedure (4.19)–(4.29).

Constraint (4.17) determines the current state of charge based on its value in the previous time period, as well as charging and discharging amounts during the current time period. This equation considers both charging and discharging efficiencies. Constraint (4.18) limits the energy storage state of charge to take values between minimum and maximum values.

Objective function of the lower-level problem (4.19) maximizes the social welfare. A DC linear approximation of the network is used to represent nodal power balance and transmission line capacity limits. Equation (4.20) enforces nodal power balance. Wind generation, energy storage discharge, generator output, and line inflows need to be balanced with the load, line outflows, and storage charge at each bus and at each time period. Power flows through the lines are calculated in (4.21) and limited in (4.22). Constraint (4.23) imposes generator offering block limits, while (4.24) limits demand-bidding blocks. Constraints (4.25) and (4.26) limit storage offering/bidding blocks, while constraint (4.27) limits the power sold by wind farms to their forecasted generation. Since wind farms are considered to offer at 0 €/MWh, their offers do not appear in the objective function of the lower-level problem. Consequently, the model will strive to use as much free wind power as possible to maximize the social welfare. Constraint (4.28) limits voltage angles for each node, and constraint (4.29) sets the voltage angle of the reference bus to zero.

This bilevel problem cannot be solved directly, so we apply Karush–Kuhn–Tucker (KKT) conditions to the lower-level problem. In order to apply KKT conditions, we define the standard form as a maximization problem with equality constraints and inequality constraints in $\leq$ form:

$$\underset{x}{\text{Maximize}} f(x) \tag{4.30}$$

$$g_i(x) \leq b_i \qquad \forall i = 1, \ldots, m \qquad \mu_i \tag{4.31}$$

$$h_j(x) = d_j \qquad \forall j = 1, \ldots, l \qquad \lambda_j \tag{4.32}$$

If a problem contains $\geq$ constraints, these can be easily converted into $\leq$ constraints by multiplying both the right- and left-hand sides of these constraints by -1. Also, a minimization problem is easily converted in a maximization problem by multiplying the objective function with -1

Lagrangian of a standard form problem is formulated as:

$$\mathcal{L}(x, \lambda, \mu) = f(x) + \sum_{i=1}^{m} \mu_i \cdot (b_i - g_i(x)) + \sum_{j=1}^{l} \lambda_j \cdot (d_j - h_j(x)) \tag{4.33}$$

The solution to the original problem has to satisfy some necessary conditions:

Stationarity conditions

$$\Delta f(x) = \sum_{i=1}^{m} \mu_i \cdot \nabla g_i(x) + \sum_{j=1}^{l} \lambda_j \cdot \nabla h_j(x) \tag{4.34}$$

Primal feasibility conditions

$$g_i(x) \le b_i \qquad \forall i = 1, \ldots, m \qquad : \mu_i \tag{4.35}$$

$$h_j(x) = d_j \qquad \forall j = 1, \ldots, l \qquad : \lambda_j \tag{4.36}$$

Dual feasibility conditions

$$\mu_i \ge 0 \qquad \forall i = 1, \ldots, m \tag{4.37}$$

Complementary slackness

$$\mu_i \quad \perp \quad b_i - g_i(x) \qquad \forall i = 1, \ldots, m \tag{4.38}$$

Applying the KKT conditions to a bilevel problem results in a mathematical problem with equilibrium constraints (MPEC). This means that the original energy storage optimal bidding problem is reformulated as:

$$\underset{\Xi^{\mathrm{UL}}}{\text{Maximize}} \sum_{t=1}^{T} \sum_{s=1}^{S} \lambda_{t,n|s} \cdot \left(p_{t,s}^{\mathrm{dis}} - p_{t,s}^{\mathrm{ch}} \right) \tag{4.39}$$

subject to:

$$\text{Upper} - \text{level constraints } (4.17) - (4.18) \tag{4.40}$$

Stationarity constraints

$$d\mathcal{L}/dg_{t,i,b} \rightarrow -\lambda_{i,b}^{\mathrm{G}} + \lambda_{t,n|i} + \gamma_{t,i,b}^{\min} - \gamma_{t,i,b}^{\max} = 0 \qquad \forall t \in T, i \in I, b \in B \tag{4.41}$$

$$d\mathcal{L}/dd_{t,n,c} \rightarrow \lambda_{n,c}^{\mathrm{D}} - \lambda_{t,n} + \sigma_{t,n,c}^{\min} - \sigma_{t,n,c}^{\max} = 0 \qquad \forall t \in T, n \in N, c \in C \tag{4.42}$$

$$d\mathcal{L}/dp_{t,s}^{dis} \rightarrow -\lambda_{t,s}^{\mathrm{DIS}} + \lambda_{t,n|s} + \phi_{t,s}^{\mathrm{dmin}} - \phi_{t,s}^{\mathrm{dmax}} = 0 \qquad \forall t \in T, s \in S \tag{4.43}$$

$$d\mathcal{L}/dp_{t,s}^{ch} \rightarrow \lambda_{t,s}^{\mathrm{CH}} - \lambda_{t,n|s} + \phi_{t,s}^{\mathrm{cmin}} - \phi_{t,s}^{\mathrm{cmax}} = 0 \qquad \forall t \in T, s \in S \tag{4.44}$$

$$d\mathcal{L}/dpf_{t,l} \rightarrow -\lambda_{t,n|l} - \beta_{t,l} + \beta_{t,l}^{\min} - \beta_{t,l}^{\max} = 0 \qquad \forall t \in T, l \in L \tag{4.45}$$

$$d\mathcal{L}/d\theta_{t,n} \rightarrow \beta_{t,l} \cdot B_{n,m} - \mu_t^{\mathrm{ref}} + \mu_{t,n}^{\min} - \mu_{t,n}^{\max} = 0 \qquad \forall t \in T, n \subset N, l \in L|N \tag{4.46}$$

Primal feasibility – equalities

$$- \sum_{w=1}^{W|N} wg_{t,w} - \sum_{s=1}^{S|N} p_{t,s}^{\mathrm{dis}} - \sum_{i=1}^{I|N} \sum_{b=1}^{B} g_{t,i,b} + \sum_{l=1}^{L|N} pf_{t,l}^{-} - \sum_{l=1}^{L|N} pf_{t,l}^{+} + \sum_{n=1}^{N} \sum_{c=1}^{C} d_{t,n,c} + \sum_{s=1}^{S|N} p_{t,s}^{\mathrm{ch}} = 0 \qquad \forall t \in T, n \in N \tag{4.47}$$

$$pf_{t,l} = B_{n,m} \cdot \left(\theta_{t,n} - \theta_{t,m} \right) \qquad \forall t \in T, \{n,m\} \in L, l \in \{n,m\} \tag{4.48}$$

$$\theta_{t,n_1} = 0 \qquad \forall t \in T \tag{4.49}$$

Primal feasibility – inequalities + dual feasibility + complementary slackness

$$0 \leq PF_l^{\max} + pf_{t,l} \quad \perp \quad \beta_{t,l}^{\min} \geq 0 \quad \forall t \in T, l \in L \tag{4.50}$$

$$0 \leq PF_l^{\max} - pf_{t,l} \quad \perp \quad \beta_{t,l}^{\max} \geq 0 \quad \forall t \in T, l \in L \tag{4.51}$$

$$0 \leq g_{t,i,b} \quad \perp \quad \gamma_{t,i,b}^{\min} \geq 0 \quad \forall t \in T, i \in I, b \in B \tag{4.52}$$

$$0 \leq G_{i,b}^{\max} - g_{t,i,b} \quad \perp \quad \gamma_{t,i,b}^{\min} \geq 0 \quad \forall t \in T, i \in I, b \in B \tag{4.53}$$

$$0 \leq d_{t,n,c} \quad \perp \quad \sigma_{t,n,c}^{\min} \geq 0 \quad \forall t \in T, n \in N, c \in C \tag{4.54}$$

$$0 \leq D_{t,n,c}^{\max} - d_{t,n,c} \quad \perp \quad \sigma_{t,n,c}^{\max} \geq 0 \quad \forall t \in T, n \in N, c \in C \tag{4.55}$$

$$0 \leq p_{t,s}^{\text{dis}} \quad \perp \quad \phi_{t,s}^{\text{dmin}} \geq 0 \quad \forall t \in T, s \in S \tag{4.56}$$

$$0 \leq p_s^{\text{dmax}} - p_{t,s}^{\text{dis}} \quad \perp \quad \phi_{t,s}^{\text{dmax}} \geq 0 \quad \forall t \in T, s \in S \tag{4.57}$$

$$0 \leq p_{t,s}^{\text{ch}} \quad \perp \quad \phi_{t,s}^{\text{cmin}} \geq 0 \quad \forall t \in T, s \in S \tag{4.58}$$

$$0 \leq p_s^{\text{cmax}} - p_{t,s}^{\text{ch}} \quad \perp \quad \phi_{t,s}^{\text{cmax}} \geq 0 \quad \forall t \in T, s \in S \tag{4.59}$$

$$0 \leq wg_{t,w} \quad \perp \quad \eta_{t,w}^{\min} \geq 0 \quad \forall t \in T, w \in W \tag{4.60}$$

$$0 \leq WF_{t,w} - wg_{t,w} \quad \perp \quad \eta_{t,w}^{\max} \geq 0 \quad \forall t \in T, w \in W \tag{4.61}$$

$$0 \leq \pi + \theta_{t,n} \quad \perp \quad \mu_{t,n}^{\min} \geq 0 \quad \forall t \in T, n \in N \tag{4.62}$$

$$0 \leq \pi - \theta_{t,n} \quad \perp \quad \mu_{t,n}^{\max} \geq 0 \quad \forall t \in T, n \in N \tag{4.63}$$

4.2.3.3 Linearization

Since the objective function (4.39) contains the product of the dual variable ($\lambda_{t,n|s}$) of the power balance constraint (4.20) and energy storage charging and discharging quantities $p_{t,s}^{\text{dis}}$ and $p_{t,s}^{\text{ch}}$, it cannot be solved directly using off-the-shelf solvers. To linearize this objective function, we use some of the KKT conditions. Using (4.43) and (4.44), the objective function is transformed to:

$$\sum_{t=1}^{T}\sum_{s=1}^{S}\left(\lambda_{t,s}^{\text{DIS}} - \phi_{t,s}^{\text{dmin}} + \phi_{t,s}^{\text{dmax}}\right) \cdot p_{t,s}^{\text{dis}} + \sum_{t=1}^{T}\sum_{s=1}^{S}\left(-\lambda_{t,h}^{\text{CH}} - \phi_{t,s}^{\text{cmin}} + \phi_{t,s}^{\text{cmax}}\right) \cdot p_{t,s}^{\text{CH}} \tag{4.64}$$

Since $\phi_{t,n}^{\text{dmin}} \cdot p_{t,s}^{\text{dis}} = 0$ and $\phi_{t,s}^{\text{cmin}} \cdot p_{t,s}^{\text{ch}} = 0$, the following holds:

$$\lambda_{t,n|s} \cdot \left(p_{t,s}^{\text{dis}} - p_{t,s}^{\text{ch}}\right) = \lambda_{t,s}^{\text{DIS}} \cdot p_{t,s}^{\text{dis}} + \phi_{t,s}^{\text{dmax}} \cdot p_{t,s}^{\text{dis}} - \lambda_{t,s}^{\text{CH}} \cdot p_{t,s}^{\text{ch}} + \phi_{t,s}^{\text{cmax}} \cdot p_{t,s}^{\text{ch}} \tag{4.65}$$

After expressing complementarity slackness constraint (4.57) as $\phi_{t,s}^{\text{dmax}} \cdot (p_s^{\text{dmax}} - p_{t,s}^{\text{dis}}) = 0$, we obtain $\phi_{t,s}^{\text{dmax}} \cdot p_{t,s}^{\text{dis}} = \phi_{t,s}^{\text{dmax}} \cdot p_s^{\text{dmax}}$. Similarly, complementarity slackness constraint (4.59) yields $\phi_{t,s}^{\text{cmax}} \cdot (p_s^{\text{cmax}} - p_{t,s}^{\text{ch}}) = 0 \rightarrow \phi_{t,s}^{\text{cmax}} \cdot p_{t,s}^{\text{ch}} = \phi_{t,s}^{\text{cmax}} \cdot p_s^{\text{cmax}}$. Therefore, the objective function can be rewritten as:

$$\text{Maximize} \sum_{t=1}^{T} \sum_{s=1}^{S} \lambda_{t,s}^{\text{DIS}} \cdot p_{t,s}^{\text{dis}} + \phi_{t,s}^{\text{dmax}} \cdot p_s^{\text{dmax}} - \lambda_{t,s}^{\text{CH}} \cdot p_{t,s}^{\text{ch}} + \phi_{t,s}^{\text{cmax}} \cdot p_s^{\text{cmax}} \tag{4.66}$$

The complementarity conditions are linearized using the big M reformulation:

$$a \perp b \quad \rightarrow \quad a \leq M \cdot x \quad \text{and} \quad b \leq M \cdot (1 - x) \tag{4.67}$$

Complementary slackness conditions are linearized at the expense of introducing binary variables. In order to avoid this, we can use strong duality equality instead.

$$\begin{aligned}
&\sum_{n=1}^{N} \sum_{c=1}^{C} \lambda_{n,c}^{\text{D}} \cdot d_{t,n,c} + \sum_{s=1}^{S} \lambda_{t,s}^{\text{CH}} \cdot p_{t,s}^{\text{ch}} - \sum_{i=1}^{I} \sum_{b=1}^{B} \lambda_{i,b}^{\text{G}} \cdot g_{t,i,b} - \sum_{s=1}^{S} \lambda_{t,s}^{\text{dis}} \cdot p_{t,s}^{\text{dis}} \\
&= \sum_{l=1}^{L} \left(\beta_{t,l}^{\min} \cdot PF_l^{\max} + \beta_{t,l}^{\min} \cdot PF_l^{\max} \right) + \sum_{i=1}^{I} \sum_{b=1}^{B} \gamma_{t,i,b}^{\max} \cdot G_{i,b}^{\max} \\
&+ \sum_{n=1}^{N} \sum_{c=1}^{C} \sigma_{t,n,c}^{\max} \cdot D_{t,n,c}^{\max} + \sum_{s=1}^{S} \left(\phi_{t,s}^{\text{dmax}} \cdot p_s^{\text{dmax}} + \phi_{t,s}^{\text{cmax}} \cdot p_s^{\text{cmax}} \right) \\
&+ \sum_{w=1}^{W} \eta_{t,w}^{\max} \cdot WF_{t,w} + \sum_{n=1}^{N} \left(\pi \cdot \mu_{t,n}^{\min} + \pi \cdot \mu_{t,n}^{\max} \right) \quad \forall_t \in T
\end{aligned} \tag{4.68}$$

After transforming the initial bilevel program (4.16)–(4.29), we obtain the following MILP:

Objective function (4.66)

subject to:

upper-level constraints (4.17)–(4.18)

stationarity constraints (4.41)–(4.46)

primal feasibility equalities (4.47)–(4.49)

primal feasibility inequalities, which are left-hand side inequalities in (4.50)–(4.63)

dual feasibility inequalities, which are right-hand side inequalities in (4.50)–(4.63)

linearized complementary slackness from (4.50)–(4.63)

Alternatively, using the duality theory and rules for transforming a primal problem to a dual, our problem is:

Objective function (4.66)

subject to:

upper-level constraints (4.17)–(4.18)

primal problem constraints (4.47)–(4.49) and left-hand side inequalities in (4.50)–(4.63)

dual problem constraints (4.41)–(4.46) and right-hand side inequalities in (4.50)–(4.63)

strong duality equality (4.68)

4.3 Energy storage in reserve market

As an ultraflexible resource, energy storage devices are suitable for providing various frequency regulation services in addition to the energy arbitrage. For the sake of convenience, in this chapter the frequency regulation is defined as active power injections to or withdrawals from the power grid to maintain the system-wide frequency at a given reference value. The most common mean for a storage device to provide frequency regulation is to follow the system operator's automatic generation control (AGC) signal, which computes the area control error from frequency deviations and interchange power imbalances. As storage technologies evolve, new types of flexibility resources emerge (e.g. new battery chemistries, flywheel storage) and, therefore, there is a need to use such AGC commands that would be feasible for all technologies.

Unlike ramp-constrained conventional generators, practically all energy storage devices have no significant ramping limits that would affect their ability to provide frequency regulation. The fast ramping rates of energy storage devices can actually reduce the total system need for regulation capacity, thus relieving a regulation burden on conventional generators. On the other hand, using energy storage devices for both energy arbitrage and frequency regulation requires coordination to avoid violating power and energy capacity limits. The following constraints are used:

$$p_t^{\text{dis}} + r_t^{\uparrow} \leq P^{\text{dmax}} \qquad \forall_t \in T \tag{4.69}$$

$$p_t^{\text{ch}} + r_t^{\downarrow} \leq P_s^{\text{cmax}} \qquad \forall_t \in T \tag{4.70}$$

$$soc_t = soc_{t-1} + \left(ch_t + r_t^{\uparrow}\right) \cdot \eta^{\text{ch}} - \left(dis_t + r_t^{\downarrow}\right) / \eta^{\text{dis}} \tag{4.71}$$

where

$r_t^{\uparrow}$ upward frequency regulation provided by energy storage (MW)
$r_t^{\downarrow}$ downward frequency regulation provided by energy storage (MW)

4.3.1 Policies enhancing energy storage flexibility

Energy storage devices can simultaneously provide upward and downward frequency regulation and there is no need to enforce complementarity condition on $r_t^{\uparrow}$ and, $r_t^{\downarrow}$ i.e. $r_t^{\uparrow} . r_t^{\downarrow}$ can be different than 0. However, to avoid depleting stored energy and to increase value of energy storage as a provider of frequency regulation, some system operators (e.g. CAISO [12], ISO New England [13], MISO [14], NYISO [15]) enforce additional policies on frequency regulation provided by storage devices [11]:

1. *Energy neutrality*: Under this policy, at the end of every time interval, energy storage would be called to charge/discharge the same amount of energy as it is charged/discharged for regulation services over that period. In other words, if energy storage is scheduled to provide $r_t^{\uparrow}$ for upward frequency regulation, exactly the same amount of energy would be charged at the end of t. In some cases, charging would occur over several subsequent time intervals to

minimize the operating cost. This policy enables energy storage devices to maintain a sufficient state of charge to meet their obligations on the day-ahead values p_t^{ch} and p_t^{dis}. Even though this policy is called energy neutral, it does not guarantee that storage devices providing frequency regulation would have the same state of charge over the course of multiple time intervals.

2. *Energy offset compensation*: Even if energy neutrality is enforced, there will still be roundtrip efficiency losses and the state of charge would gradually decrease over multiple time intervals. These losses need to be accounted for at every dispatch interval or when selling or purchasing electricity. As a result, the losses accumulated by energy storage devices providing frequency regulation are routinely offset within real-time operations. However, in case of emergency operations, the offset can be postponed until the emergency is resolved to avoid resource scarcity.
3. *Unbundled AGC signal*: To take advantage of ramping capabilities of energy storage devices and to incentivize their participation in frequency regulation, the AGC signal can be split in two: the "fast" signal is communicated to relatively flexible resources (energy storage devices or gas-fired generators), while the "slow" signal is communicated to relatively inflexible resources. Using two AGC signals helps unbundle flexible and inflexible resources and use them for frequency regulation at different regulation intervals more efficiently.
4. *Pay for performance*: In order to create an incentive for energy storage devices to provide "fast" AGC response, some system operators employ the so-called pay-for-performance policies, which additionally compensate flexible resources for accurately matching the AGC signal. To evaluate accuracy of the match, system operators compute the difference between the "mileage" of each resource over a given period and its deviation from the "mileage" of the AGC signal.

4.3.2 Generalized payment scheme

Using the aforementioned policies naturally paves the way to new compensation schemes intended to fairly compensate energy storage devices[1] for their enhanced flexibility and, thus, to create additional incentives to support their participation in frequency regulation. If generalized, these schemes usually factor in the capacity (C_t) provided by energy storage for frequency regulation at each time interval t, mileage of the AGC signal (M_t), and performance factor (ρ). Assuming that the upward and downward regulation capacity is symmetric, i.e. $C_t = r_t^{\uparrow}/\Delta t = r_t^{\downarrow}/\Delta t$, the payment for providing frequency regulation at each time interval (Λ_t^{R}) can be computed as:

$$\Lambda_t^{\text{R}} = C_t \lambda_t^{\text{C}} + C_t \rho M_t \lambda_t^{\text{M}} \tag{4.72}$$

The first term in (4.72) is structurally similar to the option payment collected by conventional generators and, therefore, λ_t^{C} can be explicitly derived as the dual

[1] For the sake of clarity and concision, this chapter discusses such schemes exclusively in the context of energy storage devices, but they are also applicable for all other flexible resources that meet the eligibility criteria.

variable of the system-wide reserve requirement constraint. The second term in (4.72) represents the "mileage" payment based on the "mileage" price (λ_t^{M}) and intends to factor in the accuracy of individual energy storage devices in following the AGC signal. For an arbitrary AGC signal given by discrete samples $\{P_1, \ldots, P_n\}$ over each time period t, the "mileage" can be computed as:

$$M_t = \sum_{i>1}^{n} |P_i - P_{i-1}|/C. \tag{4.73}$$

In other words, the "mileage" in (4.73) reflects how quickly the AGC signal changes its absolute value and serves as an indirect metric of flexibility exercised by energy storage devices. Real-valued parameter ρ is a score defined in the range [0,1], which is uniquely assigned by the system operator to each frequency regulation provider based on its historical performance in following the AGC signal. Higher values of ρ indicate better compliance in following the AGC signal. Approaches to compute ρ vary among different systems, but usually account for time delays and correlations in following the AGC signal; see [11] for more details.

4.3.3 Dilemma: Trade energy or frequency regulation?

Since providing frequency regulation offers an alternative source of revenue for energy storage owners, the objective function (4.7) needs to reflect this component:

$$\text{Maximize} \sum_{t=1}^{T}\sum_{s=1}^{S} \lambda_{t,n|s} \cdot \left(p_{t,s}^{\text{dis}} - p_{t,s}^{\text{ch}}\right) + \sum_{t=1}^{T}\sum_{s=1}^{S} \left(C_{t,s}\lambda_{t,n|s}^{\mathrm{C}} + C_{t,s}\rho M_{t,s}\lambda_{t,n|s}^{\mathrm{M}}\right) \tag{4.74}$$

This objective function makes it possible for price-making energy storage owners to estimate revenues streams and costs achieved by trading energy or frequency regulation services. These estimates are particularly important in the presence of renewable generation resources. These resources have near-zero marginal costs and, thus, create additional competitive pressure and subsequently reduce energy prices. On the other hand, renewable generation resources tend to increase the need for frequency regulation that in turn is likely to spearhead frequency regulation prices. Price-making energy storage can act in both the energy and frequency regulation markets and arbitrage between the two products. In this case, it becomes increasingly important to factor in various uncertainty factors that have bearing on (4.74) using the stochastic and robust optimization methods described in Sections 4.2.1.2–4.2.1.3. These methods unlock the risk-averse decision-making in both the the energy and frequency regulation markets.

Under high deployment rates of energy storage, these devices with the dual producer–consumer nature can arbitrage between the energy and the frequency regulation markets to gain additional market power and further increase their profits. Traditional market power mitigation strategies do not explicitly recognize the dual producer–consumer nature of energy storage and, thus, may fail in reality. Developing such market power mitigation strategies that would account for unique characteristics of energy storage is an area of ongoing research.

Appendix

Deterministic Model

```
set t time periods/t1*t24/;

scalar delta_t length of time periods in hours/1/;
scalar eta_ch storage charging efficiency/0.95/;
scalar eta_dis storage discharging efficiency/0.95/;

parameter lambda (t) market prices/

t1      25
t2      22
t3      24
t4      27
t5      29
t6      27
t7      33
t8      39
t9      34
t10     41
t11     44
t12     48
t13     45
t14     47
t15     40
t16     37
t17     38
t18     44
t19     51
t20     57
t21     55
t22     50
t23     39
t24     36
/;

scalar p_cmax charging capacity limit in MW/10/;
scalar p_dmax discharging capacity limit in MW/10/;
scalar soc_min minimum storage state of charge in pu/0.2/;
scalar soc_max maximum storage state of charge in pu/0.95/;
scalar storage_capacity storage capacity in MWh/20/;
scalar soc_0 initial storage state of charge in pu/0.5/;

positive variables p_dis(t), p_ch(t), soc(t);
free variable profit;
binary variable x(t);

equations obj_1, constr_2, constr_3, constr_4, constr_5, constr_6,
constr_7, constr_7a;
obj_1..
        profit =e= sum(t, lambda(t)*(p_dis(t)-p_ch(t)));
```

```
constr_2(t)..
        p_dis(t) =l= p_dmax*x(t);

constr_3(t)..
        p_ch(t) =l= p_cmax*(1-x(t));

constr_4(t)$(ord(t) ge 2)..
        soc(t) =e= soc(t-1) + delta_t*p_ch(t)*eta_ch - delta_t*p_dis(t)/
eta_dis;

constr_5(t)$(ord(t) eq 1)..
        soc(t)  =e=  soc_0*storage_capacity + delta_t*p_ch(t)*eta_ch -
delta_t*p_dis(t)/eta_dis;

constr_6(t)$(ord(t) = card (t))..
        soc(t) =e= soc_0*storage_capacity;

constr_7(t)..
        soc(t) =g= soc_min*storage_capacity;

constr_7a(t)..
        soc(t) =l= soc_max*storage_capacity;
model deterministic /all/;

solve deterministic maximizing profit using mip;

execute_unload "Output_deterministic.gdx" p_ch,p_dis,soc,profit
```

Stochastic Model

```
set t time periods /t1*t24/;
set s scenarios /s1*s5/;

scalar delta_t length of time periods in hours /1/;
scalar eta_ch storage charging efficiency /0.95/;
scalar eta_dis storage discharging efficiency /0.95/;

table lambda (t,s) market prices
          s1        s2        s3        s4        s5
t1        25        26        24        29        23
t2        22        23        21        28        22
t3        24        25        22        22        25
t4        27        29        28        26        30
t5        29        33        31        35        34
t6        27        26        30        31        26
t7        33        34        35        29        33
t8        39        37        38        35        37
t9        34        36        33        37        37
t10       41        38        39        40        42
t11       44        45        42        43        44
t12       48        47        49        45        47
t13       45        48        49        52        51
t14       47        49        46        52        47
```

```
t15    40    44    41    49    39
t16    37    40    38    39    40
t17    38    39    40    42    44
t18    44    41    45    42    49
t19    51    48    49    50    52
t20    57    55    59    55    58
t21    55    56    58    60    57
t22    50    52    48    47    51
t23    39    40    39    42    42
t24    36    37    40    35    37
;

parameter pi (s) scenario probabilities/
s1     0.2
s2     0.2
s3     0.2
s4     0.2
s5     0.2
/;


scalar p_cmax charging capacity limit in MW /10/;
scalar p_dmax discharging capacity limit in MW /10/;
scalar soc_min minimum storage state of charge in pu /0.2/;
scalar soc_max maximum storage state of charge in pu /0.95/;
scalar storage_capacity storage capacity in MWh /20/;
scalar soc_0 initial storage state of charge in pu /0.5/;

positive variables p_dis(t), p_ch(t), soc(t);
free variable profit;
binary variable x(t);

equations obj_1, constr_2, constr_3, constr_4, constr_5, constr_6,
constr_7, constr_7a;

obj_1..
        profit =e= sum(t,sum(s, pi(s)*lambda(t,s)*(p_dis(t)-p_ch(t))));

constr_2(t)..
         p_dis(t) =l= p_dmax*x(t);

constr_3(t)..
         p_ch(t) =l= p_cmax*(1-x(t));

constr_4(t)$(ord(t) ge 2)..
         soc(t) =e= soc(t-1) + delta_t*p_ch(t)*eta_ch - delta_t*p_dis(t)/
eta_dis;

constr_5(t)$(ord(t) eq 1)..
         soc(t) =e= soc_0*storage_capacity + delta_t*p_ch(t)*eta_ch -
delta_t*p_dis(t)/eta_dis;

constr_6(t)$(ord(t) = card(t))..
         soc(t) =e= soc_0*storage_capacity;
```

```
constr_7(t)..
         soc(t) =g= soc_min*storage_capacity;
constr_7a(t)..
         soc(t) =l= soc_max*storage_capacity;

model deterministic /all/;
solve deterministic maximizing profit using mip;
execute_unload "Output
```

Robust Model

```
set t time periods/t1*t24/;

scalar delta_t length of time periods in hours /1/;
scalar eta_ch storage charging efficiency /0.95/;
scalar eta_dis storage discharging efficiency /0.95/;

parameter lambda(t) market prices/

t1        23
t2        21
t3        22
t4        26
t5        29
t6        26
t7        29
t8        35
t9        33
t10       38
t11       42
t12       45
t13       45
t14       46
t15       39
t16       37
t17       38
t18       41
t19       48
t20       55
t21       55
t22       47
t23       39
t24       35
/;

parameter d(t) market price deviations/

t1        6
t2        7
t3        3
t4        4
t5        6
t6        5
t7        6
t8        4
t9        4
```

```
t10      4
t11      3
t12      4
t13      7
t14      6
t15      10
t16      3
t17      6
t18      8
t19      4
t20      4
t21      5
t22      5
t23      3
t24      5
/;

scalar p_cmax charging capacity limit in MW /10/;
scalar p_dmax discharging capacity limit in MW /10/;
scalar soc_min minimum storage state of charge in pu /0.2/;
scalar soc_max maximum storage state of charge in pu /0.95/;
scalar storage_capacity storage capacity in MWh /20/;
scalar soc_0 initial storage state of charge in pu /0.5/;
scalar Gamma_dis budget of uncertainty when discharging /0/;
scalar Gamma_ch budget of uncertainty when charging /4/;

positive variables p_dis(t), p_ch(t), soc(t), z_dis, q_dis(t),
        z_ch, q_ch(t), y_ch(t);
free variable profit;
binary variable x(t);

equations obj_1, constr_2, constr_3, constr_4, constr_5, constr_6,
constr_7, constr_7a,
         constr_rob_1, constr_rob_2, constr_rob_3, constr_rob_4;

obj_1..
        profit =e= sum(t, (lambda(t)+d(t))*p_dis(t)) - sum(t, (lambda(t))
*p_ch(t))
        - z_dis*Gamma_dis - sum(t, q_dis(t))
        - z_ch*Gamma_ch - sum(t, q_ch(t));

constr_2(t)..
         p_dis(t) =l= p_dmax*x(t);

constr_3(t)..
         p_ch(t) =l= p_cmax*(1-x(t));

constr_4(t)$(ord(t) ge 2)..
        soc(t) =e= soc(t-1) + delta_t*p_ch(t)*eta_ch - delta_t*p_dis(t)/
eta_dis;

constr_5(t)$(ord(t) eq 1)..
         soc(t) =e= soc_0*storage_capacity + delta_t*p_ch(t)*eta_ch -
delta_t*p_dis(t)/eta_dis;

constr_6(t)$(ord(t) = card(t))..
         soc(t) =e= soc_0*storage_capacity;
```

```
constr_7(t)..
        soc(t) =g= soc_min*storage_capacity;

constr_7a(t)..
        soc(t) =g= soc_min*storage_capacity;

constr_rob_1(t)..
        z_dis + q_dis(t) =g= d(t)*y_dis(t);

constr_rob_2(t)..
        p_dis(t) =l= y_dis(t);

constr_rob_3(t)..
        z_ch + q_ch(t) =g= d(t)*y_ch(t);

constr_rob_4(t)..
        p_ch(t) =l= y_ch(t);

option optcr=0.0;

model robust /all/;

solve robust maximizing profit using mip;
_stochastic.gdx" p_ch,p_dis,soc,profit
```

References

[1] H. Pandžić, Y. Wang, T. Qiu, Y. Dvorkin, and D. S. Kirschen, "Near-optimal method for siting and sizing of distributed storage in a transmission network," *IEEE Transactions on Power Systems*, vol. 30, no. 5, pp. 2288–2300, September 2015.

[2] A. Berrada, K. Loudiyi, and I. Zorkani, "Profitability, risk, and financial modeling of energy storage in residential and large scale applications", *Energy*, vol. 119, no. 1, pp. 94–109, 2017.

[3] A. A. Solomon, D. M. Kammen, and D. Callaway, "The role of large-scale energy storage design and dispatch in the power grid," *Applied Energy*, vol. 134, pp. 75–89, 2014.

[4] M. C. W. Kintner-Meyer, P. J. Balducci, V. V. Viswanathan, *et al.*, "Energy storage for power systems applications: A regional assessment for the Northwest Power Pool," Technical report, 2010. [Online]. Available at: http://www.pnl.gov/main/publications/external/technical_reports/PNNL–19300.pdf

[5] D. Pudjianto, M. Aunedi, P. Djapic, and G. Strbac, "Whole-systems assessment of the value of energy storage in low-carbon electricity systems," *IEEE Transactions on Smart Grid*, vol. 5, no. 2, pp. 1098–1109, March 2014.

[6] B. Dunn, H. Kamath, and J.-M. Tarascon, "Electrical energy storage for the grid: A battery of choices," *Science*, vol. 334, pp. 928–935, 2011.

[7] V. Bobanac and H. Pandžić, "Lithium-ion Batteries: Experimental research and application to battery swapping stations," *Proceedings of ENERGYCON 2018*, Limassol, Cyprus, pp. 1–6, June 2018.

[8] D. Bertsimas and M. Sim, "Robust discrete optimization and network flows," *Mathematical Programming*, vol. 98, pp. 49–71, May 2003.
[9] L. Baringo and A. Conejo, "Offering strategy via robust optimization," *IEEE Transactions on Power Systems*, vol. 26, no. 3, pp. 1418–1425, August 2011.
[10] H. Pandžić, T. Qui, Y. Wang, and D. Kirschen, "Toward cost-efficient and reliable unit commitment under uncertainty," *IEEE Transactions on Power Systems*, vol. 31, no. 2, pp. 970–982, March 2016.
[11] B. Xu, Y. Dvorkin, D. S. Kirschen, C. A. Silva-Monroy, and J. P. Watson, "A comparison of policies on the participation of storage in U.S. frequency regulation markets," in *Proceedings of the 2016 IEEE Power and Energy Society General Meeting*, Boston, MA, USA, pp. 1–5, 17–21 July 2016.
[12] CAISO, "Non-generator resource (NGR) and regulation energy management (REM)," Workshop presentation, August 2012. [Online]. Available at: http://www.caiso.com/documents/presentation-nongeneratorresourceregulation energymanagement_workshopaug30_2012.pdf
[13] ISO New England, "Description of the energy neutral dispatch algorithm," Technical report, March 2015. [Online]. Available at: https://www.iso-ne.com/static-assets/documents/2015/03/energy_neutral_dispatch_algorithms.pdf
[14] Midwest ISO, "Automatic generation control (AGC) enhancement for fast ramping resources," Workshop presentation, October 2014. [Online]. Available at: https://www.misoenergy.org/Library/Repository/Meeting%20Material/Stakeholder/MSC/2014/20141028/20141028%20MSC%20Item%2005b%20AGC%20Enhancement%20for%20Fast%20Ramping%20Resources.pdf
[15] NYISO, "Limited Energy Storage Resource (LESR) Market Integration Update," Price Responsive Load Working Group, April 2014. [Online]. Available at: http://www.nyiso.com/public/webdocs/markets_operations/committees/bic_prlwg/meeting_materials/2009-04-27/LESR_PRLWG_Presentation.pdf

Further Reading

1. H. Mohsenian-Rad, "Optimal bidding, scheduling, and deployment of battery systems in California day-ahead energy market," *IEEE Transactions on Power Systems*, vol. 31, no. 1, pp. 442–453, January 2016.
2. Y. Dvorkin, R. Fernandez-Blanco, D. S. Kirschen, H. Pandžić, J. P. Watson, and C. A. Silva-Monroy, "Ensuring profitability of energy storage," *IEEE Transactions on Power Systems*, vol. 32, no. 1, pp. 611–623, January 2017.
3. Y. Wang, Y. Dvorkin, R. Fernandez-Blanco, B. Xu, T. Qiu, and D. S. Kirschen, "Lookahead bidding strategy for energy storage," *IEEE Transactions on Sustainable Energy*, vol. 8, no. 3, pp. 1106–1117, July 2017.
4. M. Kazemi, H. Zareipour, N. Amjady, W. D. Rosehart, and M. Ehsan, "Operation scheduling of battery storage systems in joint energy and ancillary services markets," *IEEE Transactions on Sustainable Energy*, vol. 8, no. 4, October 2017.

Chapter 5

The role of storage in transmission investment deferral and management of future planning uncertainty

Ioannis Konstantelos[1] *and Goran Strbac*[1]

5.1 Introduction

Electricity systems are facing great challenges across the world to achieve the climate change mitigation targets set by governments. The transition to a decarbonized economy will entail unprecedented amounts of transmission investment due to the fact that low-carbon energy sources are usually located far from the load centres, rendering the transmission investment framework of primary importance [1]. Another big challenge to cost-efficient decarbonization will be the greater requirement for operational flexibility to deal with large and rapid changes in demand and supply [2]. In addition, the significant uncertainty that characterizes future system developments poses severe problems. Under the ownership unbundling that has taken place in many jurisdictions, planners are facing increasing uncertainty regarding the type, size and position of new connections. In a similar vein, a critical driver whose timing and extent is unknown is the long-term demand growth due to the electrification of transport and heat. Studies on the UK system have shown that the potential electrification of the heating and transport sectors could increase peak demand by up to two or three times in 2050 when compared to the present levels [3]. The possibility for low-growth scenarios may lead to asset stranding in cases of over-investment, while greater-than-expected growth may require costly piecemeal upgrades that forego economies of scale and lead to increased interim network constraint costs.

It is critical to highlight that the long lead times that characterize conventional transmission projects render them more prone to these adverse effects. In contrast, projects aimed at improving the use of the existing assets and infrastructure, such as energy storage (ES) and FACTS, have been shown to assist with interim uncertainty management and embed strategic flexibility within an investment plan [4]. The above points indicate that the ongoing decarbonization effort is altering fundamental aspects of the transmission planning process. The emerging reality involves multiple

[1]Department of Electrical and Electronic Engineering, Imperial College, London, UK

uncertainty sources and a high number of candidate technologies. As such, planners require novel approaches for navigating this setting and designing future-proof systems.

It is imperative that transmission planning is viewed as a long-term portfolio optimization problem across many different assets; each candidate asset be substantially different in terms of cost, build time, technical capability and operational flexibility. As such, investment strategies should be drawn on a comprehensive 'what-if' basis where the optimal recourse actions for minimum-cost adjustment to the unfolding reality are taken into account. The objective is to identify strategies that include an optimal mix of (i) flexibility-driven elements for interim network management (ii) large-scale commitments characterized by economies of scale, which can be deployed once uncertainty has been resolved.

5.1.1 System benefits of ES

The discovery of effective ways to store electricity, enabling its on-demand use, has remained an elusive goal since the discovery of electricity. Over the last decades, substantial advances have been achieved, presenting us with promising solution across different voltage levels. Studying the potential of ES to reduce future electricity system costs is critical in guiding policy in this area. In fact, ES presents a significant number of benefits to the electricity system [5], as outlined below.

5.1.1.1 Contribution to operational flexibility

With the advent of non-dispatchable generation, operators are increasingly facing balancing problems. ES can greatly assist in this aspect by providing a broad array of grid services across timescales such as providing frequency response and regulation, operating and ramping reserves and assisting with energy balancing [6]. Most importantly, ES can carry out these tasks by using stored zero-marginal-cost renewable energy which would otherwise have been spilled during low demand periods.

5.1.1.2 Contribution to security and adequacy

Another important advantage of ES is the ability to provide reliable system capacity, potentially replacing generation assets. Nevertheless, this ability is still not formally assessed in many jurisdictions, prohibiting ES from participating in capacity auctions. As pointed out in [7], two unique characteristics pertaining to ES (energy constraints and coupling to the upstream charging infrastructure) raise the need for new ES capacity credit methodologies.

5.1.1.3 Management of long-term uncertainty

The fact that ES can make a substantial contribution to the management of long-term uncertainty is one of the least-documented benefits and constitutes the primary focus of the present chapter. This subtle yet important aspect has in the past been studied under the name of strategic value [4], and option value (e.g. [1] and [8]). The concepts of option (or strategic) value corresponds to the value placed on the ability to utilize an asset in the future (as opposed to the value extracted from its immediate use which may be zero) and can comprise a substantial part of project's economic value, as discussed below.

In general, network planning until now has mainly been an exercise of ensuring adequate security of supply a minimum cost. However, this philosophy is no longer relevant under the uncertainty that characterizes future demand and generation developments. System planners are not able to make fully informed investment decisions since there are substantial risks related to *asset stranding*, *premature commitment* to suboptimal investment paths and *lack of adaptability* to alternative scenario realizations. These risks can ultimately lead to increased costs for consumers and ineffective, costly, or delayed decarbonization of the electricity system. The irreversible nature of capital investments combined with this increasing uncertainty regarding future developments means that attractive investment opportunities should not be identified solely in terms of net benefit, but also in terms of the *option value* that they may provide. In general, two principle features of the transmission planning process render option value of ES substantial: learning over time and irreversibility.

Learning over time signifies that some uncertainty is resolved over time. For example, generation investment is based on expected profitability and influenced by factors such as the regulatory framework, use of system charges, and investment costs. Although these are beyond the planner's immediate control, they are directly observable over time and can be used to infer possible evolution paths. Given that transmission investment is a dynamic process, it can be materially informed by the evolution of these parameters over the planning horizon. In addition, since generation investment consists of distinct stages (e.g. planning permission acquisition, construction, and commissioning), key trigger events can be identified and used in the decision process as informed indicators for subsequent scenario transitions.

Irreversibility refers to the fact that the majority of transmissions have high capital costs, long lifetimes spanning several decades and very low or zero salvage value. As such, it is critical to ensure that committed funds provide long-lasting value while avoiding assets that may eventually be stranded or under-utilized leading to severe welfare loss.

The option value concept is highly applicable when valuing the investment in ES assets. The benefit of such solutions may be because of not only the service they provide (e.g. more flexible use of resources, etc.) but also in how these can facilitate and de-risk subsequent decisions. ES can provide interim network management by increasing utilization of the existing assets, and thus 'buy time' until some uncertainty is resolved; thereafter, capital-intensive commitments can be well justified and entail reduced stranding risk. In other words, ES operational flexibility can ultimately lead to planning flexibility where planners can plan against a more certain background and reduce capital risk [4]. As such, it is imperative that future planning frameworks consider these aspects to fully capture the diverse benefits stemming from smart technologies.

5.1.2 Valuation model variants

Historically, network design had been driven by the need to meet peak demand with sufficient reliability. In systems dominated by high-capacity value thermal generators, this approach has led to economically efficient solutions. However,

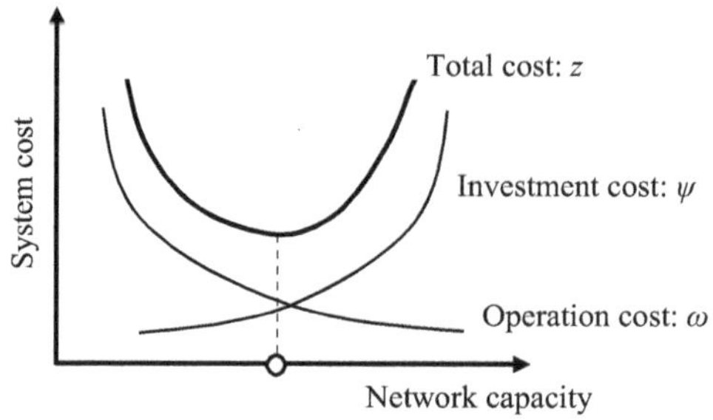

Figure 5.1 Trade-off between capital and operational costs in transmission investment

under high penetration of intermittent sources of energy, that have a much lower capacity value, accommodating peak flows during high-demand ceases to be the primary investment driver. Instead, transmission investment is undertaken on a cost-benefit basis where the solution that minimizes total costs (investment and operation cost over a target horizon) is pursued, as shown in Figure 5.1.

The investment process is usually modelled as a deterministic one-stage decision problem where the future is taken to be firmly known a priori and not influenced by exogenous sources of uncertainty that exist in a real planning setting. However, in reality, there is a wide range of uncertainties affecting the planning process. Inaccuracies in long-term load forecasting, copper price fluctuations in international commodity markets affecting transmission investment costs and ambiguous environmental constraints subject to continuous governmental reviews render the planner unable to make decisions with perfect foresight. In addition, the unbundling of the electricity sector has resulted in even more uncertainty surrounding the transmission planning process, primarily related to the future generation developments. Planning models that adopt a deterministic view of the future are no longer relevant in such a market landscape and stochastic approaches accommodating uncertainty have to be employed. In the following sections, we present the main types of valuation models that incorporate uncertainty.

5.1.2.1 Single-scenario analysis

In this approach, the optimum expansion plan is determined on the basis of a single scenario, deemed to be the most probable forecast of the uncertain parameters. A deterministic cost-benefit optimization is performed to determine the optimal investment plan with respect to that particular scenario; all other realizations are ignored. Naturally, this method has some very attractive characteristics such as ease of implementation and severely reduced problem size. However, by disregarding all other plausible scenarios, uncertainty is essentially ignored, leaving the system severely exposed to alternative realizations. Despite its shortcomings, this approach is widely used by transmission planners across the world due to its straightforward nature and clearly defined objective along the lines of transmission investment methodologies used prior to the unbundling of the electricity sector.

5.1.2.2 Multi-scenario analysis

This technique makes use of a number of alternative scenarios. As described in [9] and [10], this approach consists of two levels. Initially, the optimal transmission expansion for each individual scenario is determined, obtaining a set of optimal plans. Based on this set, a range of analytical techniques can be used to identify 'well-performing' decisions that are common across all scenarios. In the case of a risk-averse planner, this approach can be extended to quantify the impact of alternative scenario realizations and the contingent investment required to cope with unforeseen events. The basic limitation of the scenario analysis approach is that scenario-wide optimality cannot be guaranteed when utilizing decisions 'tailor-made' for different scenarios.

5.1.2.3 Stochastic planning

Multi-stage stochastic programming approaches consider the dynamics of uncertainty over successive planning periods through a multi-stage scenario tree. The number of transmission expansion models considering the dynamic evolution of uncertainty has thus far been very limited due to the large problem size entailed. Nevertheless, the topic has been gaining traction recently. Most notably, the authors in [4] study the option value of flexible solutions when facing uncertainty and extend this idea to the distribution planning problem (e.g., see [8,11–14]). Other authors have carried out similar work with a focus on western US system [15], clearly demonstrating the relevance and substantial benefit of stochastic planning tools.

5.1.2.4 Risk-constrained planning

The stochastic planning approach is risk-neutral, meaning that the planner's objective is solely the minimization of expected system costs. However, there are aspects of the optimal solution that the planner may find unattractive such as excessive costs under specific scenario paths. Using a risk-constrained formulation, the planner can immunize decisions against adverse scenarios and ensure that the strategy being followed can satisfy some risk-averse measure. This topic has received limited attention thus far due to the computational complexity it entails. Efforts have mostly focused upon the use of the conditional-value-at-risk (CVaR) due to its suitability for linear optimization programmes [16]. We indicatively mention [17] and [18], where the authors demonstrate CVaR-constrained system planning at the transmission level.

5.1.2.5 Robust planning

Another approach to embedding risk-aversity is robust planning and in particular the adoption of Wald's maximin decision model. Under this approach, the planner identifies the investment strategy that leads to the minimization of the maximum regret experienced, where regret is defined as the optimal cost achievable if perfect information was available a priori. References related to robust transmission planning include [19–21]. The regret is a measure of how early investment decisions can ill-condition the system and hinder its ability to adjust to eventual scenario realizations at least cost. Note that unlike risk-constrained planning where the level of risk aversity can be tuned by the planner, this approach yields a single optimal plan. As such it is a suitable decision criterion for cases of a highly risk-averse planner.

Note that in this work we are focusing exclusively on the risk-neutral stochastic formulation. However, the main messages regarding ES's strategic role when facing uncertainty also pertain to the risk-constrained and robust planning paradigms.

5.1.3 Motivating example

In this section, we will briefly contrast between deterministic (single-scenario and multi-scenario analysis) and stochastic planning approaches using a simple system. We aim to clearly demonstrate how they can result in fundamentally different investment decisions, highlighting the limited capability of deterministic approaches to leverage the strategic potential of ES.

Figure 5.2 illustrates a simple medium-voltage substation. It has two 15 MVA transformers, which under the $N-1$ security standard allow a maximum transfer of 15 MVA. The secondary is connected to three feeders with a total peak demand of 15 MW. Note that for reasons of simplicity we only consider active power in the study and focus on thermal network constraints.

Operation across a year is represented by 11 days; weekdays and weekends for each calendar season as defined by Elexon (winter, spring, summer, high summer and autumn) as well as a peak demand day. Data for these days have been taken from the aggregate GB demand in the year 2015 and scaled down to the study substation. Demand time series for the representative days are shown in Figure 5.3; demand of 1 p.u. represents the peak demand equal to 15 MW as stated earlier. Note that we have assumed equal distribution of the total demand across the three feeders.

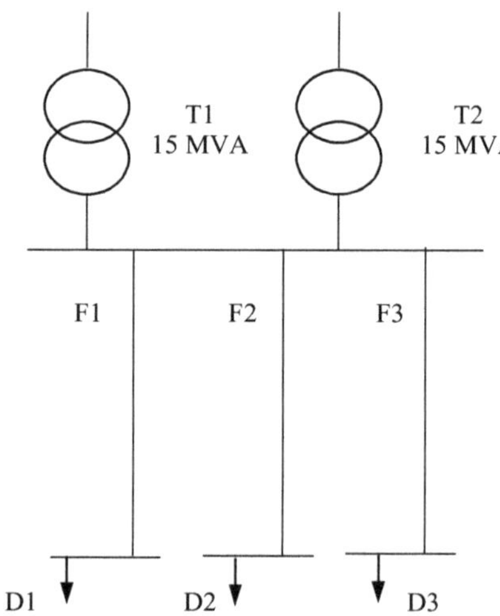

Figure 5.2 Test distribution network used in the case study

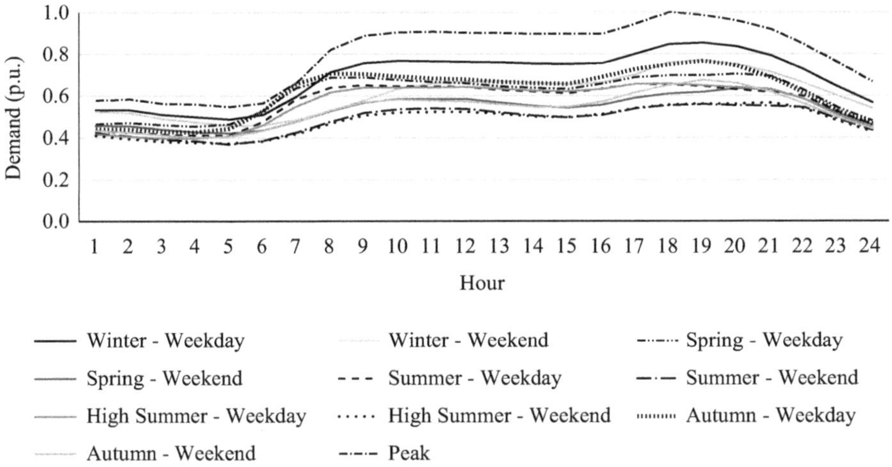

Figure 5.3 Example demand time series across different day types and seasons

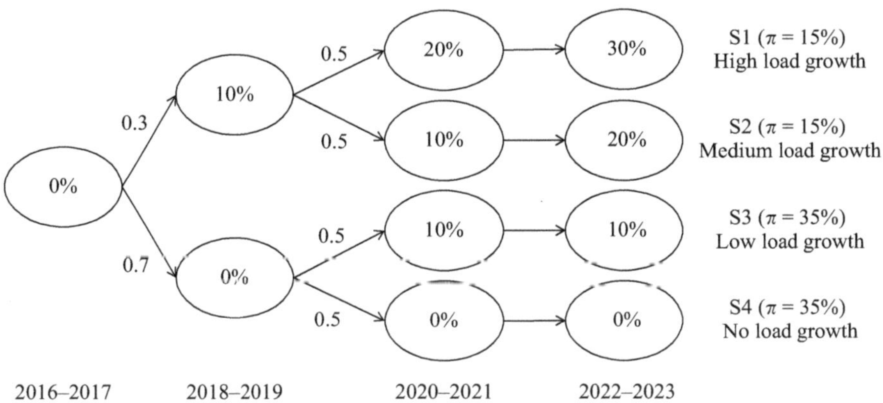

Figure 5.4 Scenario tree showing evolution of demand over the years; arrows show the corresponding transition probabilities

The four scenarios, shown in Figure 5.4, have been developed to capture potential trajectories of the system's demand evolution. Each scenario tree node shows the per cent of demand growth relative to the starting basecase of 15 MW peak demand and represents two years of operation. The scenario tree comprises four two-year stages (also referred to as epochs); 2016–2017, 2018–2019, 2020–2021 and 2022–2023. The tree's root node captures operation in the two years 2016 and 2017; during this period, there is a 0% increase of peak demand compared to the basecase of 15 MW. In the years 2018 and 2019, there can be two eventualities: a 10% increase in demand with a 30% probability or no change to demand with a higher probability of 70% and so on.We assume that the distribution network operator (DNO) has two following options for reinforcing the system and ensuring all future demand is met.

The first option is to install a 15 MVA transformer. The lump present cost of commissioning this transformer is estimated to be £2 million. Using a weighted average cost of capital (WACC) of 7% and an asset lifetime of 20 years, the equivalent annuitized capital cost is £190k. It is critical to note that the commissioning of the transformer is not considered to be instantaneous. In fact, two years are required to carry out the necessary works (civil works, modification to protection schemes, etc.) within the substation to accommodate this large piece of equipment.

The second option is to install an ES fuel cell plant with rating of 5 MW and an energy capacity of 25 MWh (i.e. 5 h at full output). The cost of this plant is £4 million. Using a WACC of 7% and an asset lifetime of 20 years results in an annuitized investment cost of £380k. Note the increased cost compared to a transformer; if we were to ascribe a cost on a £ per kW basis, the transformer and ES are, respectively, £133/kW and £800/kW. In addition, we assume that the ES plant can be deployed in the same epoch that the investment decision is made, i.e. no build time delay applies due to the ability for deployment outside of the substation premises, providing increased flexibility. In this study, we only consider the ES ability for carrying out peak shaving.

5.1.3.1 Deterministic planning

We first present the optimal investment plan obtained under a *deterministic* approach, i.e. when the DNO considers only a single scenario in isolation. The optimal investment plan for each of the four scenarios is shown in Figure 5.5. Note that no investment is undertaken in the third and fourth epochs.

In the case of scenarios 1 and 2, the optimal investment plan is to build a new transformer from the very first epoch. This is because the substation's importing capability must be upgraded by the second epoch where, as shown in the scenario

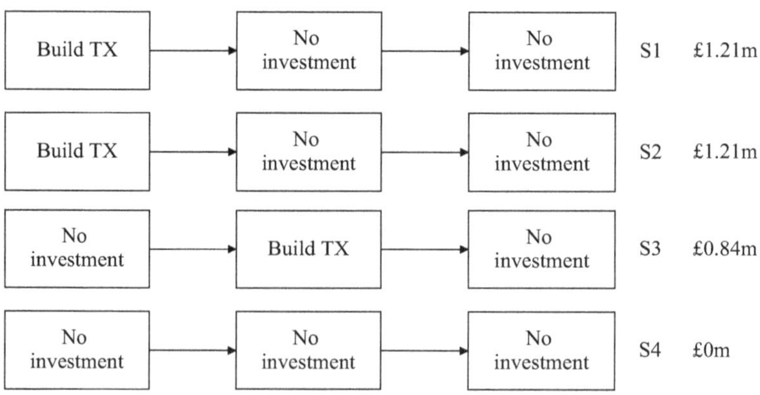

Figure 5.5 Optimal investment schedule for the four scenarios when each is examined in isolation and the planner can choose to invest in transformers and a new storage plant

tree in Figure 5.4, a 10% demand increase occurs. Due to the transformer's building time of two years, the commitment is made early in the first epoch. The total cost for both scenarios is the same; £1.21 million which corresponds to the discounted investment cost for the transformer over the case study's duration of 8 years. Scenario 3 is different in the sense that the increase of the substation's import capability is required later in the third epoch. In the case of scenario 4, no investment is needed since there is no demand increase. As such, the optimal cost for accommodating this scenario is zero.

It is of great importance to note that even though the option to build ES instead of a new transformer is available, it is never chosen as the optimal choice due to the high capital cost entailed. As mentioned earlier, ES is several times more expensive and does not present any additional benefit under full knowledge of the future system trajectory. However, as will be shown in the following section, its deployment can be advantageous in cases of long-term uncertainty. ES can constitute a valuable interim solution until uncertainty can be resolved and more informed decisions can be made.

5.1.3.2 Stochastic planning

In this section, we assume that the planner does not have perfect foresight and does not know which of the four scenarios will occur in the future. A stochastic optimization analysis is undertaken where the best investment strategy hedging against the possibility for stranded assets and under-investment is identified. Of most interest are the first-stage decisions since they are the directly implementable part of the chosen strategy.

Two variants are studied in this section. In the first variant, the planner is capable of investing solely in a new transformer; the challenge is identifying the decision points in the scenario tree where such investments would be optimal. In the second variant, the planner can invest in both storage and a new transformer. By comparing the expected cost in both cases, we are able to ascribe an expected economic benefit towards the ability to invest in ES.

We first focus on the variant that allows solely investment in transformers. The optimal strategy is shown in Figure 5.6. As can be seen, the longer transformer build time necessitates an early commitment from the very first stage, driven by the load growth under S1 and S2. As such, it is not possible to differentiate between the investments undertaken across the four scenarios. (All scenarios are subject to an investment cost of £1.21 million.) This results in earlier-than-necessary and over-investment in the case of scenarios 3 and 4, respectively. Naturally, the expected investment cost is £1.21 million.

In the second study variant where investment in both transformers and ES is allowed, the optimal investment strategy shown in Figure 5.7 is fundamentally different. Note that no investment is undertaken in the first stage; if high demand growth eventually occurs, then a ES is constructed first to deal with the increased growth experienced in the second epoch (10% increase corresponding to a peak demand of 16.5 MW). Subsequently if a scenario tree transition to S1 occurs, which entails further substantial increases in demand, a transformer is deployed to ensure

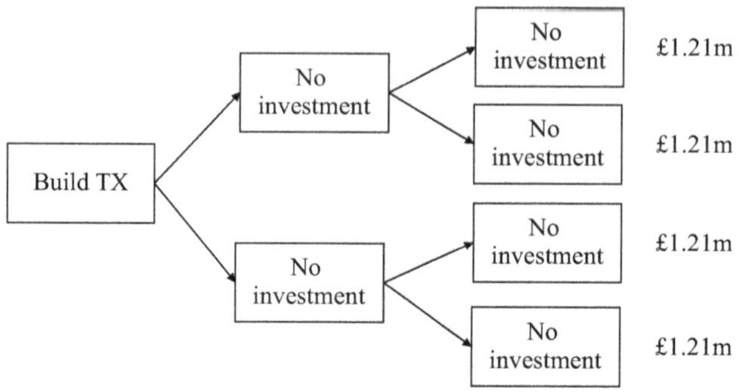

Figure 5.6 Optimal investment strategy when the planner can only build new transformers

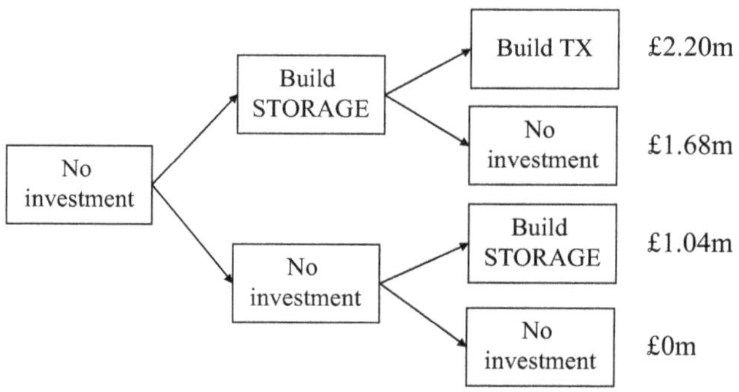

Figure 5.7 Optimal investment strategy when the planner can invest in new transformers and storage plants

peak demand requirements can be covered. On the other hand, under scenario 3 a storage plant is deployed in the third epoch. ES is preferred over building a new transformer due to the ability to differentiate between S3 and S4 and commit on a conditional basis. In other words, if a transformer was chosen to accommodate S3, it would have to be deployed in the second epoch at which point S3 and S4 have not been differentiated. This would give high stranding risk under S4 which has a substantial 30% probability of occurrence.

It is clear that the ability to swiftly deploy storage alleviates the need for a first-stage commitment. This is an important benefit since all subsequent decisions can be made with more information being available and on a conditional basis, enabling the planner to be more flexible in his/her decision-making. Although in the case of the high-growth scenarios, namely S1 and S2, the investment costs entailed are higher compared to deploying a transformer from the very first stage, this downside risk is balanced by the substantial savings made in the case that S3 and S4 occurs, which in this case are assumed to be highly probable. The expected investment cost is £0.95 million, £0.26 million lower than in the previous case study. As such, the option value of ES is a substantial £0.26 million, representing the expected benefit of having the ability to invest in ES.

In conclusion, we underline the fact that flexible investment options such as ES possess significant option value due to their ability to defer or even avoid premature commitment to capital projects. They can do this by taking advantage of the inter-temporal resolution of uncertainty (i.e. the fact that exogenous uncertainties are bound to be resolved with the passage of time). Although ES may not be the optimal investment under all uncertainty realizations, it can render 'wait-and-see' strategies viable. Note that if we were to ignore the possibility of deploying smart assets capable of cost-efficient interim network management, these 'wait-and-see' strategies would be deemed unattractive. This is aligned with recent findings published in [4] and [15] stating that although flexible investments (such as ES) may be considered suboptimal from a perfect-information/single-scenario point of view, they can be highly beneficial in the way they can manage uncertainty.

5.1.4 Chapter structure

In Section 5.2, we present the full mathematical formulation of the stochastic transmission expansion planning with storage problem. Since the computational burden can be very substantial, we also present a suitable decomposition scheme that renders the analysis of marge systems tractable. In Section 5.3, we present a large case study on the IEEE-24 busbar system where the planner identifies the optimal investment strategy while facing uncertainty with respect to future generation connections and demand growth. Our focus is computing the benefit of deploying ES to defer premature commitment to conventional reinforcement projects and manage interim uncertainty.

5.2 Stochastic transmission expansion planning with storage

Although in many jurisdictions, market unbundling rules have resulted in decoupling ES from the transmission planning process, there is growing worldwide interest for transmission system operators (TSOs) to use ES as a transmission asset for managing congestion, resource variability and uncertainty. This debate is currently under way in North America [22], Australia [23] and Europe [24], with regulators supporting the possibility for TSOs to own storage assets. The increasing relevance of ES for transmission planning gives rise to a new class of planning problems: integrated transmission and storage planning.

In this section, we first carry out an extensive literature review of the existing planning models and then proceed with providing the full mathematical formulation along with a suitable decomposition scheme.

5.2.1 Literature review

The importance of considering multiple technologies, such as ES, in expansion decision problems over a finite planning horizon has been widely recognized in the past (e.g. see [25]). The computational difficulties that arise when trying to accommodate many operating points and investment options have severely limited the development and adoption of multi-asset strategic planning models.

Recently, cost-benefit analysis of deploying ES in electricity systems has been carried out in [26–30] under different assumptions. A mixed integer-linear programming (MILP) deterministic static investment model has been proposed in [26] to reduce network investment cost. The potential of reducing network investment cost through the deployment of ES is analysed in [27] using a deterministic single-stage transmission planning model. Authors in [28] use a MILP transmission expansion planning formulation considering two 24-hour operating scenarios to show that optimal location and capacity of ES is sensitive not only to cost but also to variability and shape of demand in the network. A robust formulation of an integrated transmission and ES expansion planning problem is presented in [29]. The static model includes transmission switching and uncertainty in load demand and wind power productions through uncertainty sets. An interesting two-stage stochastic MILP planning model comprising investment in generation, transmission and ES is introduced in [30]. A case study on the standard 24-bus IEEE Reliability Test System will be used to demonstrate the ability of ES to defer investment in transmission and generation capacity. In general, although there have been numerous publications on the combined planning of transmission and storage, this has not been fully explored while considering long-term uncertainty.

5.2.2 Mathematical formulation

5.2.2.1 Nomenclature

The mathematical nomenclature used in the following sections is explained below.

Sets and indices

Ω_W	Set of available options for reinforcing a transmission corridor, indexed w
Ω_L	Set of all transmission lines
Ω_N	Set of all buses
Ω_G	Set of all generators
Ω_B	Set of all time blocks
Ω_T^b	Set of all time periods in block b
σ_b	The first period of demand block b
Ω_E	Set of epochs
ε_m	The epoch to which node m belongs
Ω_S	Set of all scenario paths
Ω_M	Set of all scenario tree nodes
Φ_m^0	An ordered set consisting of all parents of node m, including m
Φ_m^γ	An ordered set consisting of all parents of node m, from the first stage, up to stage $\varepsilon_m - \gamma$

Decision variables

$B_{m,\ l,w}$	Line upgrade decision variable (scenario node m, line l, option w)
$\tilde{F}_{m,l}$	Total extra capacity built (scenario node m, line l)
$Q_{m,\ l}$	Phase-shifter investment decision variable (scenario node m, line l)
$\tilde{Q}_{m,l}$	Auxiliary phase-shifter investment state variable (scenario node m, line l)
$H_{m,\ n}$	Storage device investment decision variable (scenario node m, bus n)
$\tilde{H}_{m,n}$	Auxiliary storage device investment state variable (scenario node m, bus n)
$p_{m,t,g}$	Power output (scenario node m, time period t, unit g)
$f_{m,t,l}$	Power flow (scenario node m, time period t, line l)
$\theta_{m,t,n}$	Bus angle (scenario node m, time period t, bus n)
$\xi_{m,t,l}$	Angle shift due to phase-shifter (scenario node m, time period t, line l)
$h_{m,t,n}$	Storage device power (scenario node m, time period t, bus n)
$\tilde{h}_{m,t,n}$	Storage device state-of-charge (scenario node m, time period t, bus n)
$d_{m,t,n}$	Demand curtailment (scenario node m, time period t, bus n).

5.2.2.2 Objective function

In Equation (5.1), we present the objective function to be minimized. It involves the total expected system cost, evaluated across the multi-stage horizon examined. Note that the expectation arises through the probability weighting where π_m is the occurrence probability of scenario tree node m. In addition, The input parameters r_e^I and r_e^O denote the cumulative discount factor for investment and operating costs, respectively, corresponding to epoch e.

In particular, the objective function comprises the probability-weighted investment costs (5.2) relating to ES devices, phase-shifting transformers (PS) and line reinforcements. Note that input parameters κ_w^F, κ^O and κ^H denote the annual cost of line upgrades (line l, option w), PS and ES, respectively. In addition, it involves expected operating cost (5.3), expressed as the summation of generation costs and the cost of load curtailment across all time periods, where each period lasts τ_t hours. Generation cost and cost of curtailment are denoted by κ_g^G and Γ, respectively. Note that we have assumed a perfectly competitive power market.

$$z^* = \min_{F,Q,H,p,d} \left\{ \sum_{\forall m} \pi_m r^I_{\varepsilon_m} \psi_m + \sum_{\forall m,b} \pi_m r^O_{\varepsilon_m} \omega_{m,b} \right\} \tag{5.1}$$

where

$$\psi_m = \sum_{\forall l} \left(B_{m,l,w} \kappa^F_{l,w} + Q_{m,l} \kappa^Q \right) + \sum_{\forall n} H_{m,n} \kappa^H, \quad \forall m \in \Omega_M, \forall b \in \Omega_B \tag{5.2}$$

$$\omega_{m,b} = \sum_{t \in \Omega_T^b} \tau_t \left[\sum_{\forall g} p_{m,t,g} \kappa_g^G + \sum_{\forall n} d_{m,t,n} \Gamma \right], \quad \forall m \in \Omega_M, b \in \Omega_L \tag{5.3}$$

5.2.2.3 Investment constraints

$$\tilde{F}_{m,l} = \sum_{w\in\Omega_W} \sum_{\phi\in\Phi_m^{\gamma_w^F}} B_{\phi,l,w}\bar{F}_w, \quad \forall m \in \Omega_M, l \in \Omega_L \tag{5.4}$$

$$\tilde{Q}_{m,l} = \sum_{\phi\in\Phi_m^{\gamma^Q}} Q_{\phi,l}, \quad \forall m \in \Omega_M, l \in \Omega_L \tag{5.5}$$

$$\tilde{H}_{m,n} = \sum_{\phi\in\Phi_m^{\gamma^H}} H_{\phi,n}, \quad \forall m \in \Omega_M, n \in \Omega_N \tag{5.6}$$

$$B_{m,l} \in \{0,1\}, \qquad \forall m \in \Omega_M, l \in \Omega_L \tag{5.7}$$

$$Q_{m,l} \in \{0,1\}, \qquad \forall m \in \Omega_M, l \in \Omega_L \tag{5.8}$$

$$H_{m,n} \in \{0,1\}, \qquad \forall m \in \Omega_M, n \in \Omega_N \tag{5.9}$$

$$\tilde{Q}_{m,l} \in \{0,1\}, \qquad \forall m \in \Omega_M, l \in \Omega_L \tag{5.10}$$

Constraint (5.4) states that the transfer capability available for line l, at node m, is a function of line reinforcement decisions across all the preceding epochs (an epoch is equivalent to stage) of the corresponding scenario path. Note that the state variable $\tilde{F}_{m,l}$ is a real-valued decision variable and depends on the input parameters $\bar{F}_w$; option's capacity addition in MW. Constraints (5.5) and (5.6) impose a similar relation for the investment in ES and PS. Note that we are explicitly considering build time γ_w^F, γ^Q and γ^H denote the building time for line upgrade w, build time for PS and ES respectively; they are integers expressed in terms of epochs. For example, γ^H corresponds to the candidate ES plant having a build time of one epoch (e.g., five years). Note that input parameters γ_w^F, constraints (5.7)–(5.9) establish the binary nature of investment decision variables.

5.2.2.4 Operation constraints

$$\sum_{\forall g\in n} p_{m,t,g} + \sum_{\forall\{l:v_l=n\}} f_{m,t,l} - \sum_{\forall\{l:u_l=n\}} f_{m,t,l} = D_{t,n} - d_{m,t,n} + h_{m,t,n}, \quad \forall t \in \Omega_T^b, , \forall n \, \text{in} \, \Omega_N \tag{5.11}$$

$$0 \le p_{m,t,g} \le \bar{p}_{m.g}, \qquad \forall t \in \Omega_T^b, \forall g \in \Omega_G \tag{5.12}$$

$$f_{m,t,l} = b_l\left(\theta_{m,t,u_l} - \theta_{m,t,v_l} + \xi_{m,t,l}\right), \quad \forall t \in \Omega_T^b, \forall n \in \Omega_N \tag{5.13}$$

$$|f_{m,t,l}| \le F_l^0 + F_{m,l}^*, \qquad \forall t \in \Omega_T^b, \forall l \in \Omega_L \tag{5.14}$$

$$|\xi_{m,t,l}| \le Q^*{}_{m,n}\bar{\xi}, \qquad \forall t \in \Omega_T^b, \forall l \in \Omega_L \tag{5.15}$$

$$|h_{m,t,n}| \le H^*{}_{m,n}\bar{h}, \qquad \forall t \in \Omega_T^b, \forall n \in \Omega_N \tag{5.16}$$

$$\tilde{h}_{m,t,n} \leq \overline{\eta}, \quad \forall t \in \Omega_T^b, \qquad \forall n \in \Omega_N \tag{5.17}$$

$$\tilde{h}_{m,t,n} = \tilde{h}_{m,t-1,n} + h_{m,t,n}\tau_t, \quad \forall t \in \Omega_T^b \backslash \{\sigma_b\}, \forall n \in \Omega_N \tag{5.18}$$

Constraints (5.11)–(5.18) describe the system's pre-fault system operation. Equation (5.11) states that at each node, the demand $D_{t,n}$ is equal to the net of incoming and outgoing power because of generation and power flows ($u_l = n$ means that line has been defined as originating from bus n while $v_l = n$ means that line has l been defined as terminating to bus n) while also considering charging/discharging of ES devices. Variable d represents demand curtailment and essentially acts as a slack variable to ensure feasibility; it is penalized by Γ, essentially representing the value of lost load. Constraint (5.12) establishes the limits on generation dispatch; power output cannot surpass the generator's maximum stable generation level $\overline{p}_{m,g}$. It is crucial to highlight that since the capacity of some generators is uncertain, this input parameter depends upon the scenario tree node m. In the case of intermittent plants, time-variable limits should be introduced to scale output capability. Equation (5.13) states the standard DC power flow equation, which describes how power flows in the network where b_l denotes the line's susceptance; note that PS capability to change bus angles is also considered. Constraints (5.14)–(5.16) are particularly important since they couple operation and investment decisions. They bound operation decision variables according to what investments have been completed at the node m being evaluated. In particular, constraint (5.14) states that the power flow on line is bounded by the sum of the initial corridor capacity F_l^0 and the total capacity upgrades deployed up until that point (i.e. scenario tree node m). In a similar manner, constraint (5.15) states that the phase-shifter is operational and can shift the bus angle up to the device's technical capability $+\overline{\xi}$ only if this asset has been commissioned following investment. Constraint (5.16) establishes charge/discharge bounds of each ES device, according to the corresponding investment decisions where $\overline{h}$ is the candidate ES power rating. Constraint (5.17) bounds ES state-of-charge (SOC), where $\overline{\eta}$ denotes the ES energy rating. Constraint (5.18) establishes that the SOC is computed as the summation of current and past charging and discharging activity. Note that an initial condition, defining SOC at the first period of each temporal block according to user input, should be included in the formulation.

$$\forall t \in \Omega_T^b, \forall c \in \Omega_c :$$

$$\sum_{\forall g \in n} p_{m,t,g} + \sum_{\forall \{l: v_l = n\}} f_{m,t,c,l}^C - \sum_{\forall \{l: u_l = n\}} f_{m,t,c,l}^C = D_{t,n} - d_{m,t,n} + h_{m,t,c,n}^C, \quad \forall n \in \Omega_N \tag{5.19}$$

$$f_{m,t,c,l}^C = \begin{cases} b_l\left(\theta_{m,t,c,u_l}^C - \theta_{m,t,c,v_l}^C\right) & \text{if } c \neq l \\ 0 \text{ if } c = l \end{cases}, \qquad \forall l \in \Omega_L \tag{5.20}$$

$$|f^C_{m,t,c,l}| \leq \begin{cases} F^0_l + \tilde{F}^*_{m,l} & \text{if } c \neq l \\ 0 \text{ if } c = l \end{cases}, \quad \forall l \in \Omega_L \tag{5.21}$$

$$|\xi^C_{m,t,c,l}| \leq \begin{cases} Q^*_{m,n}\overline{\xi} & \text{if } c \neq l \\ 0 \text{ if } c = l \end{cases}, \quad \forall l \in \Omega_L \tag{5.22}$$

$$|h^C_{m,t,c,n}| \leq H^*_{m,l}\overline{h}, \quad \forall n \in \Omega_N \tag{5.23}$$

$$h^C_{m,t,c,n} \leq H^*{}_{m,n}\overline{\eta} - h_{m,t,n}, \quad \forall n \in \Omega_N \tag{5.24}$$

Constraints (5.19)–(5.24) model the system's post-fault operation, where we consider all $N-1$ line outages. In order to describe how the system is operated after a potential $N-1$ fault (i.e. post-fault operation), we use a new set of variables. All post-fault variables are denoted by the superscript C and an extra index $c \in \Omega_L$, where c denotes the line under fault. Constraint (5.19) enforces system balance following a line fault by linking the pre-fault generation dispatch to line flows after the outage and any corrective control measures that may be available (e.g. corrective dispatch of ES and corrective operation of PS). Constraints (5.20)–(5.24) largely follow their pre-fault equation counterparts (5.13)–(5.18), while also ensuring that flow and PS-driven angle shift over the out-of-service line is zero (i.e. when $c = l$).

5.2.3 Decomposition for computational tractability

In general, multi-stage problems can be very large in size, leading to computational intractability. In addition, the consideration of ES leads to an even further increase in the computational burden due to the time-coupled constraints introduced. In response, decomposition schemes can be used to alleviate the burden and allow the optimization of realistic-sized systems with the necessary level of detail. In general, the planning-operation problem can be visualized as in Figure 5.8 below. As can be

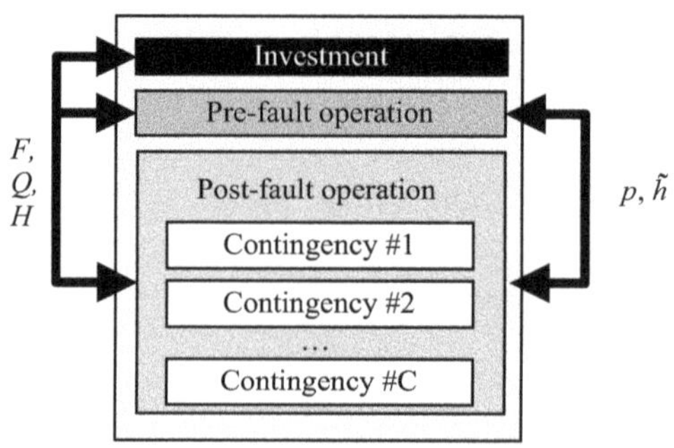

Figure 5.8 Schematic showing the levels of coupling in a single-stage planning problem

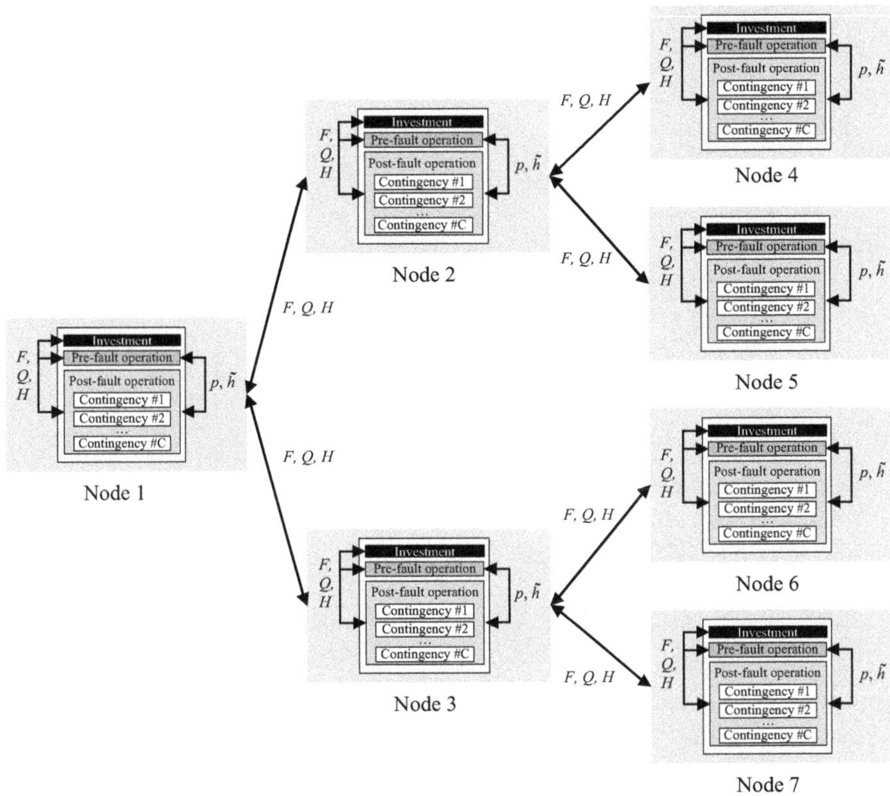

Figure 5.9 Schematic showing the levels of coupling in a multi-stage stochastic planning problem

seen it accommodates three sets of constraints and variables related to investment, pre-fault and post-fault operation. Of course these three levels are inter-related via complicating variables and coupling constraints, as shown in the figure. For example, the investment decisions F, Q and H impact upon both pre-fault and post-fault operation. In a similar vein, the pre-fault generation and storage dispatch schedule impact on what can be carried out post-fault.

Extending the above schematic to the present multi-stage stochastic planning case, the coupling structure looks like Figure 5.9, where investment decisions introduce path-dependent coupling across the different scenario tree nodes.

5.2.3.1 Hierarchical decomposition via Benders

Many problems in power system planning exhibit a structure suitable for Benders decomposition. Reference [31] is a good summary of the advent of Benders decomposition applications to problems typically faced in deregulated power systems. The principle of this technique is to take advantage of the problem structure and split the large original problem into a master and a subproblem. The master problem is solved while approximating the subproblem's optimal value. The master's optimal solution of the complicating variable constitutes a trial value that is passed to the subproblem. The subproblem is then solved with respect to the proposed trial value and the dual variable of the coupling constraint[1] is used to construct a linear constraint (also known as Benders cut) that is appended to the master problem. The set of the appended Benders cuts constitutes a linear piecewise representation of the subproblem. This process is repeated in an iterative manner where additional Benders cuts are added until the master's subproblem approximation accurately represents the subproblem.

Master problem

The master problem's objective function (5.25) approximates the original expected total system cost (5.1). However, the operational cost has been replaced with the probability-weighted estimate expressed in terms of α. This estimate is progressively informed through the appended Benders cuts until it is equal to the true optimal value. In general, the master problem consists only of investment-related variables and constraints and the impact of operation on system cost is captured solely through the approximation terms α.

$$z^* = \min_{F,Q,H} \left\{ \sum_{\forall m} \pi_m r^I_{\varepsilon m} \psi_m + \sum_{\forall m,b} \pi_m \alpha_{m,b} \right\} \tag{5.25}$$

Subject to
(5.2)–(5.5)

Subproblem

We create one operational subproblem per scenario tree node per time block. The objective function is defined as the sum of generation and demand curtailment costs. The problem pools all constraints and variables associated with system operation and considers the investment variables as input from the master problem. The latter is enforced by introducing auxiliary variables and using Equations (5.27)–(5.29) shown below, similar to the approach adopted in [32].

$$\omega^{(k)}_{m,b} = \min_{p,d} \left\{ r^o_{\varepsilon_m} \sum_{t \in \Omega^b_T} \tau_t \left[\sum_{\forall g} p_{m,t,g} \kappa^G_g + \sum_{\forall n} d_{m,t,n} \Gamma \right] \right\} \tag{5.26}$$

Subject to

$$F^*_{m,l} = \tilde{F}^{(k)}_{m,l} : \lambda^{F,k}_{m,b,l}, \qquad \forall l \in \Omega_L \tag{5.27}$$

$$Q^*_{m,l} = \tilde{Q}^{(k)}_{m,l} : \lambda^{Q,l}_{m,b,l}, \qquad \forall l \in \Omega_L \tag{5.28}$$

[1] The coupling constraint is the constraint in the subproblem that includes the complicating variable, thus coupling the trial value to the subproblem's objective function.

$$H^{*}_{m,n} = \tilde{H}^{(k)}_{m,n} : \lambda^{H,k}_{m,b,n}, \qquad \forall n \in \Omega_N \tag{5.29}$$

$$(5.11)-(5.18) \tag{5.30}$$

Finally an important differentiation of the presented scheme is that the post-fault constraints (5.19)–(5.24) presented earlier are not imposed for all contingencies Ω_c. In fact, in most cases a very small number of contingencies are usually problematic and require resource re-dispatch compared to the pre-fault case. As such, post-fault constraints related to these non-problematic cases can be considered unnecessary, resulting in a substantial reduction of problem size. A contingency screening module is used to determine which contingencies lead to post-fault demand curtailment and should be included in the formulation. The list of binding contingencies $\tilde{\Delta}_{m,t,c}$ is iteratively constructed ($\tilde{\Delta}_{m,t,c} = 1$ signifies a condition that can be potentially binding and should be included in the formulation). As such, the following constraints are added to the operation subproblem:

$$\forall c \in \Omega_L : \tilde{\Delta}_{m,t,c} = 1 \tag{5.31}$$

$$(5.19)-(5.24) \tag{5.32}$$

Benders cut

Following the multi-cut paradigm, at each iteration, the master problem increases by $|\Omega_M||\Omega_B|$ constraints. Each constraint (also known as cut) provides a lower bound estimate for $\omega_{m,b}$, i.e. the operational cost of demand block (m, b). Note that the Benders cut is formulated in terms of the optimal investment decisions made in the previous iteration $(k + 1)$. The corresponding dual variables are denoted by λ and the subproblem's optimal value is denoted by ω.

$$\alpha_{m,b} \geq \omega^{(k-1)}_{m,b} + \begin{array}{l} \sum_{\forall l} \lambda^{F,(k-1)}_{m,b,l}\left(\tilde{F}_{m,l} - \tilde{F}^{(k-1)}_{m,l}\right) +, \qquad \forall m \in \Omega_M, \forall b \in \Omega_B \\ \sum_{\forall l} \lambda^{Q,(k-1)}_{m,b,l}\left(\tilde{Q}_{m,l} - \tilde{Q}^{(k-1)}_{m,l}\right) + \\ \sum_{\forall n} \lambda^{H,(k-1)}_{m,b,n}\left(\tilde{H}_{m,n} - \tilde{H}^{(k-1)}_{m,n}\right) \end{array} \tag{5.33}$$

Criterion of convergence

The criterion shown in (5.36) is expressed as a function of the upper–lower bound distance of the objective function, as follows:

$$\frac{Z^{(k)}_{upper} - Z^{(k)}_{lower}}{Z^{(k)}_{upper}} \leq \varepsilon \tag{5.34}$$

$$Z_{upper}^{(k)} = \sum_{\forall m} \pi_m r_{\varepsilon(m)}^{l} \psi_m^{(k)} + \sum_{\forall m,b} \pi_m \alpha_{m,b} \tag{5.35}$$

$$Z_{lower}^{(k)} = \sum_{\forall m} \pi_m r_{\varepsilon(m)}^{l} \psi_m^{(k)} + \sum_{\forall m,b} \pi_m \omega_{m,b}^{(k)} \tag{5.36}$$

Note that ε, which is the threshold value below which convergence is achieved, should be chosen to be close to zero, so as to guarantee a small approximation error.

Contingency filtering

When convergence has been reached, the obtained investment strategy must result in a system that can sustain all credible contingencies without loss of load. To check if this is the case, we need to simulate each post-fault operating point (m,t,c) and characterize it as binding (results in loss of load) or non-binding (does not result in loss of load). This can be established by quantifying whether system balance following a fault has been violated (captured by slack variables $d_{m,t,c,n}^{+}$ and $d_{m,t,c,n}^{-}$) using the investment solution $\{\tilde{F}^{(k)}, \tilde{Q}^{(k)}, \tilde{H}^{(k)}\}$ and subproblem's pre-fault operation solution $\{p, f, \xi, \overline{h}\}$. An optimization problem for each point (m,t,c) can be formulated as shown below:

$$\delta_{m,t,c} = \min_{d^+, d^-} \left\{ \sum\nolimits_{\forall n} \left(d_{m,t,c,n}^{+} + d_{m,t,c,n}^{-} \right) \right\} \tag{5.37}$$

subject to constraints (5.19)–(5.24).

The node balance equation following a line fault (5.19) now incorporates two slack variables; d^+ and d^- representing shortage of demand or generation, respectively:

$$\sum_{\forall g \in n} p_{m,t,g} + \sum_{\forall \{l:v_l=n\}} f_{m,t,c,l}^{c} - \sum_{\forall \{l:u_l=n\}} f_{m,t,c,l}^{c}$$
$$= D_{t,n} + d_{m,t,c,n}^{+} - d_{m,t,c,n}^{-} + h_{m,t,c,n}^{c}, \quad n\forall \in \Omega_N \tag{5.38}$$

Each operating point can be characterized as safe or not by comparing the objective function with a user-determined value $\overline{\delta}$, appropriately set close to zero, as below:

$$\Delta_{m,t,c} = \begin{cases} 1 & \text{if } \delta_{m,t,c} \geq \overline{\delta} \\ 0 & \text{otherwise} \end{cases} \tag{5.39}$$

When $\Sigma_{\forall (m,t,c)} \Delta_{m,t,c} > 0$, it means that there are some contingencies violating post-fault balance. In this case, the list of offending contingencies must be updated and the model must be solved again. To this end, (5.40) is used:

$$\tilde{\Delta}_{m,t,c} = \max\left(\tilde{\Delta}_{m,t,c}, \Delta_{m,t,c}\right) \tag{5.40}$$

As indicated above, once a point (m,t,c) is found binding, it is always considered as potentially binding and thus included in the list, even if it is found non-binding under the current solution. Using the above update process, we make sure that any potentially offending contingencies are not taken out of the list of faults to be checked, thus preventing model oscillations between solutions.

Full algorithm

As previously stated, Benders decomposition is an iterative process. The full process is shown in Figure 5.10.

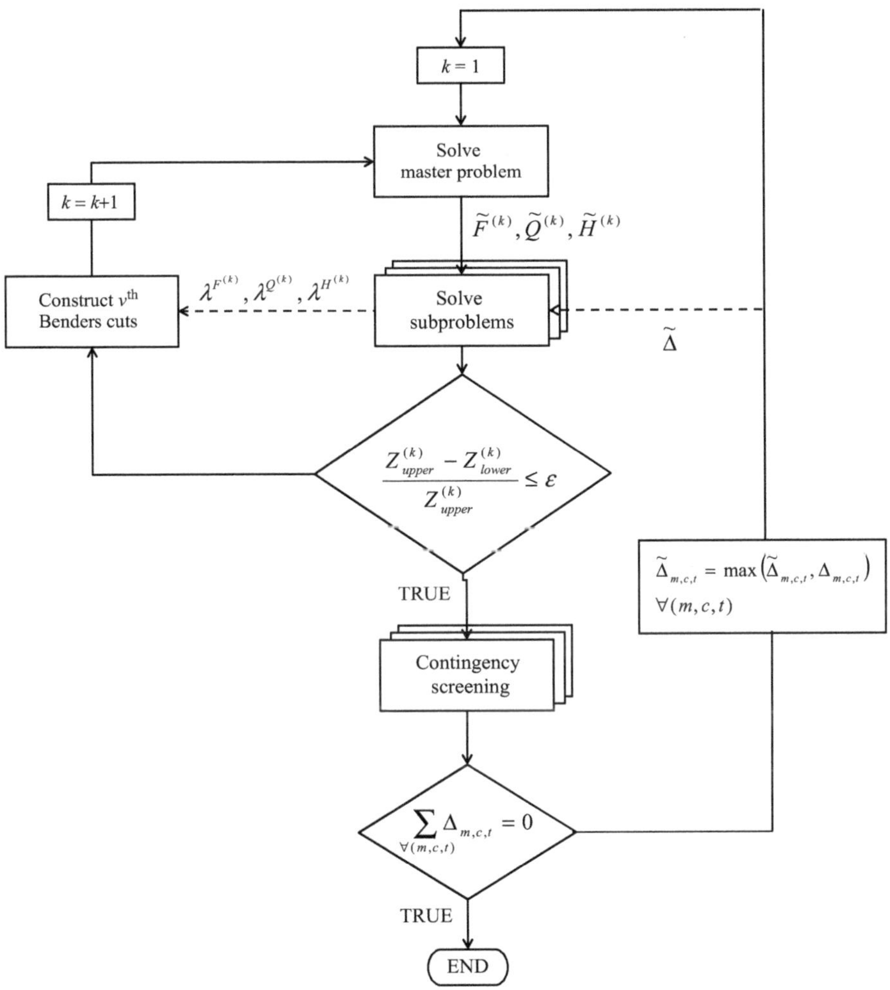

Figure 5.10 The proposed hierarchical decomposition scheme

The process is presented below.

Algorithm 1 Hierarchical decomposition via Benders

Step 1. Initialize an empty list of violating contingencies i.e. $\tilde{\Delta}_{m,t,c} = 0, \forall(m,t,c)$
Step 2. Initialize the Benders iteration index k equal to 1.
Step 3. Clear the master investment problem of any accumulated Benders cuts.
Step 4. Solve the master problem according to the formulation of the subsection 'Master Problem', while also considering all accumulated cuts (5.33).
Step 5. Solve all operation subproblems according to the formulation of the subsection 'Benders cut'.
Step 6. Evaluate convergence according to (5.34). If converged, go to **Step 7**. If not converged, build and append all relevant Benders cuts, update index $k = k + 1$ and go to **Step 3**.
Step 7. Screen all points (m,t,c) according to (5.39).
Step 8. If $\Sigma_{\forall(m,t,c)}\Delta_{m,t,c} = 0$, go to **Step 8**.
Step 9. If $\Sigma_{\forall(m,t,c)}\Delta_{m,t,c} > 0$, update $\tilde{\Delta}$ using (5.40) and return to **Step 2**.
Step 10. Termination.

5.2.3.2 Nested decomposition

It is possible to also decompose the original multistage problem on a stage-per-stage basis. This has been shown to present substantial computational benefits and is highly amenable to parallelization techniques. However such a scheme can be problematic due to (i) non-sequential state equations in the case of investment options having long building times, necessitating the introduction of additional auxiliary variables (ii) the presence of binary variables in the subproblems due to investment decisions across the different tree nodes, warranting the use of sophisticated convexification techniques. The full description of this algorithm is beyond the scope of this chapter, but interested readers are pointed to the authors' articles that contain all formulation details on this topic in [33] and [34].

5.2.4 *Operating point selection*

Under a cost-benefit planning framework, accurately capturing the plethora of operating points that can occur in a manageable subset is a highly important, yet challenging task [35]. Historical data of demand and renewables injections across the different system nodes are usually used as a starting point for the task of scenario generation. Nevertheless, the computational complexity of accommodating the entire population of thousands of operating points in a large-scale MILP planning model is highly problematic. Therefore, it is highly desirable to analyse the original data set of historical operating points and select a small set of representative scenarios that can lead to efficient planning decisions. At the high level, there are four major approaches for tackling this task, as outlined in [36]:

Clustering operating points in the *input space* (i.e. using parameters $D_{t,n}$ and $\phi_{t,n}$) where the operating points to be included in the planning model are chosen a priori on the basis of some distance or relevance metric. The obvious advantage of this approach is that it is computationally inexpensive and simple. However, proximity in the input space does not necessarily translate to similarity in the objective space (i.e. investment decisions).

Clustering in the *operational space* (i.e. using decision variables $f_{t,l}$ and $P_{t,g}$). In this case, an optimization model is solved for each candidate operating point; the single point is weighted to result in equivalent annual costs. The advantage of this method is that the clustering scheme considers the vicinity of operation points in the model's output space.

Clustering in the *investment space* (i.e. using decision variables $\tilde{F}, \tilde{Q}$ and $\tilde{H}$) can prove to be even better since investment decisions are the main focus of a planning model.

Finally, there is the hybrid *bi-level* method, where an initial clustering takes place in the investment space and a subsequent scheme clusters all operating points that do not result in investment using the respective operation state variables.

Note that clustering can be extended to time series, which are necessary when considering ES, by extending the input space dimensionality; numerous publications exist on how to cluster domestic customer profiles using standard and composite techniques (e.g. [37]).

5.3 Case study – IEEE-24 system

We proceed by presenting a study that showcases how the proposed methodology can be used to compute the benefit of different investment options under uncertainty. In particular, we investigate how the deployment of ES assists in the management of planning uncertainty and provide strategic flexibility to a transmission development plan.

5.3.1 Description

The studies have been carried out on the IEEE 24-bus reliability system (IEEE-RTS) [38]. The original bus numbering convention has been preserved, as shown in Figure 5.11. For simplification purposes, the length of all lines has been set to 50 km and each line's initial capacity rating has been set so as to just allow uncongested operation under all single line faults. In order to avoid islanding under $N-1$, an extra line has been added to connect buses 7-8. The operating costs of oil, coal and nuclear generators have been set to 150, 50 and 6£/MWh, respectively. Maximum demand is set at 2,850 MW and remains unchanged throughout the three stages studied, while initial total generation is 3,105 MW.

The study horizon has been split in three epochs, each epoch representing five years of operation. Over this horizon, an unknown amount of wind generation will be added to node 24 (initially no wind on the system exists). Figure 5.12 shows the scenario tree that describes four possible trajectories regarding the capacity of

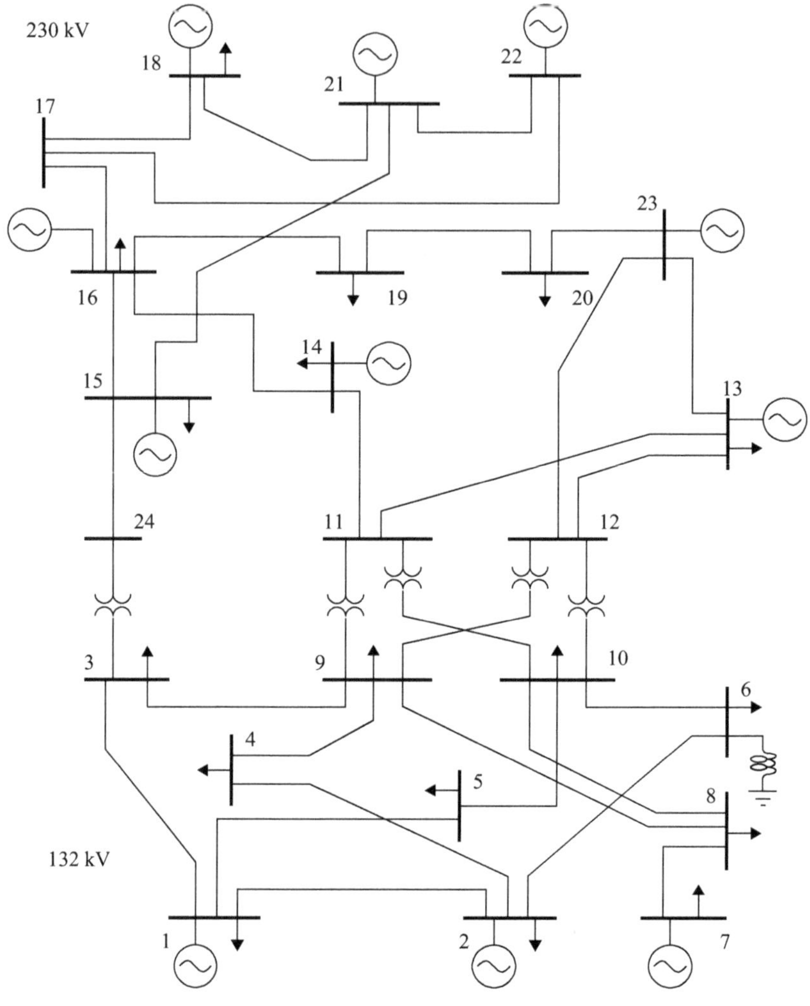

Figure 5.11 The IEEE-24 electricity transmission system

wind. For example, Scenario 1 (denoted S1) represents the build-out of 1,600 MW of wind capacity. In contrast, no wind capacity is constructed under S4. Note that our model allows for additional sources of uncertainty, but for reasons of simplicity we have limited our study to just one uncertainty source.

In order to capture the different demand and wind conditions that may arise in the network, five operation blocks are used, each block spanning 168 h. Four weekly blocks are used to represent the four calendar seasons, while an extra block corresponds to mid-winter peak demand conditions. Each temporal block is repeated 12.5 times. The block corresponding to peak loading conditions is not repeated and is assumed to occur only once. In this case study, demand and wind data from

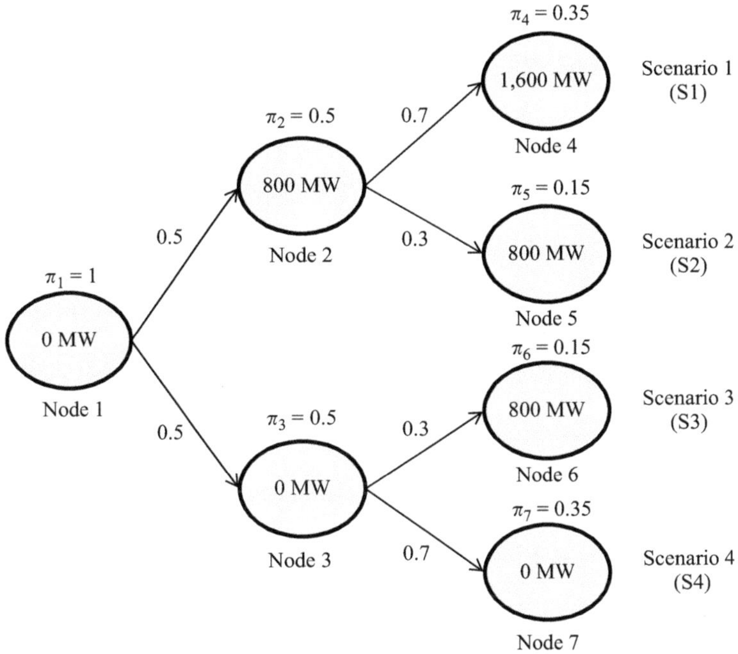

Figure 5.12 Scenario tree with node and transition probabilities

Great Britain in the year 2012 have been used. According to the 2012 historical data, mean demand and wind factors were 65.8% and 32.5%, respectively.

To accommodate the output of cheap wind power, the transmission operator can decide investing in the options shown in Table 5.1 and Table 5.2. Note that line upgrades have been modeled using an assumption of a 1 epoch (i.e. five years) build time. Conversely, ES and PS can be deployed with minimum delay as we assume that they are not subject to lengthy permission processes. ES plants have an energy rating $\overline{\eta}$ of 1,600 MWh and a power rating $\overline{h}$ of 400 MW; it can be seen as either one plant or an aggregation of multiple plants in the area. The maximum angle for phase-shifting transformers $\overline{\xi}$ has been set to 30°. A penalty value Γ of 30,000£/MWh was used for all simulations. Note that a 5% discount rate has been used in this study.

In order to further understand the impact of uncertainty and quantify the benefit of flexible assets, we have carried out studies using three different models:

- De0: Deterministic model where all assets (i.e. lines, ES and PS) are allowed.
- St1: Stochastic planning allowing investment solely in line upgrades.
- St2: Stochastic planning allowing investment in line upgrades, ES and PS devices.

For all three studies, convergence criterion $\overline{\zeta}$ was set to 0.1%.

Table 5.1 Transmission line reinforcement options

Asset type	Reinforcement capacity [MW]	Annualized capital cost [£/year]	Build time
Option A	200	1,500,000	1 epoch
Option B	400	2,500,000	1 epoch

Table 5.2 Flexible investment options

Asset type	Annualized capital cost [£/year]	Build time
Phase-shifter	600,000	0 epochs
Storage device	15,000,000	0 epochs

5.3.2 Deterministic planning

We first show results from the deterministic case study, where uncertainty is ignored and each scenario is solved in isolation.

Figure 5.13 shows the optimal planning schedule for all runs. For instance, the entry 'B(15-24)' corresponds to the upgrade of the corridor connecting buses 15 and 24 using upgrade option 'B'. ES assets are denoted as STOR while PS devices as PS. Table 5.3 shows investment, operation and total costs across all scenarios.

By ignoring uncertainty, we can identify the optimal expansion schedule while assuming perfect foresight in future developments. Under S1, which represents an eventuality of very high wind deployment, the transmission operator decided to commit to upgrading corridors (3-9), (3-24) and (15-24) from the very first stage (these are the main exporting corridors of the new wind farm); these commitments correspond to a cost of £70.8m. Further investments are undertaken in later stages, aimed at exploiting the phase shifters' ability to provide post-fault controllability. In the case of scenario 2, the main corridors are again upgraded, but now using option 'A' since there will smaller power flows to be accommodated. The same plan is followed in the case of S3, but now the planner can defer these commitments to the second stage. As expected, no investment is necessary in S4 since the system undergoes no change according to the scenario definition.

5.3.3 Stochastic planning–no storage

This study employs the full stochastic formulation presented in the previous section, but forces the planner to not invest in alternative technologies, i.e. system reinforcement is limited solely to line upgrades. Figure 5.14 displays investment decisions for each scenario, while Table 5.4 presents respective costs along with the overall expected cost.

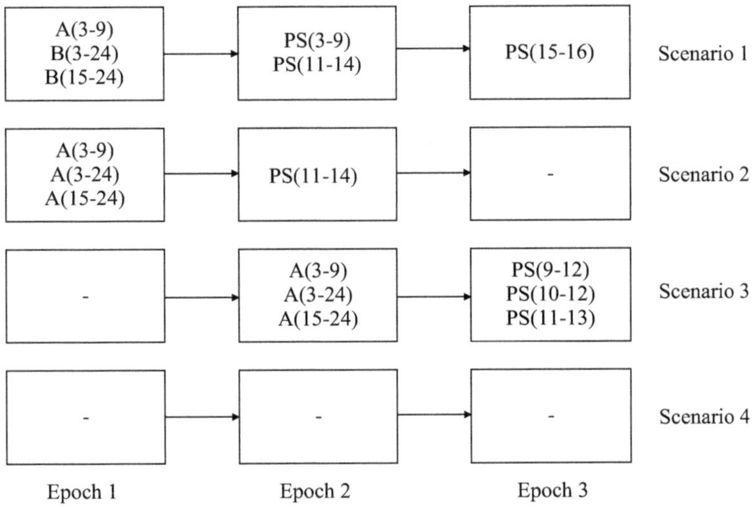

Figure 5.13 Optimal investment decisions for each different scenario under the deterministic planning paradigm (i.e. uncertainty is ignored)

Table 5.3 System costs for each scenario when ignoring uncertainty

	Investment cost	Operation cost	Total cost
Scenario 1	91.3	4,957.4	5,048.8
Scenario 2	52.9	5,267.7	5,320.6
Scenario 3	33.6	5,834.9	5,868.6
Scenario 4	0.0	6,295.1	6,295.1

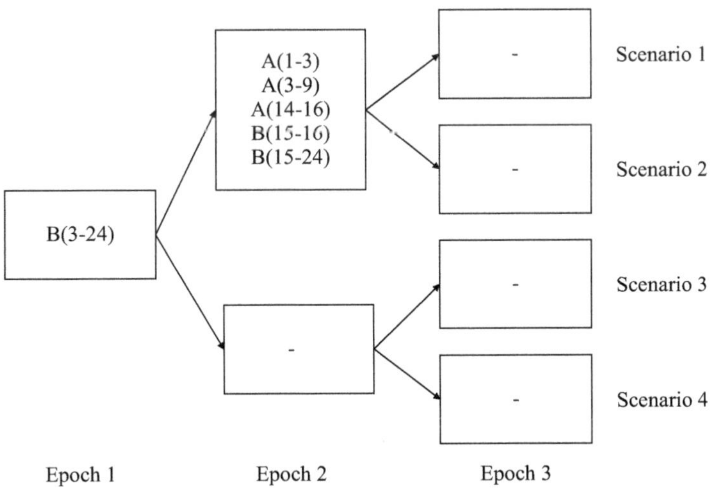

Figure 5.14 Optimal investment decisions for stochastic case study (no storage)

Table 5.4 System costs for stochastic case study (no storage)

	Investment cost	Operation cost	Total cost
Scenario 1	87.6	5,078.7	5,166.3
Scenario 2	87.6	5,336.5	5,424.1
Scenario 3	27.2	5,897.1	5,924.4
Scenario 4	27.2	6,295.1	6,322.3
Expected	57.4	5,665.9	5,723.3

As can be seen above, there is an unconditional first-stage commitment to the corridor between buses 3 (that has high demand) and 24 (wind bus) that connects the bus 3 to the incoming wind generator. Of course, this decision entails substantial risk, since it can become stranded under scenario S4 and could have been deferred to a later stage under S3. The investments implemented in the transition from the root node to node 2 involve a number of line upgrades targeted at the main exporting corridors as in the deterministic case. Note that some additional investments, compared to the deterministic case, are made in corridors (14-16) and (15-16) due to the lack of post-fault controllability that was provided by phase-shifters. No further investment is required in later stages under S3 and S4.

5.3.4 Stochastic planning with storage

This study employs the full stochastic formulation presented in the previous section, while also allowing investment in storage. Figure 5.15 displays investment decisions for each scenario, while Table 5.5 presents respective costs along with the overall expected cost.

In the case where investment in flexible assets is allowed, one important consequence is that commitments can be deferred form the first stage to the second stage. In particular, under S1, an ES device is built on the wind-exporting bus 24 so as to charge up when wind is abundant and discharge during periods of low wind and high load. This way, wind spillage can be minimized. To follow up under the high-growth trajectory, a number of line upgrades are commissioned in the third stage. PS devices are also deployed under scenarios 1 and 2, displacing the need for further security-driven reinforcements.

It is important to emphasize that the possibility for contingent deployment of ES and phase-shifting transformers allows congestion management in scenario tree node 2 without having to preemptively committing to upgrading the corridor between buses 3 and 24 as seen under the previous study, St1. In contrast, the low-growth transition (node 1 → node 3) does not warrant ES investment. Instead, a line upgrade of (3-24) is the preferable option.

We highlight the fact that investment undertaken in the first stage is much reduced compared to the deterministic study De0 for scenarios 1 and 2. This is because commitments taken in the first stage (i.e. root note), entail a large risk of asset stranding since these assets will be built under all envisioned scenario realizations and there is no possibility for strategic differentiation. In view of this, a

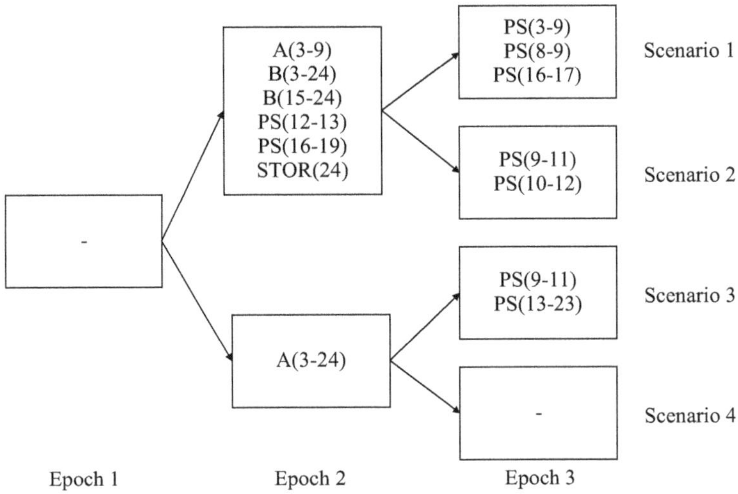

Figure 5.15 Optimal investment decisions for stochastic case study (with storage)

Table 5.5 System costs for stochastic case study (with storage)

	Investment cost	**Operation cost**	**Total cost**
Scenario 1	149.2	5,009.9	5,159.1
Scenario 2	147.6	5,253.7	5,401.3
Scenario 3	12.9	5,875.4	5,888.3
Scenario 4	9.5	6,295.1	6,304.6
Expected	79.6	5,626.1	5,705.7

strategic planner should choose to limit his first-stage commitment and try to move all decision further in time in later stages, when more information will be available. In addition, it is critical to emphasize that although expected cost of investment is higher in the latter St2 study, we see a reduction of all scenario-specific total costs. This means that no matter which scenario actually occurs, the planner is better off having the option of using flexible assets (such as ES), even if the planner may not eventually use them under some scenarios (e.g. S3 and S4). In terms of expected total cost, the option value of flexible assets is £17.6m. Even though this may seem a small benefit compared to the overall cost (which is dominated by the operational cost component, i.e. the cost of operating this large electricity system over one and a half decades), it is indeed substantial when compared to the capital cost of the ES asset considered. In fact, when viewing this benefit from the point of view of a potential ES investor, this option value could constitute a fundamental part of the overall cost-benefit business case. One of the main ideas of this chapter is that given the potential system benefits that ES can provide for managing uncertainty, suitable mechanisms should be in place to reward this ability to manage long-term uncertainty.

5.3.5 Discussion and future directions

Through the case studies discussed earlier, we have shown that by ignoring uncertainty and adopting a deterministic world view can systematically ignore flexible assets such as ES and focus solely on exploiting scale economies present in transmission investment. On the other hand, explicitly taking into account uncertainty by employing a stochastic planning model can identify attractive risk-averse investment strategies. What is critical to emphasize is that although some investment decisions (such as ES) may be not be optimal when using a deterministic world view, they can be turned out to, in fact, be valuable strategic option once uncertainty is actually considered instead of ignored. Investing in new technologies such as phase-shifting transformers and ES can enable the planner to adopt a 'wait-and-see' strategy by deferring 'here-and-now' commitments to large projects with considerable building times and limited operational flexibility provision. In the meantime, interim system operation can be facilitated and security can be ensured by deploying smart assets, such as ES that can provide a range of services to assist pre- and post-fault operation.

This highlights the increasing importance of incorporating ES in the planning process and adopting a stochastic framework where optimal solutions can be identified on the basis of future system's ability to adapt to a range of possible scenarios. For such approaches to gain ground, it is important to overcome substantial modeling and computational challenges for building useful tools as well as further studying the associated regulatory and commercial aspects of how to incentivize investments that possess substantial option value under uncertainty.

References

[1] G. Strbac, M. Pollitt, C. Vasilakos Konstantinidis, *et al.*, "Electricity transmission investment arrangements in Great Britain: Time for change?", *Energy Policy*, vol. 73, pp. 298–311, 2014.

[2] G. Strbac, I. Konstantelos, M. Pollitt, and R. Green, "Delivering future-proof energy infrastructure: Report for the National Infrastructure Commission", 2016. [Online]. Available: 24 https://www.gov.uk/government/uploads/system/uploads/attachment_data/file/507256/Future-proof_energy_infrastructure_Imp_Cam_Feb_2016.pdf

[3] D. Pudjianto, P. Djapic, M. Aunedi, *et al.*, "Smart control for minimizing distribution network reinforcement cost due to electrification", *Energy Policy*, vol. 52, pp. 76–84, 2013.

[4] I. Konstantelos and G. Strbac, "Valuation of flexible investment options under uncertainty", *IEEE Transactions on Power Systems*, vol. 30, no. 2, pp. 1047–1055, March 2015.

[5] G. Strbac, M. Aunedi, I. Konstantelos, *et al.*, "Opportunities for energy storage: Assessing whole-system economic benefits of energy storage in

future electricity systems", *IEEE PES Magazine*, vol. 15, no. 5, pp. 32–41, September 2017.

[6] NREL, Energy Storage: Possibilities for Expanding Electric Grid Flexibility, February 2016. [Online]. Available: https://www.nrel.gov/docs/fy16osti/64764.pdf

[7] I. Konstantelos, P. Djapic, G. Strbac, P. Papadopoulos, and A. Laguna, "Contribution of energy storage and demand response to security of distribution networks", *CIRED* 2017, Glasgow.

[8] S. Giannelos, I. Konstantelos, and G. Strbac, "Option valuation of storage technologies under decision-dependent innovation uncertainty", *IEEE PowerTech 2017*, Manchester.

[9] B. G. Gorenstin, N. M. Campodonico, J. P. Costa, and M.V.F. Pereira, "Power system expansion planning under uncertainty", *IEEE Transactions on Power Systems*, vol. 8, no. 1, pp. 129–136, February 1993.

[10] P. Vasquez, Z. A. Styczynski, and A. Vargas, "Flexible decision making-based framework for coping with risks existing in transmission expansion plans", *Transmission and Distribution Conference and Exposition: Latin America, 2008 IEEE/PES*, pp. 1–9, August 2008.

[11] S. Giannelos, I. Konstantelos, and G. Strbac, "Option value of soft open points in distribution networks", *IEEE PowerTech 2015,* Eindhoven.

[12] S. Giannelos, I. Konstantelos, and G. Strbac, "Stochastic optimization-based valuation of smart grid options under firm DG contracts", *IEEE EnergyCon 2016*, Leuven.

[13] S. Giannelos, I. Konstantelos, and G. Strbac, "Option value of dynamic line rating and storage", *ISGT Europe 2017*, Turin.

[14] I. Konstantelos, S. Giannelos, and G. Strbac, "Strategic valuation of smart technology options in distribution networks", *IEEE Transactions on Power Systems*, vol. 32, no. 2, pp. 1293–1303, March 2017.

[15] F. D. Munoz, B. F. Hobbs, J. L Ho, and S. Kasina, "An engineering-economic approach to transmission planning under market and regulatory uncertainties: WECC case study", *IEEE Transactions on Power Systems*, vol. 29, no. 1, pp. 307 317, 2014.

[16] R. T. Rockafellar and S. Uryasev, "Optimization of conditional value-at-risk", *Journal of Risk*, vol. 2, pp. 493–517, 2000.

[17] I. Konstantelos, "A stochastic optimization framework for anticipatory transmission investment", Ph.D. Thesis, Imperial College London, pp. 67–72, 2013.

[18] A. Inzunza, R. Moreno, A. Bernales, and H. Rudnick, "CVaR constrained planning of renewable generation with consideration of system inertial response, reserve services and demand participation", *Energy Economics*, vol. 59, pp. 104–117, 2016.

[19] I. Konstantelos, R. Moreno, and G. Strbac, "Coordination and uncertainty in strategic network investment: Case on the North Seas grid", *Energy Economics*, vol. 64, pp. 131–148, May 2017.

[20] A. Moreira, G. Strbac, R. Moreno, A. Street, and I. Konstantelos, "A five-level MILP model for flexible transmission network planning under uncertainty:

A min-max regret approach", *IEEE Transactions on Power Systems*, vol. 33, no. 1, pp. 486–501, January 2018. doi: 10.1109/TPWRS.2017.2710637
[21] B. Chen, J. Wang, L. Wang, Y. He, and Z. Wang, "Robust optimization for transmission expansion planning: Minimax cost vs. minimax regret", *IEEE Transactions on Power Systems*, vol. 29, no. 6, pp. 3069–3077, 2014.
[22] PJM, Energy Storage as a Transmission Asset, 2012. [Online]. Available: https://www.pjm.com//media/markets-ops/advancedtech-pilots/xtreme-power-storage-astransmission.ashx
[23] AEMC, Integration of Energy Storage: Regulatory Implications, October 2015. [Online]. Available: http://www.aemc.gov.au/Major-Pages/Technology-impacts/Documents/Integration-of-Storage-Discussion-Paper.aspx
[24] European Commission, "Energy storage: Which market designs and regulatory incentives are needed?" [Online]. Available: http://www.europarl.europa.eu/RegData/etudes/STUD/2015/563469/IPOL STU(2015)563469 EN.pdf
[25] S. Li and D. Tirupati, "Dynamic capacity expansion problem with multiple products: Technology selection and timing of capacity additions", *Operations Research*, vol. 42, pp. 958–976, 1994.
[26] Z. Hu, F. Zhang, and B. Li, "Transmission expansion planning considering the deployment of energy storage systems", *IEEE Power and Energy Society General Meeting*, San Diego, CA, 2012, pp. 1–6 [Online]. Available: https://ieeexplore.ieee.org/abstract/document/6344575/
[27] F. Zhang, Z. Hu, and Y. Song, "Mixed-integer linear model for transmission expansion planning with line losses and energy storage systems", *IET Generation, Transmission & Distribution*, vol. 7, pp. 919–928, 2013.
[28] C. MacRae, M. Ozlen, and A. Ernst, "Transmission expansion planning considering energy storage", *IEEE International Autumn Meeting on Power, Electronics and Computing (ROPEC), Ixtapa*, 2014, pp. 1–5 [Online]. Available: https://ieeexplore.ieee.org/abstract/document/7036327/
[29] S. Dehghan and N. Amjady, "Robust transmission and energy storage expansion planning in wind farm-integrated power systems considering transmission switching", *IEEE Transactions on Sustainable Energy*, vol. 7, pp. 765–774, 2016.
[30] R. S. Go, F. D. Munoz, and J. P. Watson, "Assessing the economic value of co-optimized grid-scale energy storage investments in supporting high renewable portfolio standards", *Applied Energy*, vol. 183, pp. 902–913, 2016.
[31] M. Shahidehpour and F. Yong, "Benders decomposition: Applying benders decomposition to power systems", *Power and Energy Magazine, IEEE*, vol. 3, no. 2, pp. 20–21, March–April 2005.
[32] N. Alguacil and A. J. Conejo, "Multiperiod optimal power flow using benders decomposition", *IEEE Transactions on Power Systems*, vol. 15, no. 1, pp. 196–201, February 2000.
[33] P. Falugi, I. Konstantelos, and G. Strbac, "Application of novel nested decomposition techniques to long-term planning problems", *PSCC 2016*, Genoa.

[34] P. Falugi, I. Konstantelos, and G. Strbac, "Planning with multiple transmission and storage investment options under uncertainty: A nested decomposition approach", *IEEE Transactions on Power Systems*, vol. 33, no. 4, pp. 3559–3572, July 2018.

[35] M. Sun, I. Konstantelos, and G. Strbac, "Transmission network expansion planning with stochastic multivariate load and wind modelling", *PMAPS 2016*, Beijing.

[36] M. Sun, T. Fei, I. Konstantelos, and G. Strbac, "An objective-based scenario selection method for transmission network expansion planning with multivariate stochasticity in load and renewable energy sources", *Energy*, vol. 145, pp. 871–885, February 2018 [Online]. Available: https://www.sciencedirect.com/science/article/pii/S0360544217321977

[37] M. Sun, I. Konstantelos, and G. Strbac, "C-Vine copula mixture model for clustering of residential electrical load pattern data", *IEEE Transactions on Power Systems*, vol. 32, no. 3, pp. 2382–2393, May 2017.

[38] Reliability Test System Task Force of the Application of Probability Methods Subcommittee, "IEEE reliability test system", *IEEE Transactions on Power Systems*, vol. 13, no. 6, pp. 2047–2054, November/December 1979.

Chapter 6

Sizing of battery energy storage for end-user applications under time of use pricing

Guido Carpinelli[1], *Pierluigi Caramia*[2], *Fabio Mottola*[2] *and Daniela Proto*[1]

6.1 Introduction

The increasing peak energy consumption in modern grids makes demand response a key solution for utilities to balance production and loads. Achieving price responsive demand is the best way for utilities to reduce peaks and avoid grid congestions. For this reason, many demand response programmes have been implemented by utilities thus giving consumers the chance to play an active role in the operation of the electric grid by shifting their electricity consumption on the basis of the electricity tariffs. This is made in response to time-based rates in which energy prices vary over time. All this is possible thanks to the use of smart meters, capable of recording electricity usage on frequent basis. Smart meters, now widely diffused in modern distribution grids, are one of the few elements contributing to the quick evolution of the grids into the new concept of smart grids. In this context, storage systems play a significant role since they allow customers to reduce the costs sustained for electricity by taking part to demand response programmes (thus shifting their demand) without impacting their comfort or the manufacturing process. Thanks to the presence of the storage system, in fact, a user can reduce the electrical energy requested from the grid during on-peak hours (when the grid needs load's curtailment) while continuing to supply its loads from the storage system. In this way, the load demand management is performed without impacting the user's comfort. This is even more important in the case of industrial customers for whom load shifting could result in changes in the manufacturing process and thus in uneconomical operation.

The reduction of the expense for the electricity by using energy storage systems is only possible if the storage system is optimally planned. Battery energy storage systems (BESSs), in fact, are still an expensive technology even if the expected technological trend presages a sensible reduction of their costs in the next future.

[1]Department of Electrical Engineering and Information Technology, University of Naples Federico II, Napoli, Italy
[2]Department of Engineering, University of Naples Parthenope, Napoli, Italy

Thus, significant benefits can be attained by their use if their optimal planning is achieved, first. In the case of BESSs, the planning is strictly related to their operation and end-users can maximize the benefits achievable in terms of electricity cost by optimally sizing their BESS once imposed its correct operation. However, the BESS optimal sizing involves the solution of a cumbersome problem, taking into account that some of the inputs of the sizing procedure are characterized by significant uncertainties. For example, load demand as well as price of energy are typical variables subject to random changes along the years and, then, they cannot be considered deterministic input quantities. At this purpose, a probabilistic approach must be adopted that treats some of the input variables as random variables.

In the relevant literature, probabilistic approaches for the sizing of batteries equipping generation systems based on renewable energy sources have been proposed [1–4]. With reference to the case of end-user applications, in some papers, the battery sizing issue was dealt with for both residential and industrial customers [5–10]. The sizing methodology proposed in [5] refers to industrial customers and is aimed at minimizing the total energy cost while facing uncertainties in energy price and load demand. In [6], the sizing problem focuses on-peak demand minimization by considering only load's uncertainties. These last have been considered also in [7] where the size's choice is based on the energy requests of groups of customers. In [8], a sizing method has been proposed for batteries used as uninterruptible power supplies with the objective of minimizing the energy costs related to power interruptions; in this case, uncertainties on load profile, occurrence and duration of outages' events are taken into account. In [9] and [10], a probabilistic battery optimal sizing is proposed based on the time of use (ToU) tariffs and aimed at cost minimization. Both [9] and [10] consider loads, energy tariffs and discount rate uncertain; in addition, [10] deals also with uncertainties on technological aspects (e.g. battery efficiency and life time duration).

This chapter focuses on the optimal sizing of BESSs in end-user applications in the frame of time-varying energy pricing structures by adopting a probabilistic approach. More in detail, the proposed procedure focuses on one of the most used time-varying tariff structures (i.e. the ToU tariff) but it can be easily extended to other structures. In this chapter, starting from the procedure proposed in [9] and [10], the probabilistic optimal sizing is performed applying the point estimate method (PEM), an algorithm that guarantees accuracy of the results with computational effort significantly lower than that implied by the Monte Carlo procedure [11–18].

The chapter is structured as follows. A brief description of the energy tariff structures is presented in Section 6.2. Then, the general formulation of the sizing procedure is described in Section 6.3. Section 6.4 describes the probabilistic approach used to size the BESS. In Section 6.5, numerical applications are proposed with reference to different typologies of end-user applications. Finally, conclusions are drawn in Section 6.6.

6.2 Energy tariff structures

The energy tariffs can be based on time-variant and non-time-variant pricing structures. Time-variant pricing is able to reflect variation over time of the energy costs. Non-time-variant pricing is characterized by the same charge independently of the ToU or season. Already implemented time-varying tariffs are (Figure 6.1): real-time pricing (RTP), ToU and critical peak pricing (CPP). Other tariff structures that are basically a further particularization of the above are the variable peak pricing (VPP) and critical peak rebate (CPR) [19]. Figure 6.1 shows the aforementioned pricing options together with a typical daily load demand. For comparison purposes, in the figure, in correspondence of each time-varying tariff structure, the standard flat rate is shown which refers to a constant tariff independent of the season or time of day. The dotted lines in the figure refer to critical load demand (e.g. during a heat wave) and the corresponding values of the tariff.

In the figure, peak periods of the day are also evidenced, by means of a grey area. The pricing options are shown through separate graphs representing the case of RTP, ToU, VPP, CPP and CPR. In all these graphs, the standard flat rate is also reported.

RTP is the tariff structure that tunes more finely the variation of the energy cost. In this tariff, the price varies during the day to reflect price variations in the wholesale electricity market. Typically, the electricity price varies hourly. ToU tariffs are probably the most frequently applied nowadays. With this tariff, the day is divided into periods of time, typically two periods that are peak period and off-peak period (in case of three periods, a mid-peak period is also considered). The ToU tariff values as well as the number of time periods in which the day is divided, typically vary with the seasons (i.e. winter and summer tariffs). VPP is a particular type of ToU (dynamic ToU tariff) where the peak price varies depending on system conditions. CPP provides for higher electricity costs on specific days of the year characterized by high demand which are typically identified as 'critical events'. CPR is a particular CPP where the utility pays the customers to reduce an amount of the electricity normally used during 'critical events'. Pricing programmes based on time-varying tariff structures allow applying demand response so achieving various benefits in terms of energy efficiency, customer's bill reduction, pollution reduction, shifting of network reinforcement costs, etc. As reported in [19], numerous pilots effected in the United States have shown that time-varying tariff structures can substantially reduce the load when the grid is particularly stressed. In case the CPP rates are implemented, the reduction of the peak load during critical events reached in some cases a value of 47 per cent.

In what follows, without loss of generality, we focus on the ToU tariff structure which is nowadays widely applied.

6.3 The cost of the storage system

The sizing procedure requires the assessment of the costs sustained by the customer for including the BESS with reference to a specified planning period. This cost is given by [9]:

$$C_{\mathrm{T}}(\mathbf{x}) = C_0 + C_{\mathrm{R}} + C_{\mathrm{D}} + C_{\mathrm{M}} + C_{\mathrm{E}} \tag{6.1}$$

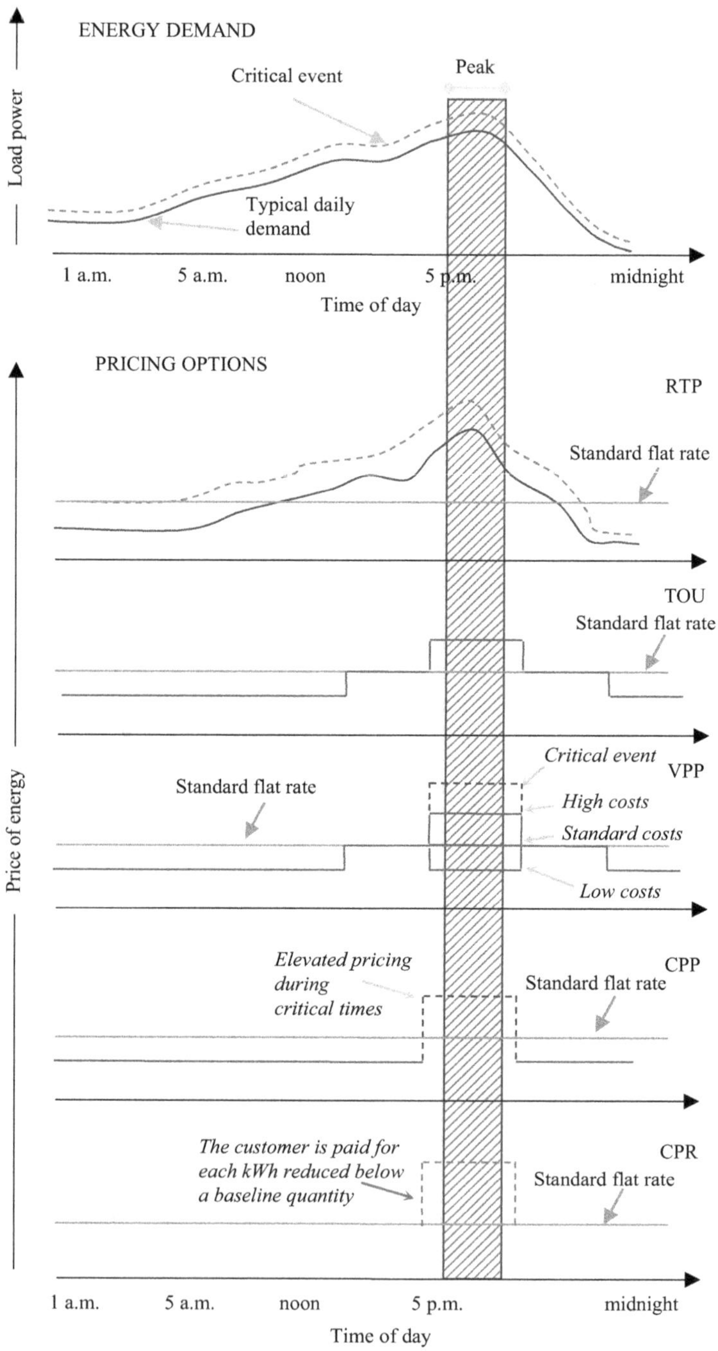

Figure 6.1 How tariff structures work [19]

with **x** the vector of parameters on which the total cost (C_T) depends that, in this chapter, are (i) ToU tariff energy prices, (ii) load power values, (iii) discount rate value for evaluating the costs' net present value and (iv) size of the battery. In (6.1), C_0 is the BESS capital cost, which includes purchase and installation costs of both the battery and the interfacing converter; C_R is the BESS replacement cost, which is related to the need of purchasing a new battery in the planning period, if needed; C_D is the BESS disposal cost, which can take into account also the benefit derived from the recycling of the battery or its secondary use in other less-demanding applications; C_M is the BESS maintenance cost; C_E is the energy cost, which is incurred by the system's owner for the consumption of energy. In the case the customer can sell energy to the grid, this term will also include the relative revenue. It should be noted that the BESS maintenance costs can be included as a percentage of capital costs; in this case, maintenance costs depend only on the BESS' size [20]. Moreover, in the evaluation of the total cost of the battery, for the sake of simplicity, it is supposed that the customer cannot sell energy to the grid. In [9], the case that the customer is allowed to sell energy to the grid is presented.

The BESS capital cost (C_0) in (6.1) includes (i) the battery capacity costs which depends on the specific technology, (ii) the power converter's costs and (iii) the balance of plant's costs which include the costs sustained for project engineering, heating/air-conditioning, protection and safety systems, etc. [21, 22].

The BESS replacement costs in (6.1) have to be considered when the considered planning period exceeds the lifetime of the battery. The battery's lifetime expressed in years (L_B) can assume different values depending on the way the battery is operated. More specifically, it depends on the total number of charging/discharging cycles (N_{cycles}) that the manufacturer provides in correspondence of a specified maximum depth of discharge (DoD):

$$L_B = \frac{N_{cycles}}{365 * v} \tag{6.2}$$

where v is the number of daily charging/discharging cycles. This last value depends on the operational strategy chosen for the BESS.

The BESS's disposal costs in (6.1) are more difficult to evaluate since they strictly depend on the expected development of recycling technology. Also, these costs depend on the country where the battery is disposed. In some cases, disposal activities could also represent a benefit rather than a cost if the possibility of a secondary use of the battery is considered [23].

The BESS operation and maintenance cost (C_M) in (6.1) refers to both the corrective maintenance and preventive maintenance. This cost item refers also to the costs sustained for the equipment to operate over its lifetime (labour, supplies, repair parts and spares, etc.) [24].

The *energy costs* in (6.1) are strictly related to the way the BESS is operated. They include both the electricity costs sustained to supply the customer's loads and those required to charge the battery. The operation strategy of the battery influences the considered costs in that significant benefits can be achieved by using the battery to supply the loads during the high price hours.

More specifically, the *energy costs*, C_E in (6.1) include three terms: the energy expense of loads, the energy expense of charging the battery and the savings related to the provision of energy to the loads by discharging the battery:

$$C_E = C_{ls} + C_{ch} - C_{dch}. \tag{6.3}$$

In (6.3), C_{ls} is the total cost required to supply the customer's loads when BESSs are not installed, C_{ch} is the cost sustained by the customer for the energy needed for charging the battery and C_{dch} is the cost avoided by the customer for the load demand satisfied by the BESS. Obviously, the greater is the value of this last term, the lower is C_E. It has to be noted that C_{ls} does not depend on the BESS size, so it does not influence the optimal sizing solution. However, to evidence the incidence of the BESS on the total cost, it has been considered in the cost analysis of the system. For the evaluation of the terms of C_E (6.3), each season of the planning period is assumed to include a specified number of typical days, each characterized by a specific ToU tariff and load profile.

The cost of the electricity requested by the loads, C_{ls}, in (6.3) for the ToU tariff can be evaluated as follows:

$$C_{ls} = \sum_{n=1}^{N} \frac{1}{(1+\alpha)^{n-1}} \left(\sum_{i=1}^{Sn} d(i) \left(\sum_{j=1}^{T(i)} E_{load}(n,i,j) Pr(n,i,j) \right) \right). \tag{6.4}$$

In (6.4), N is the planning period (years), α is the discount rate for the evaluation of the present value of future costs, Sn is the number of periods (i.e. seasons) into which each year is divided, d(i) is the number of typical days of the ith season, T(i) is the number of time periods into which the day is divided based on the ToU tariff scheme, Pr(n,i,j) is the associated price of energy and $E_{load}(n,i,j)$ is the energy consumed by the load during the jth time period of a typical day of the ith season of the nth year.

Let us refer to the case of a ToU tariff with only two price levels, i.e., off-peak periods and on-peak periods. In this case, (6.4) can be simplified as follows:

$$C_{ls} = \sum_{n=1}^{N} \frac{1}{(1+\alpha)^{n-1}} \left(\sum_{i=1}^{Sn} d(i) \left(\overline{E}_{load}(n,i)\overline{Pr}(n,i) + \overline{\overline{E}}_{load}(n,i)\overline{\overline{Pr}}(n,i) \right) \right) \tag{6.5}$$

In (6.5), two values of energy price are identified with reference to the ith season of the nth year, $\overline{Pr}(n,i)$ and $\overline{\overline{Pr}}(n,i)$, referring to the off-peak and on-peak periods, respectively. The total energy consumed by the load during the off-peak period is referred to as $\overline{E}_{load}(n,i)$, and that absorbed during the off-peak period is $\overline{\overline{E}}_{load}(n,i)$. Starting from the profile of the load demand of a typical day, the energy required by the load during the above periods can be discretized by splitting the day into n_t time intervals of duration Δt, each characterized by a constant value of active power requested by the load $p_{load}(n,i,t)$, (with $t = 1, \ldots, n_t$). This would result in a specific number of time intervals belonging to the off-peak and on-peak periods, $\overline{\mathrm{T}}(\mathrm{n,i})$ and $\overline{\overline{\mathrm{T}}}(\mathrm{n,i})$, respectively. Then, the energy requested by the load during the off-peak and on-peak periods in Eq. (6.5) is:

$$\overline{\mathrm{E}}_{\mathrm{load}}(\mathrm{n,i}) = \sum_{\mathrm{t}\in\overline{\mathrm{T}}(\mathrm{n,i})} \mathrm{p}_{\mathrm{load}}(\mathrm{n,i,t})\Delta\mathrm{t} \tag{6.6}$$

$$\overline{\overline{\mathrm{E}}}_{\mathrm{load}}(\mathrm{n,i}) = \sum_{\mathrm{t}\in\overline{\overline{\mathrm{T}}}(\mathrm{n,i})} \mathrm{p}_{\mathrm{load}}(\mathrm{n,i,t})\Delta\mathrm{t}. \tag{6.7}$$

The other two terms in (6.3) are the charging and discharging cost items (C_{ch}, C_{dch}) that can be evaluated as follows:

$$C_{\mathrm{ch}} = \sum_{n=1}^{N} \frac{1}{(1+\alpha)^{n-1}} \left(\sum_{\mathrm{i}=1}^{\mathrm{Sn}} \mathrm{d(i)} \frac{1}{\eta_{\mathrm{ch}}} e(n,i) \overline{\mathrm{Pr}}(\mathrm{n,i}) \right) \tag{6.8}$$

$$C_{\mathrm{dch}} = \sum_{n=1}^{N} \frac{1}{(1+\alpha)^{n-1}} \left(\sum_{\mathrm{i}=1}^{\mathrm{Sn}} \mathrm{d(i)} \eta_{\mathrm{dch}} e(n,i) \overline{\overline{\mathrm{Pr}}}(\mathrm{n,i}) \right) \tag{6.9}$$

where $e(n,i)$ is the energy charged and discharged by the BESS (always with reference to the typical day of the ith season of the nth year), and η_{ch} (η_{dch}) is the BESS charging (discharging) efficiency. By substituting (6.5)–(6.9) in (6.3), the energy cost C_{E} can be written as:

$$C_{\mathrm{E}} - \sum_{n=1}^{N} \frac{1}{(1+\alpha)^{n-1}} \cdot \sum_{\mathrm{i}=1}^{\mathrm{Sn}} \mathrm{d(i)}$$
$$\left[\left(\overline{\mathrm{E}}_{\mathrm{load}}(\mathrm{n,i}) \overline{\mathrm{Pr}}(\mathrm{n,i}) + \overline{\overline{\mathrm{E}}}_{\mathrm{load}}(\mathrm{n,i}) \overline{\overline{\mathrm{Pr}}}(\mathrm{n,i}) \right) + \mathrm{e(n,i)} \left(\frac{\overline{\mathrm{Pr}}(\mathrm{n,i})}{\eta_{\mathrm{ch}}} - \overline{\overline{\mathrm{Pr}}}(\mathrm{n,i}) \eta_{\mathrm{dch}} \right) \right]. \tag{6.10}$$

Note that, by using (6.5) for the evaluation of C_{ls} in (6.10), we made the simplifying assumption of a ToU tariff with only two price levels. The general case of more than two levels can be easily derived by implementing formula (6.4) instead of (6.5).

It has to be noted also that, given a specified battery size (S), with the aim of minimizing C_{E} in (6.10), the charged/discharged energy e(n,i) should be maximized. In fact, being $\overline{\overline{\mathrm{Pr}}}(\mathrm{n,i})$ significantly greater than $\overline{\mathrm{Pr}}(\mathrm{n,i})$ and taking into account the high battery efficiencies values, the quantity $\overline{\mathrm{Pr}}(\mathrm{n,i})/\eta_{\mathrm{ch}}$ is less than $\overline{\overline{\mathrm{Pr}}}(\mathrm{n,i})/\eta_{\mathrm{dch}}$ (thus, the more the energy discharged in the peak periods for supplying the loads, the less the expense sustained by the customer for the energy provision). Then, the BESS operation strategy must be able to maximize the charged/discharged energy (and, then, to minimize the total costs). This objective must be reached subject to technical constraints related to the preservation of the battery's lifetime as well as to constraints related to the assumed customer's/utility's contractual agreements. More specifically, the following constraints were assumed:

1. *Preservation of the battery's lifetime*:
   - the battery's DoD cannot exceed a specified maximum value;
   - the number of charging/discharging cycles per day cannot be larger than a maximum value imposed by (6.2), once the expected lifetime of the battery has been specified;
2. *Contractual agreements*:
   - a maximum value is imposed on the total power absorbed from the grid by the customer;
   - the customer cannot supply energy to the grid.

With reference to the constraints related to the preservation of the battery's lifetime, the first depends on the battery's technology for which the maximum value of DoD can be fixed equal to the value that corresponds to the desired lifetime of the battery. Obviously, if the DoD is chosen so to obtain a lifetime equal to the planning period, the replacement costs will be neglected. Regarding the number of daily charging/discharging cycles, in the case of the ToU tariff made by two price levels, it is naturally fixed equal to one per day, since the battery is allowed to charge during the off-peak price hours and to discharge during the on-peak price hours. This condition can be still valid in case of ToU tariffs made by three price levels i.e. the tariffs that provide a different price in the mid-peak hours. In fact, typically, mid-peak periods refer to a few hours, thus a benefit can be achieved if the battery is idle during these hours [25].

The constraints related to the contractual agreements, instead, limit the maximum energy to be charged/discharged by the battery during a day. In case of the ToU tariff with two price levels, the maximum value of charged/discharged energy during the days and, then, the value of the minimum total costs, can be evaluated analytically on the basis of the following considerations:

- during the on-peak periods ($\overline{\overline{T}}(n,i)$), the maximum energy that can be discharged by the battery $e^{dch}_{max}(n,i)$ is limited by the constraint that the customer cannot sell energy to the grid;
- during the off-peak periods ($\overline{T}(n,i)$), the maximum energy that can be stored in the battery $e^{ch}_{max}(n,i)$ is limited by the constraint on the maximum power that can be absorbed from the grid.

The time intervals identified for the on-peak and off-peak periods, $\overline{\overline{T}}(n,i)$ and $\overline{T}(n,i)$, can be further split into sub-periods. In more detail, the on-peak period (i.e. the discharging period of the day) includes two sub-periods:

- $\overline{\overline{T}}_1(n,i)$, including the set of time intervals in which the maximum value of the power that can be supplied by the BESS ($p_{B,max}$), is greater than the power requested from the load: $\overline{\overline{T}}_1(n,i) = \{t : p_{B,max} > p_L(n,i,t)\}$.

- $\overline{\overline{T}}_2(n, i)$, including all the time intervals in which the power requested from the load, is greater than or equal to the maximum value of the power that can be supplied by the BESS: $\overline{\overline{T}}_2(n, i) = \{t : p_{B,max} \le p_{load}(n, i, t)\}$.

During $\overline{\overline{T}}_1(n, i)$, the maximum energy that can be discharged is:

$$e_1(n, i) = 1/\eta_{dch} \sum_{t \in \overline{\overline{T}}_1(n,i)} p_L(n, i, t)\Delta t. \tag{6.11}$$

During $\overline{\overline{T}}_2(n, i)$, the maximum value of the energy that can be discharged by the BESS is

$$e_2(n, i) = 1/\eta_{dch} \sum_{t \in \overline{\overline{T}}_2(n,i)} p_{B,max}\Delta t. \tag{6.12}$$

During the on-peak period, the maximum energy that can be discharged is given by

$$e_{max}^{dch}(n, i) = e_1(n, i) + e_2(n, i) \tag{6.13}$$

Similarly, the off-peak period (i.e. charging period of the day) can be split into two sub-periods:

- $\overline{T}_3(n, i)$, including all the time intervals in which the sum of the load power and the maximum value of the power that can be used by the BESS for charging, is lower than the maximum value of the power that can be imported from the grid, which is imposed by the contractual agreement (P_{max}^{ca}) : $\overline{T}_3(n, i) = \{t : (p_{B,max} + p_{load}(n, i, t)) < P_{max}^{ca}\}$.
- $\overline{T}_4(n, i)$, including all the time intervals in which the sum of the load power and the maximum value of the power that can be used by the BESS for charging, is greater than or equal to the maximum value imposed by the contractual agreement: $\overline{T}_4(n, i) = \{t : (p_{B,max} + p_{load}(n, i, t)) \ge P_{max}^{ca}\}$.

During $\overline{T}_3(n, i)$, the maximum value of the energy that can be charged is:

$$e_3(n, i) = \eta_{ch} \sum_{t \in \overline{T}_3(n,i)} p_{B,max}\Delta t \tag{6.14}$$

During $\overline{T}_4(n, i)$, the maximum value of the energy that can be charged is:

$$e_4(n, i) = \eta_{ch} \sum_{t \in \overline{T}_4(n,i)} \left(P_{max}^{ca} - p_{load}(n, i, k)\right)\Delta t \tag{6.15}$$

Hence, during the off-peak period, the maximum value of the energy that can be charged is given by:

$$e_{max}^{ch}(n, i) = e_2(n, i) + e_4(n, i). \tag{6.16}$$

Besides the above limits, a further constraint has to be added that poses limitations to the maximum daily energy that can be exchanged between the BESS and the grid, which is related to the maximum DoD. This constraint can be expressed as:

$$e_{max} = s\frac{\delta}{100} \tag{6.17}$$

where δ is the value of the maximum DoD of the battery [26].

In conclusion, based on the above constraints, the actual energy exchanged in a typical day of the ith season of the nth year is given by:

$$\mathrm{e}(\mathrm{n},\mathrm{i}) = \min\{\mathrm{e}_{\max}, \mathrm{e}_{\max}^{\mathrm{dch}}(\mathrm{n},\mathrm{i}), \mathrm{e}_{\max}^{\mathrm{ch}}(\mathrm{n},\mathrm{i})\}. \tag{6.18}$$

Eventually, by assuming the maintenance and disposal costs as a specific percentage of capital costs, the total costs (6.3) will depend only on the size of the battery (s) and it will assume the form:

$$C_{\mathrm{T}}(\mathrm{s}) = C_0(\mathrm{s}) + \sum_{\mathrm{n}=1}^{\mathrm{N}} \frac{1}{(1+\alpha)^{\mathrm{n}-1}} \sum_{\mathrm{i}=1}^{\mathrm{Sn}} \mathrm{d}(\mathrm{i}) \left[\overline{\mathrm{E}}_{\mathrm{load}}(\mathrm{n},\mathrm{i})\overline{\mathrm{Pr}}(\mathrm{n},\mathrm{i}) + \overline{\overline{\mathrm{E}}}_{\mathrm{load}}(\mathrm{n},\mathrm{i})\overline{\overline{\mathrm{Pr}}}(\mathrm{n},\mathrm{i}) + \mathrm{e}(\mathrm{n},\mathrm{i})\left(\frac{\overline{\mathrm{Pr}}(\mathrm{n},\mathrm{i})}{\eta_{\mathrm{ch}}} - \overline{\overline{\mathrm{Pr}}}(\mathrm{n},\mathrm{i})\eta_{\mathrm{dch}}\right)\right]. \tag{6.19}$$

Note that, in (6.19), e(n, i) assumes the constrained value given by (6.18). The term e(n, i) given by (6.18) in fact includes all of the constraints discussed above.

6.4 Probabilistic approach for sizing battery systems

In the case of deterministic scenarios, the inputs of the total cost in (6.19) are all known a priori. The optimal size of the BESS can be obtained by assessing the value of the total cost function for a finite number of sizes chosen among those commercially available; then, the optimal size is that corresponding to the lowest total cost. However, some of the input variables that appear in the total cost function in (6.19) are characterized by high levels of uncertainty. This requires that a probabilistic approach be used to solve the BESS sizing problem by treating such quantities as random input variables. In the frame of the probabilistic approach, the total cost in (6.19) is thus the output random variable whose average value is evaluated for each of the considered storage system sizes. The optimal value of the BESS size is that corresponding to the minimum average value of the total cost.

In this chapter, the evaluation of the average value of the total cost is effected by means of the $2m+1$ scheme of the PEM, whose application requires the statistical data provided by the first four central moments of the input random variables. The variables considered as random inputs in (6.19) are (i) the load demand, (ii) the price of energy and (iii) the discount rate for the actualization of the costs[3].

[3] Also other input variables could be considered as random variables, such as the parameters related to the battery technology (e.g. lifetime duration, efficiency). A deep discussion on their impact on the sizing procedure can be found in [10].

6.4.1 Brief background on PEM algorithm

In the most general case, the PEM allows calculating the moments of a random variable y_i that is a function g of m random variables x_i:

$$y_i = g(x_1, x_2, \ldots \ldots ., x_m) \tag{6.20}$$

through the solution of a few deterministic problems respect to the massive number of trials required when the Monte Carlo simulation procedure is applied.[4]

In the frame of the PEM approach, it is possible to apply different schemes. Each of them requires solving a different number of deterministic problems. As previously mentioned, in this chapter, the $2m+1$ scheme has been considered with m random input variables. In the case of the $2m+1$ scheme, for each input random variable x_r, the function g is evaluated at two points whose coordinates are given by:

$$\left(\mu_{x_1}, \mu_{x_2}, \ldots \ldots \ldots ., \hat{x}_{r,1}, \ldots \ldots . . \mu_{x_m}\right) \tag{6.21}$$

$$\left(\mu_{x_1}, \mu_{x_2}, \ldots \ldots \ldots ., \hat{x}_{r,2}, \ldots \ldots . . \mu_{x_m}\right) \tag{6.22}$$

More specifically, the coordinates of the two points include:

- specified values referred to as *locations* ($\hat{x}_{r,1}$ and $\hat{x}_{r,2}$) that are calculated from their statistical moments;
- the mean values of the remaining $m-1$ random input variables, μ_{x_i} with $i = 1, \ldots, r-1, r+1, \ldots, m$.

Also, it is required that an additional run of g at the point characterized by the mean values of all of the input random variables be effected:

$$\left(\mu_{x_1}, \mu_{x_2}, \ldots \ldots \ldots ., \mu_{x_r}, \ldots \ldots . . \mu_{x_m}\right). \tag{6.23}$$

Once the solutions of the $2m+1$ deterministic problems are known, the kth moment of the output random variables can be derived by weighting the kth power of the solution through a weighting factor. The value of this last depends once again on the third and fourth standard central moment of the input random variables [27].

6.4.2 Applications of PEM for the BESS sizing procedure

In the optimal BESS probabilistic sizing procedure proposed in this chapter, we are interested in calculating only the first-order moment of the total cost for each of the considered BESS sizes. If the total cost in (6.19) is expressed as a function g of the m random input variables, x_t (i.e. the hourly load demand, the price of energy and the discount rate):

$$C_T = g(x_1, x_2, \ldots \ldots \ldots . . x_m) = g(\mathbf{X}), \tag{6.24}$$

[4] It has to be noted that the PEM requires that the input random variables be uncorrelated. In case of correlation among input random variables, a rotational transformation has to be used. The transformation is based on the eigenvector of the covariance matrix to transform the set of correlated random variables into a set of uncorrelated random variables [27, 28, 29].

The average value of the total costs can be easily obtained applying the $2m+1$ scheme of PEM as follows:

1. Evaluate the mean value, the standard deviation, the third and fourth standard central moments of each input random variable x_t
2. Set the mean value of the total cost equal to zero: $E(C_T) = 0$
3. Set $t = 1$
4. Calculate the locations $\hat{x}_{t,1}$ and $\hat{x}_{t,2}$:

$$\hat{x}_{t,1} = \mu_{x_t,} + \xi_{t,1}\sigma_{x_t}; \tag{6.25}$$

$$\hat{x}_{t,2} = \mu_{x_t,} + \xi_{t,2}\sigma_{x_t}. \tag{6.26}$$

with μ_{x_t} and σ_{x_t} the mean value and the standard deviation of the input random variable x_t, respectively;

$$\xi_{t,1} = \frac{\lambda_{t,3}}{2} + \sqrt{\lambda_{t,4} - \frac{3}{4}\lambda_{t,3}^2}; \tag{6.27}$$

$$\xi_{t,2} = \frac{\lambda_{t,3}}{2} - \sqrt{\lambda_{t,4} - \frac{3}{4}\lambda_{t,3}^2}. \tag{6.28}$$

being $\lambda_{t,3}$ and $\lambda_{t,4}$ the third and fourth standard central moment of the input random variable x_t

5. Evaluate the total costs by Eq. (6.24) at the points:

$$\boldsymbol{X}_1 = \left[\mu_{x_1,}, \mu_{x_2,}, \ldots \ \ldots \ \ldots, \hat{x}_{t,1}, \ldots \ \ldots \mu_{x_m,}\right]; \tag{6.29}$$

$$\boldsymbol{X}_2 = \left[\mu_{x_1,}, \mu_{x_2,}, \ldots \ \ldots \ \ldots, \hat{x}_{t,2}, \ldots \ \ldots \mu_{x_m,}\right]. \tag{6.30}$$

6. Update $E(C_T)$:

$$E(C_T) = E(C_T) + \sum_{k=1}^{2} w_{t,k}\, g(\mathbf{X}_k) \tag{6.31}$$

with the weighting factors equal to:

$$w_{t,1} = \frac{1}{\xi_{t,1}\left(\xi_{t,1} - \xi_{t,2}\right)}; \tag{6.32}$$

$$w_{t,2} = -\frac{1}{\xi_{t,2}\left(\xi_{t,1} - \xi_{t,2}\right)}. \tag{6.33}$$

7. Repeat steps from 3 to 6 for $t = t+1$ until the list of random input variables is finished (i.e. $t = m$).
8. Evaluate the total cost by Eq. (6.24) at the point consisting of the mean values of the input random variables:

$$\mathbf{X}_\mu = \left[\mu_{x_1,}, \mu_{x_2,}, \ldots \ \ldots, \mu_{x_t,}, \ldots \ \ldots \mu_{x_m,}\right] \tag{6.34}$$

9. Update $E(C_T)$:

$$E(C_T) = E(C_T) + w_o g(X_\mu) =$$

$$= \sum_{t=1}^{m}\sum_{i=1}^{2} f(\mu_{x_1}, \mu_{x_2}, \ldots \ldots, x_{t,i}, \ldots \ldots \mu_{x_m}) + w_o\, g(\mathbf{X}_\mu) \quad (6.35)$$

where the weight factor w_o is:

$$w_o = 1 - \sum_{t=1}^{m} \frac{1}{\left(\lambda_{t,4} - \lambda_{t,3}^2\right)}. \quad (6.36)$$

The procedure is repeated for each considered BESS size to calculate the total cost average value, $E(C_{T,si})$ for each BESS size. The optimal size is that corresponding to the minimum value of the average total cost (Figure 6.2). The PEM method application requires the statistical information derived by the first four central moments of the input random variables which are linked to their probability density functions (pdfs).

In this chapter, the load demand pdfs were assumed to be normal distributions. This distribution is the most frequently used even though in several probabilistic studies, the load characteristics could be assumed having binomial, discrete or other distributions. With reference to probabilistic balanced power flows, the authors of [30] and [31] reported a wide list of references on this subject. Recently, efforts have been made to improve probabilistic electric load forecasting by adding residuals simulated with normal distributions. In [32], this assumption was investigated, and it was concluded that the normality hypothesis for residual simulation allows significant improvement in the load forecast only in the case of deficient probabilistic models, while the improvement is negligible in the more accurate models.

Regarding the prices for the energy bought, the choice of a pdf is a very difficult task due to the challenges of associating a specific statistical behaviour with the energy prices. Then, given the lack of either data or specific studies in the relevant literature, the uniform distribution seemed to be the best choice; its parameters can be chosen so that there is an equal probability that prices may increase, decrease or remain constant over the years.

With reference to the uncertainty of the discount rate, both continuous and discrete probability distributions were assumed in the relevant literature [33, 34]. In [33], the discount rate was assumed having discrete probability distribution:

$$p(\alpha = \alpha_i) = p_i \quad i = 1, \ldots . N_t \quad (6.37)$$

Thus, the discount rate α has N_t possible values, each characterized by a probability p_i. This probability distribution can be estimated from recorded historical data. In [34], the probability distribution of the discount rate was made by a triangular function whose parameters reflect its probable range of variation. In case

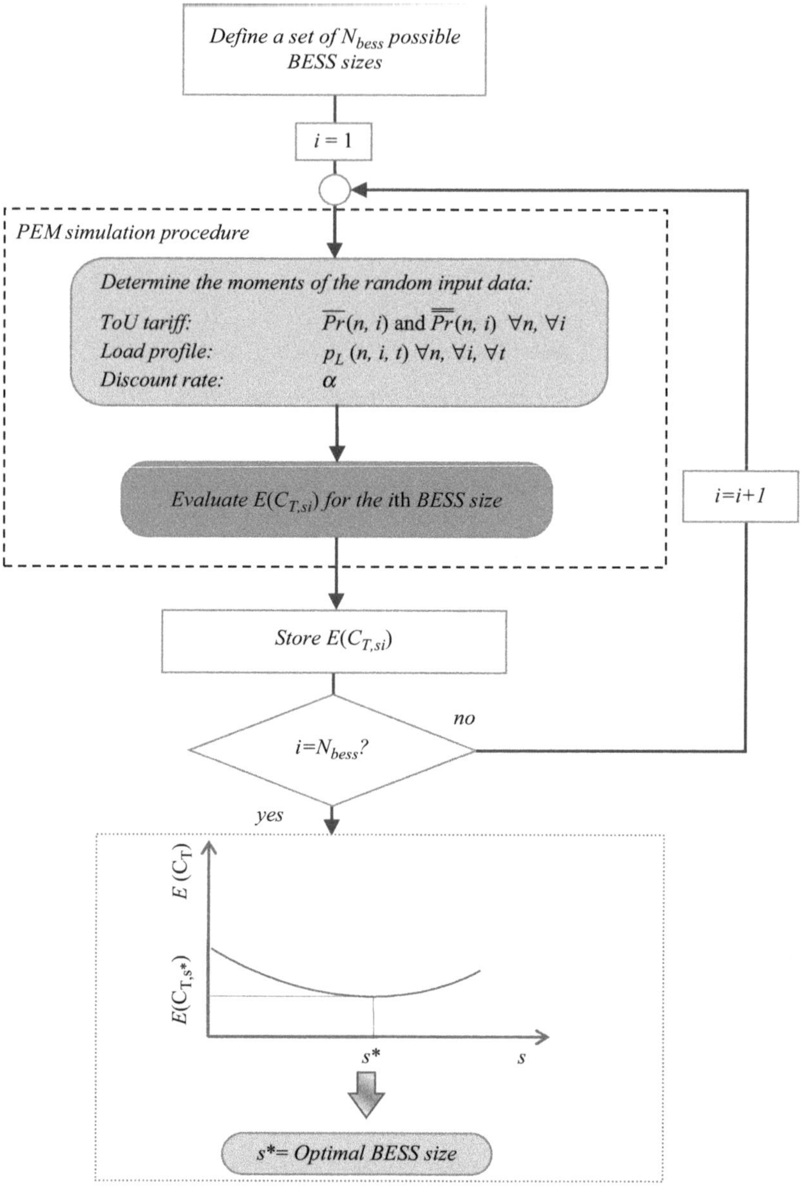

Figure 6.2 Flow chart of the probabilistic sizing procedure

of no information regarding the shape of a distribution, a uniform distribution could be used so giving equal weight to the probability of all the input values. Anyway, when the range of parameters is known, the symmetrical triangular distribution has the additional benefit of not giving advantage to a specific alternative based on the influence of lower or higher values of discount rates. It has to be noted, in fact, that lower values of discount rate typically lead to higher initial costs and lower future costs whereas higher values of discount rate usually lead to lower initial costs and higher future costs. In this chapter, the approach proposed in [33] was used.

6.5 Numerical applications

In this section, three scenarios are discussed applying the proposed approach to the load profiles of a residential customer, a commercial customer and an industrial customer. Two seasonal profiles were implemented for each of them. The profiles were obtained from the data provided by the Cigrè benchmark networks [35]. The hourly load values reported in [35] were assumed as mean of the Gaussian distribution associated at each time interval of the procedure (i.e. at each hour). The standard deviation at each time interval was assumed as 10 per cent of the corresponding mean value. For the load powers and with reference to all the time intervals into which the days were divided, different possible trends of yearly load growth were considered, by varying the mean values of the Gaussian distributions through the years by +1 per cent, +3 per cent and +5 per cent. Figure 6.3 shows the hourly mean values of the daily load demand of the industrial, commercial and residential customer, respectively, referring to the typical summer day of the first year.

The parameters of the uniform pdf of the electrical energy tariff at the first year were chosen as ±10 per cent variation around the tariff applied by an existing utility [36]. This tariff, whose electrical energy prices are reported in Table 6.1, is applicable to all of the considered typlogies of customers. It is interesting to note how different are the three typologies of load profiles while the same tariff is applied to all of them. Annual variations of +1 per cent, +3 per cent and +5 per cent were assumed for the tariffs, too. A discrete distribution was used for the

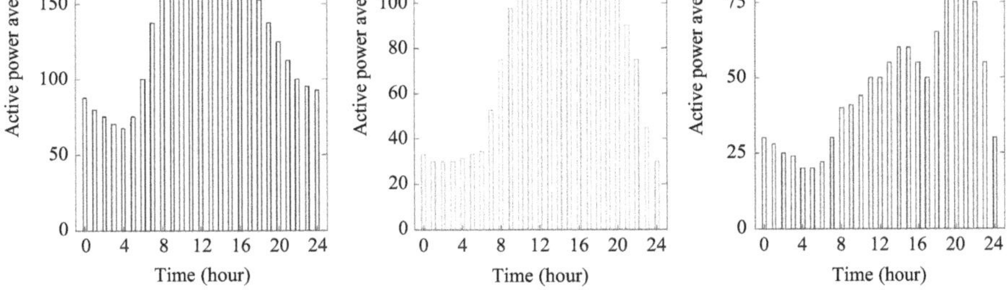

Figure 6.3 Load demand profiles for the industrial, commercial and residential loads

Table 6.1 TOU tariff

TOU levels	Period	Price ($/MWh)
	Summer tariff	
On peak	12:00 noon to 6:00 p.m.	542.04
Part peak	8:30 a.m. to 12:00 noon and 6:00 p.m. to 9:30 pm	252.90
Off peak	9:30 p.m. to 8:30 a.m.	142.54
	Winter tariff	
On peak	8:30 a.m. to 9:30 p.m.	161.96
Off peak	9:30 p.m. to 8:30 a.m.	132.54

discount rate, α, which is based on the historical data of the US discount rate [37]. The values of α used as inputs of each iteration of the sizing procedure were obtained by means of extractions from the discrete distributions of these data.

The BESS includes a lithium-ion battery connected to the power grid by a PWM-controlled DC/AC converter. The charging and discharging efficiencies of the BESS were assumed to be 0.95 and 0.98, respectively. For the installation cost, a value of 600 $/kWh was assumed. This cost includes also the cost item related to maintenance which has been considered as 2 per cent of the installation cost. The benefit deriving from the disposal of the battery has been disregarded. The planning period considered is 12 years during which no battery replacement is required. In fact, one charging/discharging cycle per day is imposed, so the life cycle of the Li-ion battery usually considered (4,000 cycles) covers the whole planning period.

Regarding the application of the PEM, since the number of the input random variables is 625, the $(2m + 1)$ scheme required 1,251 trials.

6.5.1 Industrial load

In this case study, the procedure was applied to a BESS to be included into an industrial facility. The maximum power that can be absorbed by this customer has been assumed equal to 350 kW in the first year. A yearly growth was supposed for this value with the same percentage as the load yearly growth. The range of BESS candidate sizes is 0–2,500 kWh with step sizes of 10 kWh. The results of the application of the procedure to this case are reported in Table 6.2 with reference to all of the possible loads and tariff growth trends. The results in the table are reported in terms of optimal sizes of the BESS for all of the considered scenarios. They clearly show that, when the yearly percentage variations of energy tariff and load demand increase, the size of the BESS that minimizes the average total cost increases as well. An exception appears in the case of tariff growth of 1 per cent, where the same size is found in both the scenarios implying load growth of 3 per cent and 5 per cent. This is due to the fact that, when the tariff growth is low, the marginal cost corresponding to the step size of 10 kWh is not rewarded by the avoided cost related to the load growth from 3 per cent to 5 per cent. Thus, it can be concluded that the average total cost is mainly affected by the tariff growth rather than by the load growth. Because of that, the largest size resulted from the sizing

Table 6.2 Industrial case: BESS optimal sizes for different scenarios

Tariff trend (%)	**Load trend (%)**		
	1	**3**	**5**
1	1,720 kWh	1,800 kWh	1,800 kWh
3	1,740 kWh	1,870 kWh	1,970 kWh
5	1,810 kWh	1,960 kWh	2,080 kWh

Table 6.3 Industrial case: minimum average cost for different scenarios

Tariff trend (%)	**Load trend (%)**		
	1	**3**	**5**
1	4.14 M$	4.65 M$	5.21 M$
3	4.51 M$	5.08 M$	5.74 M$
5	4.93 M$	5.58 M$	6.34 M$

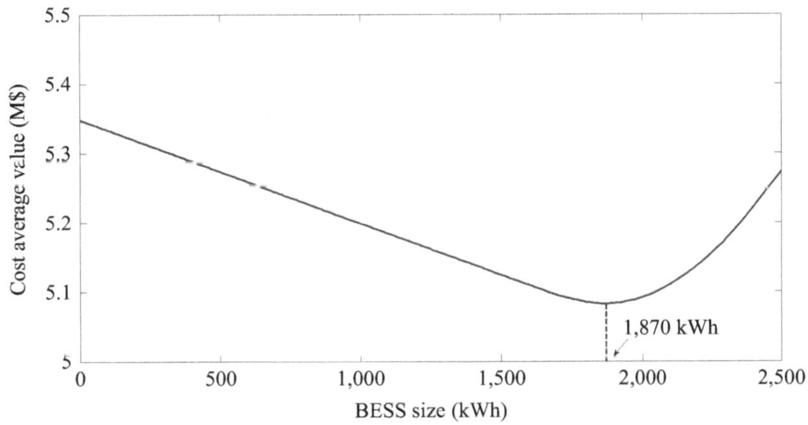

Figure 6.4 Average total costs evaluated for all of the BESSs' sizes

procedure is 2,080 kWh, obtained in the case of +5 per cent annual variation of both tariff and load demand. The total cost corresponding to the BESS sizes of Table 6.2 is reported in Table 6.3.

With reference to the scenario load variation 3 per cent – tariff variation 3 per cent, Figure 6.4 reports the values of the average total costs evaluated for all of the BESSs' sizes. More specifically, in the figure, the size that corresponded to the minimum average total cost is shown together with the value of the minimum average total cost.

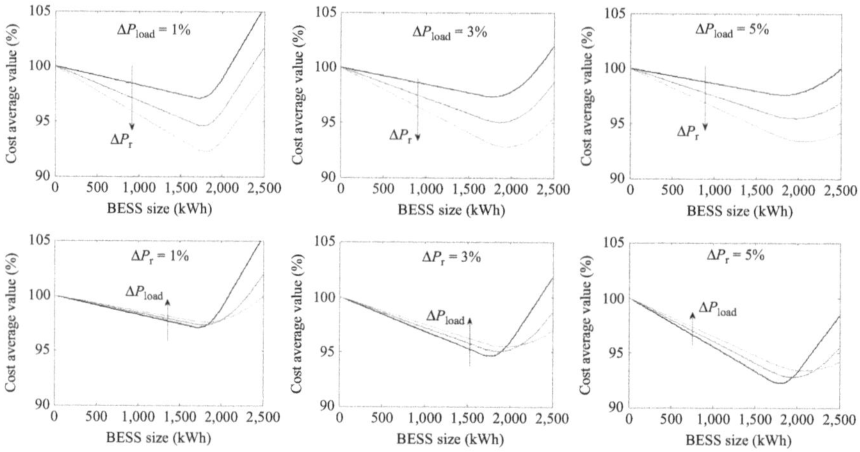

Figure 6.5 Profiles of the average value of the total costs corresponding to different combinations of load demand (ΔP_{load}) and tariff (ΔP_r) growths

The figure shows that, starting from the minimum size value (corresponding to no installation of BESS), the average cost assumes a decreasing trend with the increase of the BESS size up to the minimum average value (1,870 kWh). After that point, the average value of the total cost starts increasing as the size increases.

The average value of the total costs corresponding to different combinations of load demand (ΔP_{load}) and tariff (ΔP_{r}) growths are reported in Figure 6.5. In the figure, the average costs are reported in percentage values of the cost without BESS related to each scenario.

From the analysis of the figure, it appears that for an assigned yearly load growth of the load demand, different values of energy tariff yearly growths significantly modify the percentage value of the total average cost showing the greatest reduction in correspondence of 5 per cent yearly variation of energy prices. On the other hand, the percentage value of the total average cost slightly varies with different values of load demand yearly growths.

With reference to the scenario load variation 3 per cent – tariff variation 3 per cent, Figure 6.6 shows the average value of the total cost corresponding to different values of the installation costs. In more detail, five different values of the installation costs were considered within the range 300–800 \$/kWh. As expected, the cost reduction corresponding to the use of different BESS sizes decreases while the BESS cost increases; the cost of 700 \$/kWh corresponds to the highest installation cost for which benefit can be derived from the installation of the BESS.

6.5.2 Commercial load

In the case of the commercial load, the maximum power that can be absorbed by the customer has been assumed equal to 250 kW in the first year. Also in this case, a yearly growth was supposed for this value with the same percentage as the load

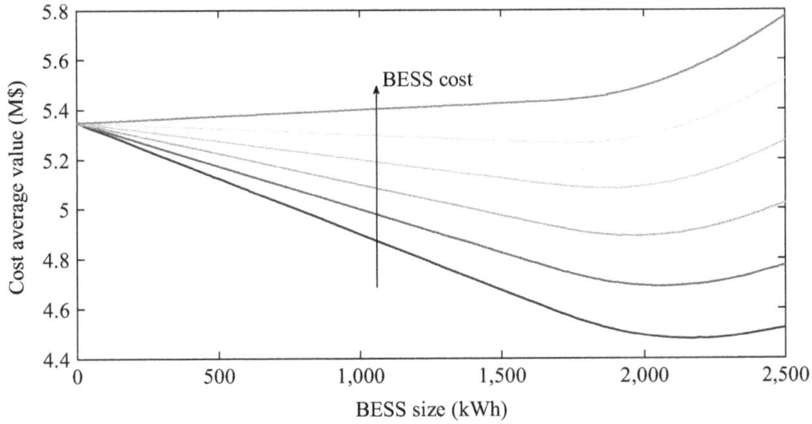

Figure 6.6 Average value of the total cost corresponding to different values of the installation costs (scenario load variation 3% - tariff variation 3%)

Table 6.4 Commercial case: BESS optimal sizes for different scenarios

Tariff trend (%)		**Load trend (%)**	
	1	**3**	**5**
1	1,010 kWh	1,060 kWh	1,060 kWh
3	1,020 kWh	1,090 kWh	1,160 kWh
5	1,060 kWh	1,150 kWh	1,220 kWh

Table 6.5 Commercial case: minimum average cost for different scenarios

Tariff trend (%)		**Load trend (%)**	
	1	**3**	**5**
1	2.40 M$	2.68 M$	3.01 M$
3	2.61 M$	2.94 M$	3.32 M$
5	2.84 M$	3.22 M$	3.66 M$

yearly growth. The range of BESS candidate sizes is 0–1,500 kWh. The BESS sizes resulted from the procedure are reported in Table 6.4 with reference to all of the possible loads and tariff growth trends and the corresponding total costs are shown in Table 6.5. The data reported in the tables, which refer to all of the considered scenarios, again show that the increase of the percentage of yearly variations of both energy tariff and load demand results in a larger size for the BESS corresponding to the minimum average total cost. The largest size obtained by the sizing procedure was 1,220 kWh, still resulted in the case of +5 per cent annual variation of both tariff and load demand.

Figure 6.7 shows the values of the average total costs evaluated for all of the BESSs' sizes in the scenario load variation 3 per cent – tariff variation 3 per cent. Also in this case, it clearly appears that, starting from the minimum size value (corresponding to no installation of BESS), the average cost assumes a decreasing trend with the increase of the BESS size. This trend continues up to the minimum value of the total cost corresponding to the BESS size 1,090 kWh after which the average value of the total cost starts increasing with the increase of the size.

The profiles of the average value of the total costs corresponding to different combinations of load demand (ΔP_{load}) and tariff (ΔP_r) growths are reported in Figure 6.8,

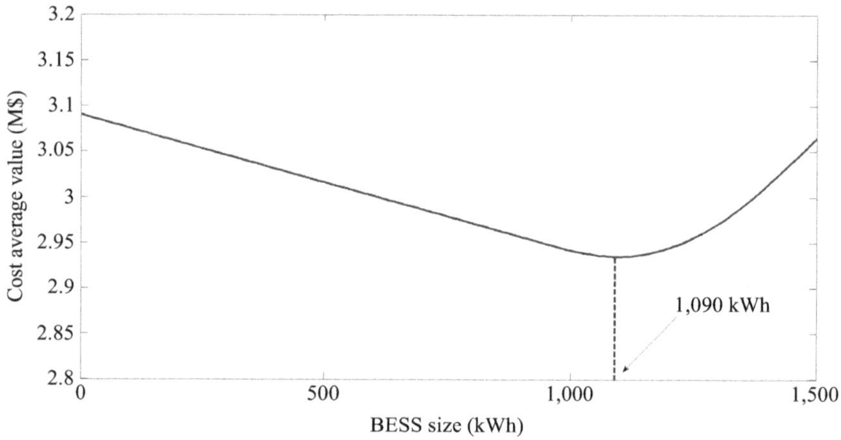

Figure 6.7 Average total costs evaluated for all of the BESSs' sizes in the scenario load variation 3% – tariff variation 3%

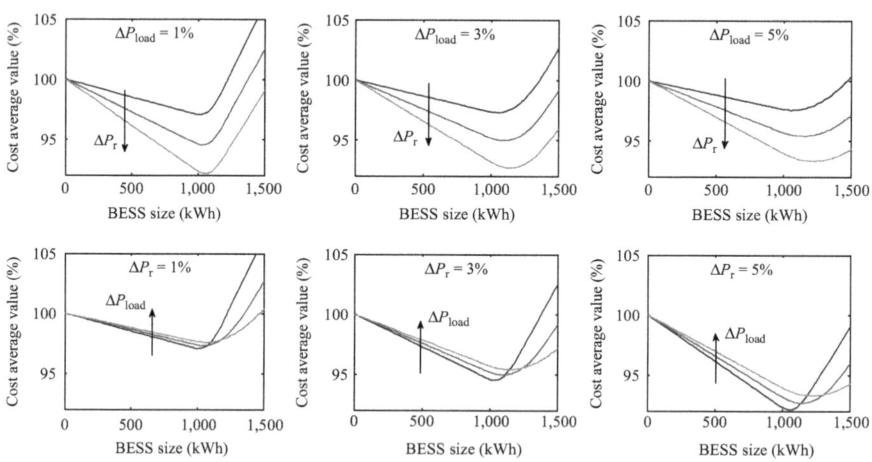

Figure 6.8 Profiles of the average value of the total costs corresponding to different combinations of load demand (ΔP_{load}) and tariff (ΔP_r) growths

with the same assumption made with reference to Figure 6.5. From the analysis of the figure, similar considerations as in the industrial case arise.

6.5.3 Residential load

In the case of the residential load, the maximum power that can be absorbed by the customer has been assumed equal to 150 kW in the first year whose yearly growth was still supposed of the same percentage as the load yearly growth. The range of BESS candidate sizes was 0–800 kWh. The BESS sizes resulted from the procedure are reported in Table 6.6, and the corresponding total costs are shown in Table 6.7. The largest size obtained by the sizing procedure was 500 kWh, again obtained in the case of +5 per cent annual variation of both tariff and load demand. Figure 6.9

Table 6.6 Residential case: BESS optimal sizes for different scenarios

Tariff trend (%)	Load trend (%)		
	1	**3**	**5**
1	410 kWh	430 kWh	430 kWh
3	420 kWh	450 kWh	470 kWh
5	430 kWh	470 kWh	500 kWh

Table 6.7 Residential case: minimum average cost for different scenarios

Tariff trend (%)	Load trend (%)		
	1	**3**	**5**
1	1.23 M$	1.37 M$	1.54 M$
3	1.34 M$	1.51 M$	1.71 M$
5	1.47 M$	1.67 M$	1.89 M$

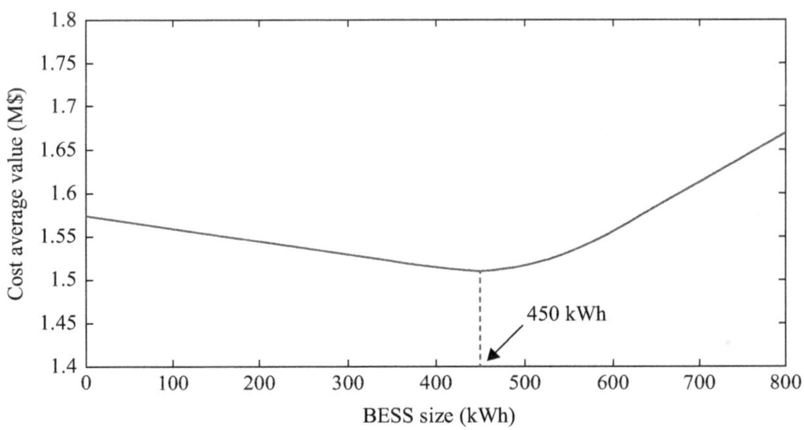

Figure 6.9 Average total costs evaluated for all of the BESSs' sizes in the scenario load variation 3% – tariff variation 3%

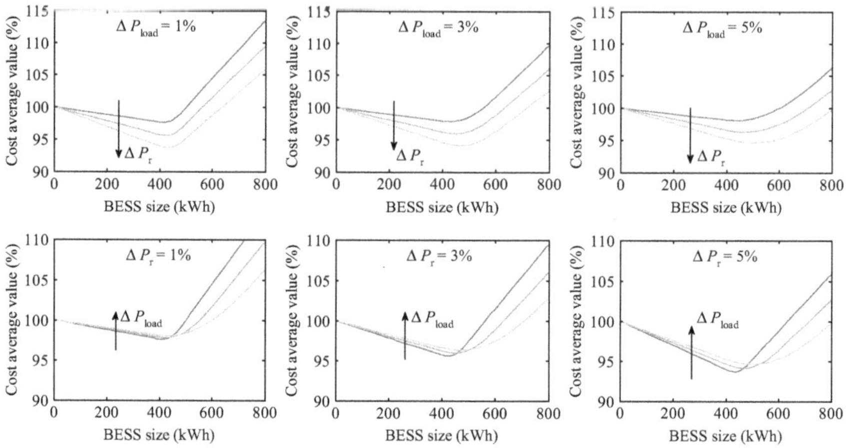

Figure 6.10 Profiles of the average value of the total costs corresponding to different combinations of load demand (ΔP_{load}) and tariff (ΔP_r) growths

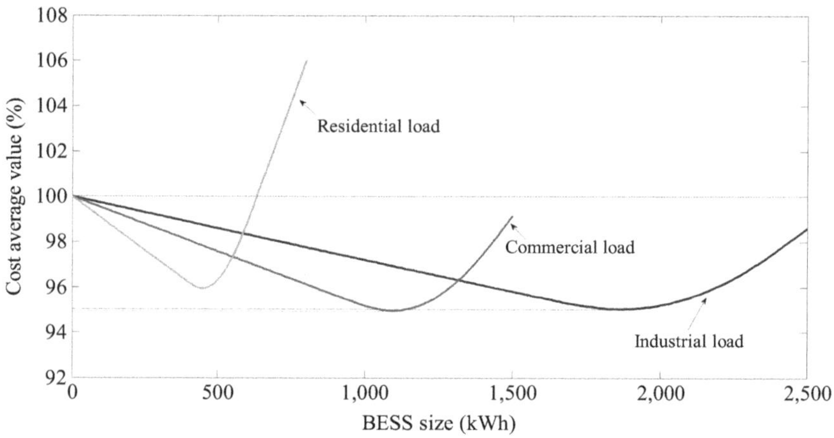

Figure 6.11 Values of the average total costs, evaluated for all of the BESSs' sizes in the scenario load variation 3% – tariff variation 3% for each customer typology

shows the values of the average total costs evaluated for all of the considered BESSs' sizes in the scenario load variation 3 per cent – tariff variation 3 per cent.

Figure 6.10 shows the profiles of the average value of the total costs corresponding to different combinations of load demand and tariff growths. From the analysis of the figures and the tables, considerations similar to those emerged in the previous case studies arise.

To compare the impact of the sizing procedure in different cases considered in this application, in Figure 6.11, the values of the average total costs are reported, evaluated for all of the BESSs' sizes in the scenario load variation 3 per cent – tariff variation 3 per cent for each customer typology.

Compared to the previous cases, the BESS size minimizing the total cost in the residential case is lower than that encountered in the other cases, due to the lower peak power. Also, a slightly lower percentage cost reduction can be observed in the case of residential customer. In fact, in the cases of the industrial and commercial loads, the percentage cost reduction is about 5 per cent whereas it is 4 per cent in the case of residential case. This is because the highest power values are confined only in a few hours of the day (see Figure 6.3).

Figure 6.11 shows how, for the residential loads considered in the chapter, the choice of sizes different from those chosen by the proposed procedure would dramatically increase the expected total costs. In the commercial and industrial loads, the increase of costs in correspondence of sizes different from that minimizing total costs is less evident (i.e. the curves of costs have a smaller slope).

6.6 Conclusions

This chapter dealt with the sizing of BESSs applied to end-user applications, in the case of ToU tariffs. More specifically, a procedure has been proposed, aimed at optimally sizing the battery storage system with the objective of reducing the user's electricity bill costs by shifting his load demand during the periods of the day characterized by low energy costs. Moreover, the battery storage system allows end-users to participate to the demand response programmes without impacting their comfort. Due to the unavoidable uncertainties characterizing some of the input variables, the sizing procedure used a probabilistic approach based on the PEM. The result of its application to the typical load demands of residential, commercial and industrial customers demonstrated its effectiveness in terms of reduction of the costs sustained by the customer for the electricity bill. Moreover, it clearly appears that significant advantages can be achieved by the inclusion of the battery energy storage in the customer's system thanks to the reduction of the electrical energy requested from the grid during on-peak hours. The comparison of the results obtained for different case studies evidenced the importance of applying an optimal sizing procedure able to capture both the peculiarities of the typology of customers and the effect of uncertain scenarios.

References

[1] Li, J., Wei, W., and Xiang, J. 'A Simple Sizing Algorithm for Stand-Alone PV/Wind/Battery Hybrid Microgrids'. *Energies*, 2012, vol. 5, pp. 5307–23.

[2] Bludszuweit, H. and Dominguez-Navarro, J. A. 'A Probabilistic Method for Energy Storage Sizing Based on Wind Power Forecast Uncertainty'. *IEEE Transactions on Power Systems*, 2011, vol. 26, pp.1651–58.

[3] Wu, J., Zhang, Li, B. H., Li, Z., Chen, Y., and Miao, X. 'Statistical Distribution for Wind Power Forecast Error and Its Application to Determine

Optimal Size of Energy Storage System'. *International Journal of Electrical Power & Energy Systems*, 2014, vol. 55, pp. 100–7.
[4] Chen, W. Z., Li, Q. B., Shi, L., *et al.* 'Energy Storage Sizing for Dispatchability of Wind Farm'. *Proceedings of the 11th International Conference on Environment and Electrical Engineering (EEEIC)*, Venice, Italy, 2012, pp. 1–6.
[5] Carpinelli, G., Di Fazio, A. R., Khormali, S., and Mottola, F. 'Optimal Sizing of Battery Storage Systems for Industrial Applications When Uncertainties Exist'. *Energies*, 2014, vol. 7. pp.130–49.
[6] Lu, C., Xu, H., Pan, X., and Song, J. 'Optimal Sizing and Control of Battery Energy Storage System for Peak Load Shaving'. *Energies*, 2014, vol. 7, pp. 8396–410.
[7] Bayram, I. S., Abdallah, M., Tajer, A., and Qaraqe, K. A. 'A Stochastic Sizing Approach for Sharing-Based Energy Storage Applications'. *IEEE Transactions on Smart Grid*, 2017, vol. 8, no. 3, pp. 1075–84.
[8] Tan, C. W., Green, T. C., and Hernandez-Aramburo, C. A. 'A Stochastic Simulation of Battery Sizing for Demand Shifting and Uninterruptible Power Supply Facility'. *Proceedings of the IEEE Power Electronics Specialists Conference (PESC)*, Orlando (Florida), USA, 2007, pp. 2607–13.
[9] Carpinelli G., Mottola F., and Proto D. 'Probabilistic Sizing of Battery Energy Storage When Time-of-Use Pricing Is Applied'. *Electric Power Systems Research*, 2016, vol. 141, pp.73–83.
[10] Carpinelli, G., Mottola, F., and Proto, D. 'Addressing Technology Uncertainties in Battery Energy Storage Sizing Procedures'. *International Journal of Emerging Electric Power Systems*, 2017, vol. 18 (2), pp. 1–16.
[11] Rosenblueth, E. 'Point Estimate for Probability Moments'. *Proceedings of National Academy of Science of United States of America*, 1975, vol. 72, pp. 3812–14.
[12] Rosenblueth, E. 'Two Point Estimate in Probabilities', *Applied Mathematical Modelling*, 1981, vol. 33, pp. 329–35.
[13] Chang C. H., Tung Y. K., and Yang J. C. 'Evaluation of Probability Point Estimate Methods'. *Applied Mathematical Modelling*, 1995, vol. 19, pp. 95–105.
[14] Verbic, G. and Canizares, C. A. 'Probabilistic Optimal Power Flow in Electricity Markets based on a Two Point Estimate Method'. *IEEE Transactions on Power Systems*, 2006, vol. 21(4), pp. 1883–93.
[15] Su, C-L. 'Probabilistic Load Flow Computation Using Point Estimate Method'. *IEEE Transactions on Power Systems*, 2005, vol. 20, pp. 1843–51.
[16] Morales, J. M. and Perez-Ruiz, J. 'Point Estimate Schemes to Solve the Probabilistic Power Flow'. *IEEE Transactions on Power Systems*, 2007, vol. 22(4), pp. 1594–601.
[17] Chang, C. H., Tung, Y. K., and Yang, J. C. 'Evaluation of Probability Point Estimate Methods'. *Applied Mathematical Modelling*, 1995, vol. 19, pp. 95–105.
[18] Caramia, P., Carpinelli, G., and Varilone, P. 'Point Estimate Schemes for Probabilistic Three-Phase Load Flow'. *Electric Power Systems Research*, 2010, vol. 80, pp. 168–75.

[19] Badtke-Berkow, M., Centore, M., Mohlin, K., and Spiller, B. 'Making the Most of Time-Variant Electricity Pricing'. *Environmental Defense Fund*, 2015. Available from https://www.edf.org/energy/electricity-pricing

[20] Akhil, A. A., Huff, G., Currier, A. B., *et al. DOE/EPRI 2013 Electricity Storage Handbook*, 2013, Sandia National Laboratories, Albuquerque, NM, USA.

[21] Zakeri, B. and Syri S. 'Electrical Energy Storage Systems: A Comparative Life Cycle Cost Analysis'. *Renewable and Sustainable Energy Reviews*, 2015, vol. 42, pp. 569–96.

[22] Anderson, M. D. and Carr, D. S. 'Battery Energy Storage Technologies'. *Proceedings of the IEEE*, 1993, vol. 81, pp. 475–79.

[23] Zheng, Y., Dong, Z. Y., Xu, Y., Meng, K., Zhao, J. H., and Qiu, J. 'Electric Vehicle Battery Charging/Swap Stations in Distribution Systems: Comparison Study and Optimal Planning'. *IEEE Transactions on Power Systems*, 2014, vol. 29(1), pp. 221–29.

[24] Swaminathan, S., Miller, N. F., and Sen, R. K. *Battery Energy Storage Systems Life Cycle Costs Case Studies*, 1998, Sandia National Laboratories Albuquerque, New Mexico and Livermore, California.

[25] Borenstein, S., Jaske, M., and Rosenfeld, A. 'Dynamic Pricing, Advanced Metering, and Demand Response in Electricity Markets'. *Center for the Study of Energy Markets, University of California Energy Institute*, 2002.

[26] Divya, K. C. and Østergaard, J. 'Battery Energy Storage Technology for Power Systems: An Overview'. *Electric Power Systems Research*, 2009, vol. 79, pp. 511–20.

[27] Hong, H. P. 'An Efficient Point Estimate Method for Probabilistic Analysis'. *Reliability Engineering and System Safety*, 1998, vol. 59, pp. 261–67.

[28] Fukunaga, L., *Introduction to Statistical Pattern Recognition (Second Edition)*, Boston, MA: Academic Press, 1972.

[29] Mardia, K., Kent, J. T., and Bibby, J.-M. *Multivariate Analysis.* London: Academic Press, 1997.

[30] Schilling, M., Leite da Silva, A. M., Billington, R., El-Kady M. A. 'Bibliography on Power System Probabilistic Analysis'. *IEEE Transactions on Power Systems*, 1990, vol. 5(1), pp. 1–11.

[31] Chen, P., Chen, Z., and Bak-Jensen, B. 'Probabilistic Load Flow: A Review'. *Proceedings of Third International Conference on Electric Utility Deregulation and Restructuring and Power Technologies*, Nanjing, China, 2008, pp. 1586–91.

[32] Xie, J., Hong, T., Laing, T., and Kang, C. 'On Normality Assumption in Residual Simulation for Probabilistic Load Forecasting'. *IEEE Transactions on Smart Grid*, 2017, vol. 8(3), pp. 1046–53.

[33] Wenyuan, L. *Probabilistic Transmission System Planning*, New York: John Wiley & Sons, 2011.

[34] Walls, J. and Smith, M. 'Life-Cycle Cost Analysis in Pavement Design-In Search of Better Investment'. *Publication No. FHWA-SA-98–079, Federal Highway Administration*, 1998.

[35] Papathanassiou, S., Hatziargyriou, N., and Strunz, K. 'A Benchmark Low Voltage Microgrid Network'. *CIGRE Symposium 'Power Systems with Dispersed Generation: Technologies, Impacts on Development, Operation and Performances'*, Athens, Greece, 2005, pp. 1–8.
[36] Rates—Pacific Gas and Electric Company. Available from http://www.pge.com/tariffs/
[37] International Monetary Fund, Interest Rates, Discount Rate for United States©[INTDSRUSM193N]. Retrieved from FRED, Federal Reserve Bank of St. Louis, September 2015.

Chapter 7
Assessment and optimization of energy storage benefits in distribution networks

Kalpesh Joshi[1], Naran Pindoriya[2] and Anurag Srivastava[3]

7.1 Introduction

The distributed generators (DGs) – mostly solar and wind-based – bring along the intermittency and uncertainty of weather conditions into the operations of power systems. Erstwhile power systems were designed to absorb the daily and seasonal demand fluctuations. However, the volatility of wind-based generation, and to a considerable extent with photovoltaic generation as well, is so high that the normal operations cannot be continued without curtailing the power evacuated from these DGs. The increasing penetration of RES-based DGs worsens the situation where, on one hand, the generation from conventional generators has to be curtailed and on the other hand, the same generation might be required to standby for absorbing the intermittency and volatility. In this context, the use of energy storage systems as a buffer agent for absorbing the problems caused by increasing penetration of DGs is tempting despite its high cost. Despite the intermittency and uncertainty of RES-based DGs, battery energy storage systems (BESS) can be used to make PV generation and wind power more dispatchable [1]. While conventional generators have operating costs which are higher with reduced profits, bulk energy storage – with their fast ramping capability – can become more cost-competitive with increased penetration levels of RES-based DGs [2]. Bulk energy storage can also allow high amount of renewable energy differing the cost of upgrade of existing infrastructure. Energy storage devices can also be used to demonstrate a positive effect on transient stability of typical distribution feeder as well [3–4].

In the above context, this chapter focuses on the analysis and assessment of multiple technical benefits of BESS in improving daily network operations of distribution systems. While this kind of analysis needs to be based on modelling and simulation of distribution networks including BESS, the time-series power flow (TSPF) simulations are also required at first in the development of dispatch

[1]Department of Electrical and Computer Engineering, University of Saskatchewan, Saskatoon, Canada
[2]Department of Electrical Engineering, Indian Institute of Technology, Gandhinagar, India
[3]Department of Electrical Engineering and Computer Science, Washington State University, WA, USA

schedule for BESS and at later stages to assess the efficacy of the scheduling algorithm over longer periods including different case scenarios. Moving up from requirements of modelling to multi-objective optimization, this chapter builds up progressively in the following steps:

- Specific requirements of modelling and simulation for the application of BESS in distribution systems – supplemented by a crisp survey of available tools
- Need for time-series analysis and its role in the development of BESS dispatch algorithms and evaluation of these algorithms under different scenarios
- Benefit analysis exercises including data preprocessor, test networks and evaluation of benefits with rule-based single-objective optimization approach for four different objectives such as peak shaving, feeder voltage profile improvement and avoiding the ANSI voltage limit violations, minimizing feeder losses and reducing the tap-changer operations for voltage regulators (VRs)
- Problem formulation for multi-objective optimization with Fuzzy systems followed by simulation results and discussion for two distribution feeders
- Handling high volatility due to local PVDGs by introducing Fuzzy-PSO-based approach for multi-objective optimization

Starting from modelling requirements, Section 7.2 describes the modelling and simulation tools and also includes the framework for time-series analysis. Assessment and optimization of BESS benefits are presented in later sections with problem formulation and results for two distribution network including a utility feeder in Northwest USA.

7.2 Modelling and simulation

In order to exploit the benefits of BESS such as performance improvement in distribution system operations, deferred investments in network upgrades, reliability enhancements, among other benefits, the planners and operators need to have powerful simulation tools that truthfully and efficiently represent the new state of distribution networks [5]. The tools should be able to perform the conventional analysis with introduction of new converter-based intermittent DGs whereas these should also be able to simulate new control algorithms that work specifically for DGs, BESS and voltage-regulating equipment in active distribution networks (ADNs) [6–7]. The near-term needs are upgradation of load-flow tools and fault-current calculation tools to accurately and adequately represent the mix of DGs present in the ADNs, whereas the long-term requirements include the capability of undertaking islanded operation studies with clusters of DGs, PV flicker analysis and advanced load-flow tools for reliability evaluation [8]. The erstwhile simulation tools were simply based on electrical parameters of the network components, inclusion of geographic information system can greatly aid in the consideration of actual economic, environmental and geo-technical aspects in the planning and operational exercises [9]. Modelling and simulation of RES-based resources such as photovoltaics is also gaining importance and need to be treated exclusively for inclusion in the analysis of ADNs [10].

7.2.1 Modelling requirements

Distribution networks are typically unbalanced due to two reasons – one, topological unbalance and two, temporal unbalance. Non-availability of one or two phases in a section of a distribution networks and inadequate phase-transposition are the prime reasons for topological unbalance. This type of phase-unbalancing remains unchanged but it is very much known, whereas temporal unbalance is caused by unpredictable behaviour of electricity customers, especially the scattered residences and sometimes commercial outlets. In order to model the existing unbalanced network, various configurations of distribution transformers, large single-phase loads and phase-unbalanced compensation by shunt capacitors also need to be accounted for. Phase-domain models of all these components along with the feeders are therefore essential to faithfully reproduce the distribution networks to perform realistic simulations. Power flow programme also needs to be replaced by backward forward sweep (BFS) algorithm or other such algorithms suitable for unbalanced networks. These algorithms are designed to handle phase-domain modelling of all major components with 3×3 (or larger) matrices.

Historically, transmission line networks are represented by single-line equivalent networks. Similar approach is also used more or less for distribution networks as well. However, with the increasing need to model distribution networks carefully, different phase-domain models of all major components have also been developed and recommended for unbalanced distribution networks [11]. Phase-domain models for voltage regulating equipment such as on load tap changer (OLTC), line drop compensator (LDC) and shunt capacitors (SCs) are now exceedingly used. Proximity of distributed generation to load centres is another reason that they perceive the effect of unbalance all the more. Also, single-phase DGs from households increase the unbalancing effects which warrants the phase-domain analysis. Collectively, all the above factors need to be accounted for in planning exercise as well as day-to-day operations of distribution networks [5].

7.2.2 Tools for operations analysis

A mix of simulation tools can be used as an advantage to get the best of available features to serve the overall objectives of operations analysis. This is achieved in [12] by combining the electrical simulation model of wind turbines from MATLAB/Simulink with the gear-train models from FAST. Similar approach is also used in [5] where the PV array modelling is based on two-stage modelling exercise in which weather parameters are used to model the module temperature in one stage [13] and the second stage uses module temperature and other parameters to obtain PV array output [14]. Some higher-level decision support tools including techno-economic analysis for DGs in distribution systems are: (1) Wind Deployment Systems (WinDS), [15], (2) Clean Energy Technology Economic and Emissions Model (CETEEM) [16], (3) DIStributed Power Economic Rationale SElection (DISPERSE) [17]. Specialized simulation programmes for PV system design and analyses are available; few of these are ATSUNPV, PV*SOL, PVGRID, PVWATSS, GridPV, PV-DesignPro, PV F-CHART, PV-FORM, PVSYST, NSOL and SolarPro. Exhaustive list of such

Table 7.1 Features, applications and intended users of reviewed simulation tools

Simulation tool	Features and applications
Hybrid2 [20]	Can be used for simulating long-term performance of hybrid power system. Also facilitates economic analysis of hybrid power systems. Useful for system planners and designers
HOMER [21]	Named as hybrid optimization of multiple energy resources, it is specially aimed at facilitating design of hybrid renewable micro grids. Useful for technical designers as well as non-technical decision-makers
OpenDSS [22]	Can be used for frequency-domain analyses for power delivery networks, can also be used for grid modernization and renewable energy research. Mainly used by researchers in smart grid and renewable energy
FAST with MATLAB [12]	Designed to accurately model the behaviour of wind turbine generators using the MATLAB Simulink environment with FAST. Useful for wind turbine designers and researchers
DISPERSE [17]	DIStributed Power Economic Rationale SElection (DISPERSE) can be used to assess the site-specific market potential for RES-based DG. It is used by manufacturers, electric utilities, research laboratories and government agencies
PVsyst [24]	Preliminary design tool for detailed design of PV system and economic evaluation of the same. Useful for PV system planners, architects and designers
GridLAB-D [25]	It allows creation of custom models and objects to represent the existing and new control and power equipment in distribution systems. Allows time-series simulations useful for system operators, distribution system planners and researchers

packages can be found in [18]. Table 7.1 lists the features and applications of some major software packages.

7.2.3 Tools for planning studies

The planning tools for distribution system analysis needs to consist of: (1) electrical performance simulators, (2) reliability analysis and (3) decision support tool for selection from the available options [19].

Hybrid2 can simulate several types of wind turbines, photovoltaic modules, diesel-generator sets, electrical loads, BESS and also provides a model to calculate the economic viability of the project [20]. HOMER's optimization and sensitivity analysis algorithms allows users to evaluate technical feasibility and economic viability from among variety of options and to account for changes in costs of different technologies, electric loads and availability of energy resources [21]. The OpenDSS is a power system simulation programme primarily for use by electric utility for their distribution networks. OpenDSS can perform nearly all commonly used frequency domain (sinusoidal steady state) analyses by distribution systems' operators. It can also simulate novel and recently developed analyses approaches that are designed to respond to the future needs related to grid modernization, smart grid and renewable energy research [22]. GridLAB-D is a tool under development with a platform that is capable to interact with a third-party simulation tool and automation models for distribution systems. It is being developed to include modules that can perform unbalanced power flow, incorporate phase-domain models for distribution network components, equipment and controls and simulate energy market models, energy operations and models of Supervisory Control and Data Acquisition (SCADA). It may also have its new graphical user interface to integrate input models and for the execution and control of the simulation [23].

7.2.4 Time-series analysis

In order to assess the worst-case operating conditions, power flow studies are carried out for typical worst-case scenarios such as peak load condition, minimum/maximum load/generation conditions and generation outage or line outage cases. Obviously, the selected cases are the ones in which system components are stressed in one or the other way. Such an approach has an underlying assumption that the worst-case scenarios can be deterministically represented and that the results obtained can be accurate enough to carry out the system operations reliably. Such a study, known as snapshot power flow analysis, has been found sufficient to carry out system operations as well as system planning studies.

However, with the ingress of renewable energy-based distributed generation in distribution system and sub-transmission systems, the nature of variables has changed from deterministic time-invariant to uncertain, time variant. The conventional thermal-generating units can be scheduled to generate any amount of real power within its capacity almost continuously. Thus, the real-power generation is known and invariant, whereas the real-power generation from renewable resources is not only time variant but also intermittent, uncertain and difficult to predict. The loads, on the other hand, approximated on the bases of transformer rating that feeds them in transmission system studies, play an important role in network operations in distribution system studies. Not only that the load profiles need to be approximated correctly, but also the nature of load also needs care and attention in modelling for accurate representation. The variable nature of real-power generation and significant impact of load profile and load composition in distribution system analysis renders the identification of worst-case scenarios very difficult and error prone. Moreover, the studies related to VR operations, reactive resource allocation and energy loss evaluations among others need to consider diurnal variability of renewable generation and load alike.

Probabilistic approach is used extensively to represent variability in transmission system studies and some attempts are also made to implement similar approach to distribution system studies with distributed energy resources (DERs). Authors have used Monte-Carlo simulation to analyse the impact of DERs on voltage limit violations in low-voltage networks in [26] and stochastic power flow in distribution network with DERs to estimate energy losses in [27]. Many similar attempts with difference in objective can be found in literature. However, one serious limitation of these studies against the analysis based on recorded measurements is that the probabilistic studies may not necessarily generate the frequent combinations of actual generation and load profiles in a given specific case. Hence, the probabilistic studies are more generic and less applicable to a specific case, whereas the distribution systems with DERs and BESS exhibit unique combination of load and generation profiles with varying nature of load composition and diurnal variability in generation throughout the year.

Given the above reasons and increasing complexities in distribution system operations in the presence of communication layer between load and generation points, TSPF studies are recommended in [28–29]. Few attempts have been made to use the set of time-series data of DGs and demand in distribution networks. A 3-φ balanced power flow simulation with approximated load profile in a residential area, approximated PV generation data based on solar irradiance and hypothetical DGs connected in distribution network, has been presented in [18] to assess the impact of DG penetration. A more rigorous attempt is demonstrated in [19] with quasi-static TSPF simulations at an interval of 1 min for an IEEE 34 node system with extrapolated load profile (from 1 h to 1 min) and assumed PV generation approximated based on the recorded solar radiation database. Both these efforts – [30] and [31] – are based on approximated load and generation data for one year, whereas the analysis presented in [32] is partially based on recorded measurements for three years in an actual distribution network and approximated wind generation data based on wind density measurements over the period of one year. The exercise in [32] leads to conclusion regarding maximum allowable wind power penetration (capacity addition) in a specific distribution network. TSPF has also been used in transmission system analysis with sizable penetration of DERs to assess its impact on voltage stability in [33].

Following observations are made from the survey of literature on the applications of TSPF:

- All these studies have been based on approximated data either on generation side or load side or both.
- All above applications of TSPF – whether it is at distribution system level or at transmission system level – have used either single line equivalent power flow or 3-φ balanced case distribution power flow.

The first observation implies that the analysis is a close approximation of the possible real test case and that reflects the probable scenarios more realistically than the stochastic or probabilistic approach. Nonetheless, the test cases lack the authenticity of recorded measurements for actual distribution networks. Second observation indicates a potential limitation of the analysis in which the essence of distribution system analysis is lost by eliminating the unbalance in distribution networks. The actual voltage profile in each phase, conductor overloading, energy losses, tap-changing operations (TCOs) and reactive resource allocation are some of the applications where 3-φ unbalanced distribution power flow study is essential and unavoidable.

7.2.5 Sequential time simulations approach

The sequential time simulation (STS) approach is used in this study with one standard test case (IEEE 37 node distribution feeder) and with another existing feeder in

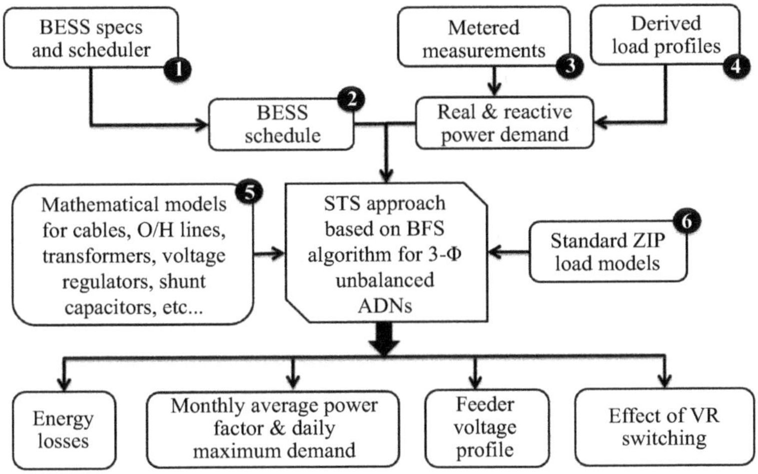

Figure 7.1 Framework for sequential time simulations approach

Northwest USA. The analysis is based on full 3-φ unbalanced modelling of all network components including the possibility of phase-sensitive dispatch of BESS. The overall framework of STS approach is depicted in Figure 7.1. Phase-domain models of all the network components are supplied to the BFS algorithm to solve three-phase unbalanced power flow simulations from block 5 in Figure 7.1. The BESS scheduler (block 1, Figure 7.1) supplies the BESS schedule to block 2 which is further used by the simulator. The load profiles are either derived from publically available data sets (such as [34]) or recorded measurements are used to supply real and reactive power demand in the form of time-series data strings. The standard load models are used from block 6 to perform BFS simulations recursively for the time-series data. Details of phase-domain modelling and application of BFS algorithm can be found in [11] and [35].

In the next and following sections, details on the development of BESS scheduler (block 1, Figure 7.1) are elaborated with benefit analysis and simulation results of multi-objective optimization.

7.3 Benefit analysis for BESS

Applications of grid-scale BESS are wide ranging; refer to Table 7.2. Depending on the application domain for BESS, the specific applications can include frequency regulation, black start and deferred infrastructure upgradation for both transmission and distribution networks and many others. However, given the context of the increasing activity in distribution systems for increasing penetration of DGs, the focus of the exercise for benefit analysis in this chapter is low-voltage distribution networks. The evaluation of benefits of BESS in low-voltage distribution networks is undertaken here for the objectives such as (a) peak demand shaving, (b) improving feeder voltage profile, (c) minimizing feeder losses and (d) reducing VR adjustment activity. The framework for analysis is however flexible to incorporate

Table 7.2 Energy storage services [36]

Broad application domain	Specific applications
Customer EMS	• Demand charge management
	• Retail electric energy time-shift
	• Power reliability
	• Power quality
Transmission infrastructure services	• Voltage support
	• System upgrade deferral
Distribution infrastructure services	• Congestion relief
	• System upgrade deferral
Ancillary services	• Black start
	• Voltage support
	• Reserves – spinning, non-spinning and supplemental
	• Frequency regulation and real power balancing
Bulk energy services	• Electric supply capacity
	• Electric energy time-shift (Arbitrage)

other objectives as well. The proposed approach can particularly be extended to assess the benefit of reactive power dispatch from BESS for objectives stated above and others. This section covers the details of the approach based on a data pre-processing exercise followed by test networks for benefit evaluation and the single-objective rule-based optimization and results.

7.3.1 DLP analyser and rule-based optimization

Daily load profile (DLP) analyser identifies 'similarity' among all the DLPs by finding and quantifying the recurrent part of DLPs among the data set of annual load profiles (ALPs). Two different similarity indices are developed to resemble similarities among DLPs and classify these profiles as characteristic DLPs (CDLP). To this end, (7.1) defines the *DD* matrix for each day δ. Closest integer value given by (7.2) is used to determine the total number of demand blocks -α_0 such that $\alpha_s \leq \gamma_{SP}$ (where α_S designates the demand block's size (in p.u.) and γ_{SP} is used to represent size of energy storage block (power)).

$$DD^{\delta}_{\alpha,\tau} = \begin{cases} 1 & \text{for } \forall P^{\delta}_{\tau} \in [\alpha_S \cdot n, \alpha_S \cdot (n+1)] \\ 0 & \text{for } \forall P^{\delta}_{\tau} \notin [\alpha_S \cdot n, \alpha_S \cdot (n+1)] \end{cases} \tag{7.1}$$

$$\alpha_0 = \frac{AMD}{\alpha_S} \tag{7.2}$$

$$\sum_{\delta=1}^{\delta_0} \sum_{\tau=1}^{\tau_0} \sum_{\alpha=1}^{\alpha_0} DD^{\delta}_{\alpha,\tau} = \delta_0 \cdot \tau_0 \tag{7.3}$$

where:

α	Real-power demand blocks ($\alpha = 1,2,\ldots,\alpha_0$)
δ_0	Days in total
$\delta,\delta\delta$	Day of a Year (DoY: $1,2,\ldots\delta_0$)
τ_0	Number of time intervals in a day
τ	Time slots ($\tau = 1,2,\ldots\tau_0$)
n	number $0,1,2,\ldots,(\alpha_0-1)$
P_τ^δ	Active power (p.u.) for the interval τ of day δ
AMD	Annual maximum demand

Element-to-element mapping of *DD* matrices for different days is used to enumerate *ESI* in (7.4). For $\forall DD_{\alpha,\tau}^{\delta} \neq 0, \forall DD_{\alpha,\tau}^{\delta\delta} \neq 0$:

$$ESI_\delta^{\delta\delta} = \frac{1}{\tau_0}\sum_{\tau=1}^{\tau_0}\theta_\tau \tag{7.4}$$

$$\theta_\tau = \begin{cases} 1 & \text{for } \forall\, \mathrm{DD}_{\alpha,\tau}^{\delta} = DD_{\alpha,\tau}^{\delta\delta} \\ 0 & \text{for } \forall\, \mathrm{DD}_{\alpha,\tau}^{\delta} \neq DD_{\alpha,\tau}^{\delta\delta} \end{cases} \tag{7.5}$$

where:

$ESI_\delta^{\delta\delta}$	Exact Similarity Index matrix between days δ and $\delta\delta$
θ_τ	Element of *ESI* matrix for interval τ

ESI can attain a value anywhere between 0 and 1 for the day. These values are shown in Figure 7.2 as a grey map. It shows: (1) clusters of group affinity based on seasonal variations and (2) recurring break in cluster patterns. The recurring dis-continuities in patterns are due to weekly variations in load demand. Therefore, the

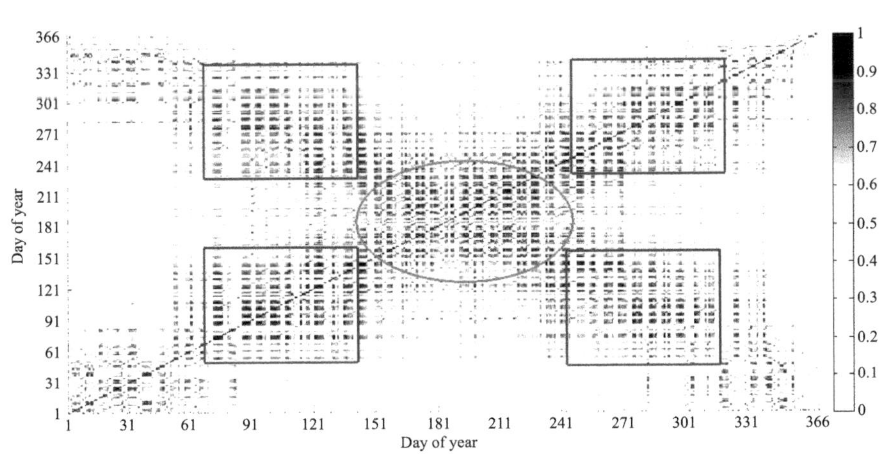

Figure 7.2 Grey-map based on ESI, similarity among DLPs is highlighted [7]

algorithm chooses DLPs of Saturday and Sundays as CDLP to account for weekend discontinuities. The clusters are highlighted by rectangles and a circle in Figure 7.2. In order to compare the neighbourhood proximity of nonzero elements in *DD* matrix, (7.6) and (7.7) together define *SI*. Naturally, the *SI* is more tolerant in determining similarities among DLS than *ESI*.

$$SI_{\delta}^{\delta\delta} = \frac{1}{\tau_0}\sum_{\tau=1}^{\tau_0}\prod_{\tau} \tag{7.6}$$

$$\prod_{\tau} = \begin{cases} 1 & \text{for } \forall\, DD_{\alpha,\tau}^{\delta} = DD_{\alpha,\tau}^{\delta\delta} \\ 0.5 & \text{for } \forall\, \mathrm{DD}_{\alpha,\tau}^{\delta} = DD_{\alpha\pm1,\tau}^{\delta\delta} \\ 0 & \text{for } \forall\, DD_{\alpha,\tau}^{\delta} \neq DD_{\alpha,\tau}^{\delta\delta} \end{cases} \tag{7.7}$$

Equation (7.9) quantifies the number of days that have their load profiles in close agreement with the CDLP. It helps identifying the group of CDLPs that efficiently can represent most days of a year.

$$ESI_{\delta}^{\delta\delta}, SI_{\delta}^{\delta\delta} = \begin{cases} 1 & \text{for } \forall\delta = \delta\delta \\ \in [0,1] & \text{for } \forall\delta \neq \delta\delta \end{cases} \tag{7.8}$$

$$\lambda_{\delta} = \sum_{\delta\delta=1}^{\delta_0} \eta_{\delta}^{\delta\delta} \quad \text{where } \eta_{\delta}^{\delta\delta} = \begin{cases} 1 & \text{for } \forall \mathrm{SI}_{\delta}^{\delta\delta} \geq \varepsilon \\ 0 & \text{for } \forall SI_{\delta}^{\delta\delta} < \varepsilon \end{cases} \tag{7.9}$$

Equation (7.9) also quantifies the obtained characterization by this algorithm. The amount of dissimilarity allowed among the group of DLPs represented by a single CDLP is quantified by ε in (7.9). CDLPs can be derived using (7.1)–(7.9) by following a set of steps elaborated in [7]. Since the CDLPs are made up of most recurrent parts of DLPs based on data sets of ALPs, all the CDLPs collectively represent the whole data set and BESS scheduler tested for CDLPs can effectively handle any and every variation found in ALPs. In a way, the BESS scheduler is tuned to perform at its best when aided by an optimization algorithm that can produce optimum results for these CDLPs. Rather than searching for the optimum solution for any and every possibility of demand-generation scenario (based on probabilistic methods), the search space for optimization algorithm now is reduced and exhaustive search for the optimum solution becomes feasible within the reduced search space made available by CDLPs. A typical CDLP is shown in Figure 7.3 wherein black colour represents high probability for load demand occurrence (based on *ESI*), whereas grey shade indicates somewhat inferior probability for load demand occurrence – based on *SI*.

7.3.2 Test networks

7.3.2.1 IEEE 37 node feeder

An urban unbalance distribution network is represented by a modified IEEE 37 node test network [36] with a hypothetical BESS in this analysis. ALP of the

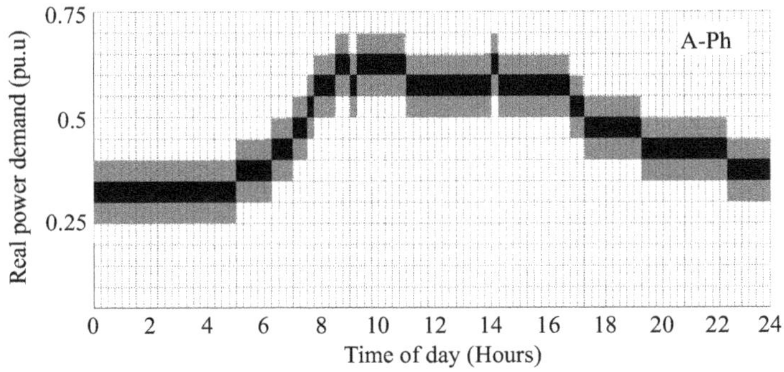

Figure 7.3 Figure 7.3 A typical CDLP [7]

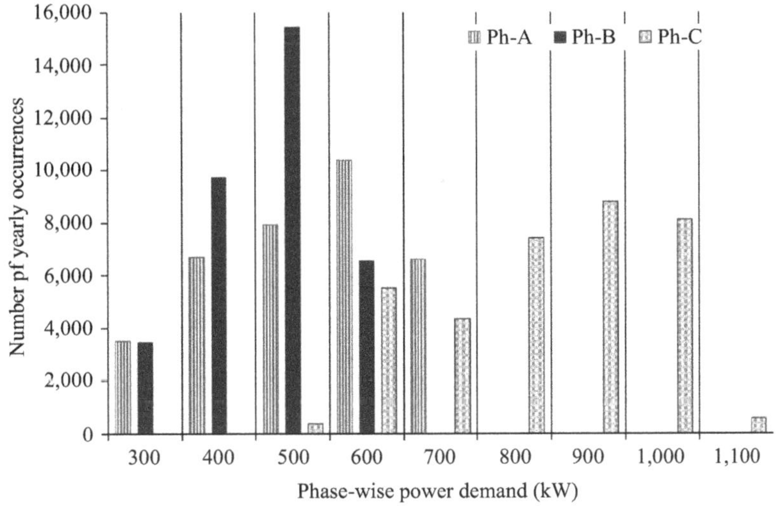

Figure 7.4 Histogram of load demand in each phase of IEEE 37 node network at substation node [7]

network forms the data sets for this exercise. A data set for the year 2012 available in [37] including industrial, residential and commercial daily profiles are used to obtain ALP for each node of the IEEE 37 node network.

The underlying phase-unbalancing cannot be represented by net power demand at sub-station node. Therefore, a histogram is shown in Figure 7.4 for the phase-wise active power load demand at source node. High unbalancing among the three phases is apparent in Figure 7.4. Range of load demand for two phases of the network stays within 300 to 600 kW whereas third phase witnesses the range of 700 to 1,000 kW, as shown in Figure 7.4. It should be noted that the histogram shown in Figure 7.4 is obtained by retaining the connected load at all the nodes of IEEE 37 node standard test feeder. The ALPs at each node never exceed this connected load.

7.3.2.2 Utility feeder in Northwest USA

A utility feeder with an interconnected BESS in Northwest USA is also considered for the application of proposed approach. The two feeders shown in Figure 7.5 have collectively ~1,500 nodes with ~450 load points. Details of the feeders are not shown due to confidential reasons. Feeder voltage profile stays healthy with 1,500 kVAr of reactive support in each phase augmented with one VR at substation node for each feeder. The BESS is connected at remote end of feeder for maximum benefits as shown in Figure 7.5.

7.3.3 Benefit evaluation

Evaluation of benefits is first done based on a rule-based single objective optimization for peak shaving and similar exercise is performed for each objective. A simple exhaustive search-based algorithm is used as a scheduler to obtain the most benefit of using BESS. For each CDLP, discharge sweep in Table 7.3 collects the information about candidate demand blocks for the objective of peak load shaving. It takes into account each α and every time interval τ to compute the potential for peak load shaving given the constraints of BESS capacity of power and energy. Recharge sweep (Table 7.3) also accounts for each interval of time and every demand block to assess the per-phase potential for absorbing active power to

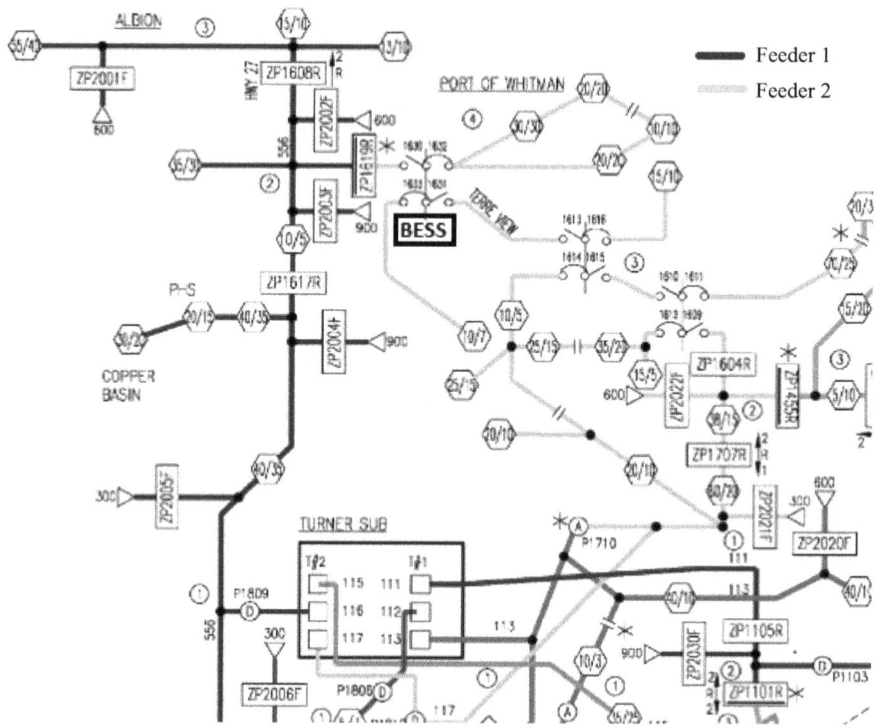

Figure 7.5 Feeder 1 and Feeder 2 in Northwest

Table 7.3 Framework for sequential time simulations approach

Discharge sweep	**Recharge sweep**
Step 1: Start with a CDLP separately for each phase	Step 1: Start with a CDLP separately for each phase
Step 2: Choose the candidate block for BESS discharge with given objective.	Step 2: Choose the candidate block for BESS recharge with given objective.
Step 3: Power and energy limits of BESS reached? If yes, dispatch γ; else stop.	Step 3: Power and energy limits of BESS reached? If yes, dispatch γ; else stop.
Step 4: Repeat from step 2 for each α	Step 4: Repeat from step 2 for each α
Step 5: Repeat from step 2 for each τ	Step 5: Repeat from step 2 for each τ
Equilibrium of charge for each CDLP	
Step 1: Estimate the potential of BESS for phase-wise discharge–recharge	
Step 2: Compute the maximum possible objective satisfaction for each α and sort it in downward order, separately for all the phases	
Step 3: Compute the maximum possible recharge potential for each α and sort them in downward order, separately for all the phases	
Step 4: Based on all the permutations of Steps 2 and 3, obtain the highest satisfaction of an objective such that the recharge potential stays higher than discharge potential consistently	
Step 5: Schedule BESS discharge	

recharge BESS. Per-phase potential in each time slot for discharge–recharge is estimated after the CDLPs are selected. This estimate also takes into account the power and energy capacity of the hypothetical BESS system. BESS scheduling is then completed with a constraint of single discharge–recharge cycle daily. Similar steps are followed to obtain benefit for all the objectives. The four objectives considered are (a) peak shaving, (b) voltage profile improvement, (c) minimizing feeder losses and (d) reducing tap-changer activity of VR.

IEEE 37 node standard test feeder [36] is modified with an assumed BESS of 500 kW, 2 MWh connected at node 775 in this analysis. Additionally, utility feeder with BESS is also used for validation of developed algorithm. Data for DLP analyser are ALPs of the network for the year 2012. Preserving the topological phase-unbalancing and keeping the maximum value of connected load unchanged at nodes in the standard test feeder, combinations of different load profiles are used to populate each node with an ALP having time interval of 15 min. The exhaustive data set is processed by a DLP analyser to obtain CDLPs. It is observed that phase C witnesses high demand as compared to phases A and B. Inherent phase-unbalancing in connected loads specified in the standard test feeder [36] is preserved while populating it with ALPs.

7.3.3.1 Peak shaving

Peak shaving or reducing the maximum demand at a substation helps in reducing the overall reserve capacity of generation to meet the extreme demands. It is also a very

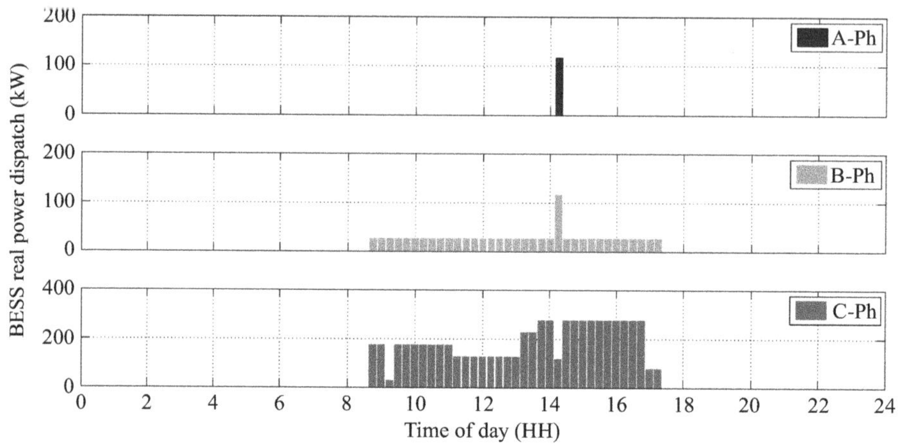

Figure 7.6 Phase-wise real power dispatch for peak-shaving objective

critical factor in small and partially independent microgrids where local generation is comparable to local demand. The use of BESS can be critical in shaving the maximum demand within a certain level. The rule-based optimization proposed here works on an unbalanced demand in three phases and works on each phase individually to assess the potential of maximum benefit of BESS real-power dispatch. The results of single objective optimization are tabulated along with results for all other objectives (refer to Table 7.4). Figure 7.6 shows the BESS dispatch in each phase as a function of time of day (ToD). These results emphasize the fact that it is important to dispatch real power in appropriate phase as and when necessary rather than having balanced dispatch for all objectives, all the time. It also ensures more efficiency compared to phase-balanced dispatch as the later will need more amount of real power in appropriate phase to achieve the same effect. Figure 7.6 displays the selection of phases for real-power dispatch from BESS as a function of ToD. It can be noticed that either phase c or phases bc are selected for BESS real-power dispatch. The same fact is also apparent from Figure 7.6.

7.3.3.2 Conservation voltage regulation and feeder voltage profile

To keep the node voltages within the specified ANSI limits, [38] is an obligation of distribution system operator, whereas improving the feeder voltage profile can be very helpful in conservation voltage regulation (CVR). The improved voltage profile can provide sufficient margins for the deliberate reduction in system voltages to conserve energy and peak power during high demand hours of operation. However, the first priority for the system operator is to maintain the node voltages within the specified range (normally within ±5 % of nominal voltage).

In case of IEEE 37 node feeder, it is observed that the node voltages happen to go below 0.95 p.u. in certain cases. Therefore, the objective framed is that of maintaining node voltages within the ±5 % of nominal voltage. It should be noted that this is the objective that requires most real-power dispatch (250 kW) but not

Table 7.4 Single objective optimization results for IEEE 37 node feeder

Objective	P_{max} (kW)	E_{max} (kWh)	Objective satisfaction
Peak shaving	350	1,925	287 kW peak shaving
Loss min.	100	2,000	1,033.4 kWh reduced to 962.7 kWh
Voltage profile	250	1,325	Improved from 0.944 p.u. to 0.95 p.u.
TCO min.	200	1,350	TCOs reduced from 18 to 12

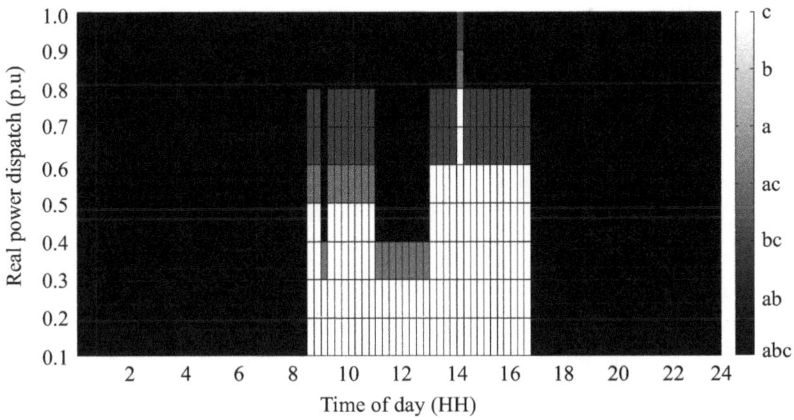

Figure 7.7 Phase selection for BESS real power dispatch to improve feeder voltage profile

the most energy from BESS (1,325 kWh) as compared to other objectives, refer to Table 7.4. Figure 7.7 shows the selection of phases for real-power dispatch from BESS to improve the voltage profile. It should be noted that the phase selection includes the combination of phase c, phase ac and phase bc at different power levels.

7.3.3.3 Feeder loss minimization

Feeder loss minimization is another important objective for system operator and it is also the objective that can demand most energy from BESS, as can be observed from Table 7.4. Figure 7.8 shows the phase selection at different power levels of BESS dispatch. When both the results – real-power dispatch from BESS (100 kW for loss minimization) and phase selection at this power level (0.2 p.u.) – it is clear that for loss minimization, all the real power is dispatched to phase c. Unless the real-power dispatch is more than 250 kW, only phase c continues to be at priority to reduce feeder losses.

7.3.3.4 Regulator adjustment activity

Tap changer operations become frequent and lead to high wear and tear especially during the large fluctuations in demand. VR's TCOs can be reduced by absorbing the demand fluctuations – partially – by BESS. This benefit is also evaluated here

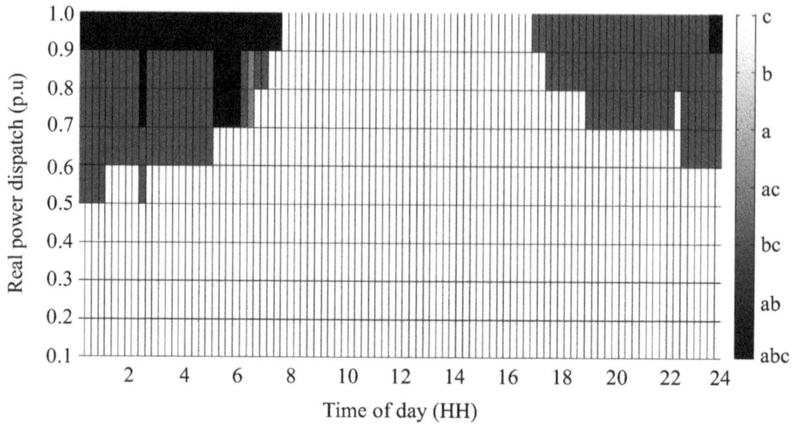

Figure 7.8 Phase selection for BESS real power dispatch to minimize feeder losses

to assess how much support from BESS can be effective in reducing TCOs of VRs. It is all the more important in the grid where the presence of RES-based DGs is increasing introducing high level of volatility in demand and generation. It is observed from the results (Table 7.4) that the benefit of BESS dispatch in an unbalanced network can be as much as 30% reduction in TCOs.

7.4 Multi-objective optimization and BESS scheduling

7.4.1 Fuzzy multi-objective optimization approach

The optimization problem needs an approximate or expected DLP to start with. This is obtained by a local intelligent predictor described in the preceding section which processes phase-wise historical demand data and produces CDLPs. These CDLPs are then forwarded to fuzzy multi-objective optimization (FMOO) algorithm. These are processed by FMOO module with updated state of charge (SoC) obtained in real time at every 15-min interval. FMOO delivers the day-ahead dispatch of real power for the day represented by given CDLPs. The algorithm can also update the scheduled dispatch based on updated SoC or operator preferences through real-time data interface.

The FMOO uses all the CDLPs to quantify the effect of incremental real-power dispatch from BESS for all the objectives. It should be noted that every incremental real-power dispatch is considered for phase-balanced and phase-unbalanced dispatch options (ϕ). This results in a repository that stores network response (in terms of indices for all objectives) of incremental BESS dispatch in each phase option for all the CDLPs.

Fuzzy systems are particularly suitable for imprecise and directly incomparable objectives. It allows to compare the different benefits on a uniform scale while also providing a human-friendly linguistic interface, which is used as a benefit in the proposed algorithm for operator preferences. Approximate reasoning can be well represented by Fuzzy systems and enables it to deal with reality [39]. Flexibility in defining limiting values of Fuzzy membership functions makes it adaptable to diverse network responses to similar stimulation. Numerous applications are found in literature with Fuzzy theory-based optimization in energy systems and distribution networks [39, 40].

7.4.1.1 Multi-objective problem formulation

In this study, following objectives have been considered for ESS optimization: (a) peak shaving, (b) voltage profile improvement, (c) loss minimization, (d) minimizing TCOs for VRs and (e) minimizing power and energy dispatch. These five objectives are defined by Equations (7.10)–(7.14) as outlined below. It should be noted that phase-balancing is recognized as an effective means for reduction in power losses in distribution network [41].

7.4.1.2 Peak shaving

Peak shaving reduces the maximum demand at source node by dispatching power during peak hours which helps in the overall flattening of DLP.

$$f_{peak} = Min\, \mathrm{P}_{peak} \tag{7.10}$$

where $P_{peak} = \max(P_\tau), \tau \in [1, 2, \ldots, \tau_0]$ and τ_0 is number of time slots in a day

7.4.1.3 Voltage profile improvement

Feeder voltage profile improvement ensures that each node voltage in feeder stays within the stipulated limits throughout the ToD. Here, the objective is framed to minimize the deviation of minimum feeder voltage from nominal feeder voltage as the feeder is observed to have some minimum node voltage limit violations.

$$f_V = \mathrm{Min}\left\{ \left| V_{\min,\tau} - V_n \right| \right\} \tag{7.11}$$

where $V_{\min,\tau} = \min(V_\tau^k)$, $k \in [1, 2, \ldots,$ number of nodes] and V_n is nominal feeder voltage.

7.4.1.4 Loss minimization

Loss minimization in distribution systems should be considered in two aspects: power loss and energy loss. While power loss minimization over a period of time results in energy loss minimization, variation in demand and limited availability of ESS makes it imperative to optimize power loss minimization such that energy losses are also minimized.

$$f_{\mathrm{Loss}} = Min \left\langle \frac{P_{Loss,\tau}}{\sum_{\tau=1}^{\tau_0} P_{Loss,\tau}} \right. \tag{7.12}$$

$P_{Loss,\tau}$ is the total real-power loss of the feeder. This objective is intended to minimize real-power loss for individual time slots as well as energy loss over a day. Considering hourly intervals for dispatch, $\sum_{\tau=1}^{\tau_0} P_{Loss,\tau}$ in (7.3) effectively represents energy saved over a period of 24 hours.

7.4.1.5 TCO minimization

VRs' function is to maintain healthy voltage profile of the feeder against diurnal variation in demand. High and frequent TCOs lead to increased wear and tear of the VRs. Minimization of TCOs extend the life span of VRs and decreases maintenance requirements.

$$f_{TCO} = Min\sum_{\tau=1}^{\tau_0} N_{TCO,\tau} \tag{7.13}$$

$N_{TCO,\tau}$ is the number of tap-change operations in VRs within one-time interval.

7.4.1.6 Minimizing power and energy dispatch

The pricey ESS has to be dispatched such that its availability is maximized while optimum benefits are harnessed by the scheduled dispatch. This objective stands in direct conflict with all other objectives.

$$f_{BESS} = Min\left\langle \frac{P_{BESS}}{\sum_{\tau=1}^{\tau_0} P_{BESS,\tau}} \right. \tag{7.14}$$

where $P_{BESS,\tau}$ is the real-power dispatch from BESS for interval τ

The objective for energy dispatch minimization (objective e) is included for two reasons. One, it is necessary to avoid undesired energy output without intended benefit. It reduces the total number of discharge–recharge cycles and thus extends the BESS life span in long run. Second, all objectives except TCO minimization have a tendency to attain high satisfaction at high-power dispatch (high discharge rate of BESS), which results in high benefit for few time slots in a day. Whereas, to maximize benefits over a span of 24 hours, it is necessary to minimize the discharge rate of BESS in each time slot thereby attaining minimum energy dispatch as well as maximizing availability of BESS over a day. This objective is obviously in conflict with all other objectives in two ways – one, it restricts the maximum rate of discharge in each time slot and two, it also minimizes the total energy dispatched over a day.

While the above described objectives – (a), (b), (c) and (d) are not necessarily conflicting when considered for perfectly balanced operations, their different sensitivity to unbalanced operation of distribution networks can be exploited as a benefit in optimizing the use of pricey ESS. Any effort to mitigate the effect of unbalanced operations is responded by these objectives differently. Also, the energy capacity of ESS imposes a great limit on exploiting the benefits of its operations and each objective competes for the limited resource. It should also be considered that each objective requests dispatch of power at different ToD for maximum benefit. Thus, maximum dispatch of ESS during peak hours does not maximize the overall benefits. Phase-unbalanced dispatch needs to optimize the benefits in network operations – objectives (a) to (d) – while minimizing the dispatch of power and energy.

All the five objectives are subject to the operational constraints as well as the constraints of BESS for its rating and depth of discharge. It should be noted here that the steady-state analysis need not incorporate the ramp-rate constraints of

BESS as the response of BESS is quite fast as compared to the planning interval for BESS. Equations (7.15–7.19) define the constraints.

Power and energy balance constraint:

$$\left.\begin{aligned} For \quad \forall\tau, \left(P_{Loss,\tau} + P_{Load,\tau}\right) = \left(P_{ss,\tau} + P_{BESS,\tau}\right) \\ \sum\nolimits_{\tau=1}^{\tau_0} \left(P_{Loss,\tau} + P_{Load,\tau}\right) = \sum\nolimits_{\tau=1}^{\tau_0} \left(P_{ss,\tau} + P_{BESS,\tau}\right) \end{aligned}\right\rangle \tag{7.15}$$

where $P_{Loss,\tau}$ and $P_{Load,\pi}$ are real-power losses and real-power demand, whereas $P_{SS,\tau}$ and $P_{BESS,\tau}$ are real power supplied at substation node and by BESS, respectively.

Branch current limits:

$$i_{\min}^{j} \leq i_{\tau}^{j} \leq i_{\max}^{j} \tag{7.16}$$

j represents line sections with line current limits given by $i_{\min}$ and $i_{\max}$ for respective line sections.

Node voltage limits:

$$V_{\min}^{k} \leq V_{\tau}^{k} \leq V_{\max}^{k} \tag{7.17}$$

k represents the node number with node voltage limits specified by V_{min} and V_{max}.

VR tap limits:

$$Tap_{\min} \leq Tap_{\tau}^{k} \leq Tap_{\max} \tag{7.18}$$

$Tap_{\min}$ and $Tap_{\max}$ are extreme positions of taps in VRs. Tap_{τ}^{k} represents tap position of VR located at node k during time slot τ. Normally, -16 and +16 are extreme tap positions in VRs.

BESS Constraints:

$$\left.\begin{aligned} P_{BESS,\tau} \leq P_{\max} \\ E_{BESS,\tau} \leq E_{\max} \\ SoC_{\min} \leq SoC_{\tau} \leq SoC_{\max} \end{aligned}\right\rangle \tag{7.19}$$

where $P_{\max}$ is the maximum specified rate at which BESS can supply energy and $E_{\max}$ is the maximum capacity of BESS to store electrical energy. $SoC_{\min}$ and $SoC_{\max}$ represent minimum and maximum SoC for BESS, respectively.

Fuzzy indices are defined and appropriate fuzzy membership functions are then associated with each objective. Effect of incremental BESS dispatch is quantified by defining appropriate indices to represent the benefit of BESS dispatch for each objective. Fuzzy membership functions are then selected that fuzzify the level of objective satisfaction.

7.4.2 Optimization with fuzzy max-average composition

Among Fuzzy compositions, Max–Min composition is commonly used in multi-objective optimization to select the maximum benefit value among the set of minimum satisfaction of objective functions. It ascertains the minimum level of satisfaction for all the objectives. It is a preferred composition when all the objectives can be satisfied to a more or less extent simultaneously. However, Max–Min composition is not used in the proposed FMOO approach for the following reasons: (1) all the included objectives in this problem formulation do not necessarily have a non-zero satisfaction value for all the time intervals in a day, (2) some objectives (2 and 5) are optimized for individual time slots as well as over a day, therefore their selection cannot merely be based on Max–Min principle and (3) Max–Min composition ascertains minimum benefit for all objectives whereas the proposed problem formulation cannot ignore the high benefits for some objectives while others might have no benefit. Different operations can be used in place of Min in the Max–Min composition – represented in (7.20) – while still performing maximization. This type of composition is known as Max–star or Max-*composition, given by (7.21). The integral sign in (7.21) is replaced by summation when the product is discrete [42].

$$\mu_{R^1 \odot R^2}(x \times z) = V_y[\mu_{R^1}(x,y) \wedge \mu_{R^2}(y,z)]/(x,z) \tag{7.20}$$

$$R^1 * R^2 \equiv \int_{X \times Z} \check{y}[\mu_{R^1}(x,y) * \mu_{R^2}(y,z)]/(x,z) \tag{7.21}$$

Max-* composition is used in the form of Max-Product composition to obtain modified set of Fuzzy membership functions for benefit objectives that incorporate the effect of cost objective (μ_{BESS}) value to penalize the use of energy as well as high rate of discharge from BESS. The benefit objectives are then maximized by using Max-Average composition given in (7.22 and 7.23). It selects the maximum among the averaged values of benefit objectives.

$$R^1 \langle + \rangle R^2 \equiv \int_{X \times Z} \check{y}\left[\frac{1}{2}\{\mu_{R^1}(x,y) + \mu_{R^2}(y,z)\}\right]/(x,y) \tag{7.22}$$

$$\mathbb{OS} = \vee_y\left[R^1 \langle + \rangle R^2 \langle + \rangle R^3 \langle + \rangle R^4\right]_{X \times Z} \tag{7.23}$$

7.4.3 Simulation results for utility feeder

Multi-objective and single-objective results obtained with FMOO approach are listed in Table 7.5. The undesired effects of high unbalancing in distribution networks can be mitigated by phase-sensitive real-power dispatch based on the

Table 7.5 FMOO results for utility feeders in Northwest USA

Objective	**Optimization**	**Objective satisfaction**
Peak shaving	Multi-obj.	565 kW peak shaving
	Single-obj.	643 kW peak shaving
Loss minimization	Multi-obj.	1,842 kWh reduced to 1,687 kWh
	Single-obj.	1,842 kWh reduced to 1,576 kWh
Voltage profile	Multi-obj.	Stays within ±5% of nominal voltage
	Single-obj.	Stays within ±5% of nominal voltage
TCO minimization	Multi-obj.	TCOs reduced from 46 to 34
	Single-obj.	TCOs reduced from 46 to 30

proposed two-stage approach in this chapter. It ensures better utilization of BESS capacity while extending most benefits to improve daily network operations.

7.4.4 Fuzzy-PSO approach

With a view to representing uncertainty of PV generation and also to exploit the benefit of human-friendly linguistic interface, Fuzzy systems are used along with PSO module to ensure global or near global optimum solution for different objectives. The use of PSO is imperative in the context of many-fold expansion of search space for optimization. The representation of uncertainty in demand and generation both results in nine different levels probabilistic load demand (including generation from PV systems). In addition to this, the step-size of incremental BESS dispatch needs to be smaller in order to respond to the temporal variations in PVDG. Apart from the fact that, the search space for optimization has expanded, the smaller step-size in BESS dispatch makes numerical comparisons (and decisions) difficult with Fuzzy Max-* compositions for final selection of BESS dispatch options. PSO is therefore used to overcome the limitations in the final selection of BESS dispatch levels, whereas benefits of Fuzzy systems are retained in initial stage of problem formulation as discussed earlier.

7.4.4.1 Inclusion of DG variability

The overall approach is depicted in Figure 7.9. Solar energy profiler (SEP) collects historical data of solar incident energy at hourly or smaller intervals. It uses a custom-made algorithm to identify characteristic days. It is similar to DLP Identifier that identifies characteristic daily load profiles. DLP analyser is developed in [7]. The SEP and DLP analysers are processed to obtain hourly forecasted load profile. This is fuzzified to represent the uncertainty of both PV generation and demand data. The fuzzified load profiles are then derived based on combinations of three levels of expectancy (low, medium and high) in both demand and PV generation. It should be noted that these levels are customizable for different CDLPs.

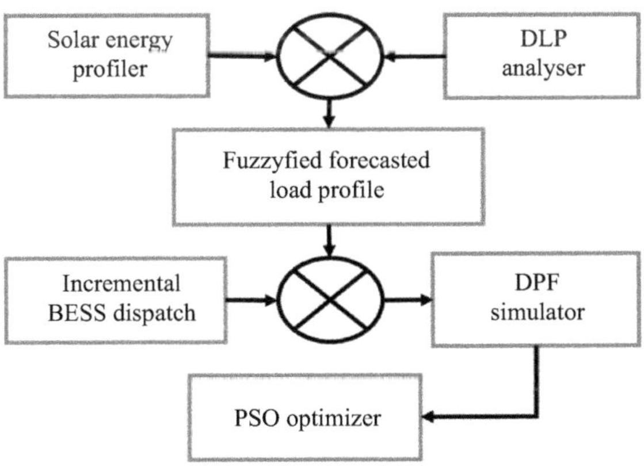

Figure 7.9 Fuzzy-PSO approach for day-ahead dispatch of BESS

This representation of uncertainty increases the search-space by many folds. In addition to this, the step size of BESS dispatch needs to be finer and smaller so as to cope up with the temporal variations in PVDG. The fuzzified load profiles are processed in DPF simulator with incremental dispatch options. DPF simulator also helps in evaluating the fuzzy indices to quantify the benefits of each incremental BESS dispatch. Different objectives considered for single-objective optimization are same as those described in the preceding section except the fifth objective which is incorporated as a constraint.

Fuzzy Max-* compositions are not used here for multi-period optimization. Rather, PSO is applied to obtain the optimized dispatch for each time interval and also to obtain the modified overall dispatch for a period of one day. With PSO, it becomes feasible to refine the step size of incremental dispatch which is reduced to as low as 0.005 p.u. whereas in FMOO approach, the step size was limited to 0.1 p.u. In addition to this, uncertainty in demand and generation is represented in three levels against a two-level representation of the same in an earlier approach. The final dispatch schedule – after a tentative dispatch for each time interval – is based on overall maximum satisfaction achievable by PSO. This exercise has used hourly time interval, but it can be used for as small as 5-min intervals also.

7.4.5 Simulation results

A single day's load profile is provided as an input to the DPF simulator that simulates the hourly load profile with incremental dispatch options from BESS. The PSO optimizer is then used to obtain the most optimum BESS dispatch option from among the set of incremental dispatch options at each hour. The single objective optimization results are obtained and presented in Table 7.6.

Optimization results for an actual feeder in Northwest USA are presented in Table 7.7. It should be noted here that the assumed PVDG plant is connected right next to the BESS. The maximum demand in this feeder is observed to occur during night hours (i.e. not typically during sun-shine hours) which results in almost same results in peak-shaving objective with FMOO and Fuzzy-PSO approach. However, other results should not be compared as these are affected by the presence of PVDG. It should be noted that a considerable rise in TCOs is effectively reduced by BESS dispatch. Similar observation is applicable in voltage profile improvement.

Table 7.6 Optimization results for IEEE 37 node feeder

Objective	**BESS dispatch**		**Objective satisfaction**
	Max. real power	**Total dispatched energy**	
Peak shaving	450 kW	3.2 MWh	Peak reduced from 2.3 MW to 1.84 MW
Losses	350 kW	4 MWh	Energy losses reduced by 4%
TCOs	600 kW	3.6 MWh	Reduced from 34 to 26
Voltage profile	600 kW	2.6 MWh	Two voltage limit violations reduced

Table 7.7 Optimization results for utility feeders in Northwest USA

Objective	BESS Dispatch		Objective satisfaction
	Max. real power	Total dispatched energy	
Peak shaving	735 kW	2..9 MWh	Peak shaving of 645 kW
Losses	285 kW	4 MWh	Energy losses reduced by 12%
TCOs	585 kW	3.4 MWh	Reduced from 53 to 28
Voltage profile	465 kW	3.6 MWh	Five voltage limit violations omitted

7.5 Summary

Low-voltage distribution networks are now witnessing increased activity in terms of new local DG-based generation, inclusion of BESS and other storage solutions, new control and monitoring infrastructure and development of semi-independent micro-grids. All these aspects need revised attention from researchers and solution developers. This chapter has focused on evaluating the technical benefits of integrating BESS in low-voltage unbalanced distribution networks with simulations and results of two distribution feeder including an actual feeder in Northwest USA. Rather than accepting the highly generic forecasts of load profile for preparing a day-ahead dispatch schedule for BESS with optimum benefits, a DLP analyser is proposed as a data pre-processor. This analyser feeds the optimization modules with most probable scenarios to be managed by the BESS scheduler. A FMOO platform is introduced that attempts to balance all the benefits from the limited capability of real-power dispatch from BESS. Finally, a Fuzzy-PSO-based extension is also included to incorporate the increased volatility in demand-generation scenarios due to photovoltaic DGs. While the analysis presented in this chapter has only considered the technical benefits of real-power dispatch from BESS, it is observed that the BESS scheduling deserves multi-fold analysis and that the same should at least include the aspect of reactive power availability from BESS as well as more general economic aspects.

Acknowledgments

This research work was supported by Indian Institute of Technology Gandhinagar, India in the form of Overseas Research Exposure grant to Dr Kalpesh Joshi. Authors are also grateful for the support of AVISTA Utilities and the US Department of Energy, USA.

References

1. Teleke S, Baran ME, Bhattacharya S, and Huang AQ. Rule-based control of battery energy storage for dispatching intermittent renewable sources. IEEE

Trans Sustain Energy [Internet]. 2010 Oct [cited 2017 Jul 11];1(3): 117–24. Available from: http://ieeexplore.ieee.org/document/5545425/

2. Li N and Hedman KW. Economic assessment of energy storage in systems with high levels of renewable resources. IEEE Trans Sustain Energy [Internet]. 2015 Jul [cited 2017 Jul 11];6(3): 1103–11. Available from: http://ieeexplore.ieee.org/document/6874522/
3. Srivastava AK, Kumar AA, and Schulz NN. Impact of distributed generations with energy storage devices on the electric grid. IEEE Syst J [Internet]. 2012 Mar [cited 2016 Jun 20];6(1): 110–7. Available from: http://ieeexplore.ieee.org/lpdocs/epic03/wrapper.htm?arnumber=6140538
4. Srivastava AK, Zamora R, and Bowman D. Impact of distributed generation with storage on electric grid stability. 2011 IEEE PES Gen Meet. 2011;pp. 1–5.
5. Joshi KA and Pindoriya NM. Case-specificity and its implications in distribution networks with high penetration of photovoltaic generation. CSEE J Power Energy Syst. 2017;3(1): 101–13.
6. Martinez JA, de León F. and Dinavahi V. Simulation tools for analysis of distribution systems with distributed resources. Present and future trends. IEEE PES Gen Meet [Internet]. 2010;1–7. Available from: http://ieeexplore.ieee.org/xpl/freeabs_all.jsp?arnumber=5589904
7. Joshi KA and Pindoriya NM. Day-ahead dispatch of battery energy storage system for peak load shaving and load leveling in low voltage unbalance distribution networks. In: IEEE PES Gen Meet. Denver: IEEE; 2015. pp. 1–5.
8. Ortmeyer T, Dugan R, Crudele D, Key T, and Barker P. Renewable systems interconnection study: Utility models, analysis, and simulation tools [Internet]. Sandia National Laboratory. 2008 [cited 2015 Sep 20]. Available from: https://www.eecbg.energy.gov/solar/pdfs/utility_models_analysis_simulation.pdf
9. Ramirez-Rosado IJ, Fernandez-Jimenez LA, Monteiro C, Miranda V, Garcia-Garrido E, and Zorzano-Santamaria PJ. Powerful planning tools. IEEE Power Energy Mag [Internet]. 2005 Mar [cited 2015 Sep 14];3(2): 56–63. Available from: http://ieeexplore.ieee.org/lpdocs/epic03/wrapper.htm?arnumber=1405871
10. Solanki CS. Solar photovoltaics: Fundamentals, technologies and applications [Internet]. PHI Learning Pvt. Ltd.; 2011 [cited 2015 Sep 10]. 512 p. Available from: https://books.google.co.in/books/about/Solar_Photovoltaics_Fundamentals_Technol.html?id=shsGEahT9wMC&pgis=1
11. Kersting WH. Distribution system modeling and analysis, Third Edition [Internet]. CRC Press; 2012 [cited 2015 Sep 10]. 455 p. Available from: https://books.google.com/books?id=aMY5JLUTKYC&pgis=1
12. Singh M, Muljadi E, Jonkman J, Gevorgian V, Girsang I, and Dhupia J. Simulation for wind turbine generators—With FAST and MATLAB-simulink modules [Internet]. National Renewable Energy Laboratory. 2014 [cited 2015 Sep 20]. Available from: http://www.nrel.gov/docs/fy14osti/59195.pdf
13. Faiman D. Assessing the outdoor operating temperature of photovoltaic modules. Prog Photovolt Res Appl. 2008;16: 307–15.
14. Villalva MG, Gazoli JR, and Filho ER. Comprehensive approach to modeling and simulation of photovoltaic arrays. IEEE Trans Power Electron [Internet].

2009 May [cited 2015 Jun 17];24(5): 1198–208. Available from: http://ieeexplore.ieee.org/lpdocs/epic03/wrapper.htm?arnumber=4806084
15. Short W, Blair N, and Heimiller D. Modeling the market potential of hydrogen from wind and competing sources: Preprint. [cited 2015 Sep 20]; Available from: http://www.researchgate.net/publication/255200820_Modeling_the_Market_Potential_of_Hydrogen_from_Wind_and_Competing_Sources_Preprint
16. Lipman TE, Edwards JL, and Kammen DM. Fuel cell system economics: Comparing the costs of generating power with stationary and motor vehicle PEM fuel cell systems. Energy Policy. 2004;32: 101–25.
17. DISPERSE Model [Internet]. 2005. Available from: http://www.distributed-generation.com/disperse.htm
18. Turcotte D, Ross M, and Sheriff F. Photovoltaic hybrid systems sizing and simulation tools status and needs. In: PV Horizon: Workshop on Photovoltaic Hybrid Systems [Internet]. 2001 [cited 2015 Sep 14]. pp. 1–10. Available from: https://www.nrcan.gc.ca/sites/www.nrcan.gc.ca/files/canmetenergy/files/pubs/2001–119_e.pdf
19. Willis HL. Power distribution planning reference book, Second Edition [Internet]. CRC Press; 2004 [cited 2015 Sep 10]. 1244 p. Available from: https://books.google.com/books?id=9EShPwTRnoUC&pgis=1
20. Baring-Gould EI. Hybrid2: The hybrid system simulation model, Version 1.0; User Manual [Internet]. National Renewable Energy Laboratory. 1996 [cited 2015 Sep 20]. Available from: http://www.nrel.gov/docs/legosti/old/21272.pdf
21. NREL, Energy H. Homer microgrid software [Internet]. 2004 [cited 2015 Apr 1]. Available from: http://www.homerenergy.com/software.html
22. OpenDSS [Internet]. Electric power research institute. 1997. Available from: http://smartgrid.epri.com/SimulationTool.aspx
23. Schneider KP, Chassin D, Chen Y, and Fuller JC. Distribution power flow for smart grid technologies. In: 2009 IEEE/PES Power Systems Conference and Exposition [Internet]. IEEE; 2009 [cited 2015 Sep 11]. pp. 1–7. Available from: http://ieeexplore.ieee.org/lpdocs/epic03/wrapper.htm?arnumber=4840078
24. Mermoud A. PVsyst [Internet]. PVsyst; 2007. Available from: http://www.pvsyst.com/en/
25. DoE, PNNL. GridLAB-D [Internet]. 2014. Available from: http://www.gridlabd.org/
26. Demailly F, Ninet O, and Even A. Numerical tools and models for Monte Carlo studies of the influence on embedded generation on voltage limits in LV grids. IEEE Trans Power Del. 2005;20: 2343–50.
27. Marinopoulos AG, Alexiadis MC, and Dokopoulos PS. Energy losses in a distribution line with distributed generation based on stochastic power flow. Electr Power Syst Res [Internet]. 2011 Oct [cited 2016 May 12];81(10): 1986–94. Available from: http://www.sciencedirect.com/science/article/pii/S037877961100143X
28. Kersting WH and Dugan RC. Recommended practices for distribution system analysis. Power Systems Conference and Exposition. 2006. pp. 499–504.

29. Arritt RF and Dugan RC. Distribution system analysis and the future smart grid. IEEE Trans Ind Appl. 2011 Nov;47(6): 2343–50.
30. Thomson M, and Infield DG. Network power-flow analysis for a high penetration of distributed generation. IEEE Trans Power Syst. 2007;22(3): 1157–62.
31. Mather BA. Quasi-static time-series test feeder for PV integration analysis on distribution systems. IEEE PES Gen Meet. 2012. pp. 1–8.
32. Boehme T, Wallace AR, and Harrison GP. Applying time series to power flow analysis in networks with high wind penetration. IEEE Trans Power Syst. 2007;22(3): 951–7.
33. Vittal E, O'Malley M, and Keane A. A steady-state voltage stability analysis of power systems with high penetrations of wind. IEEE PES Gen Meet. 2010. pp. 1–8.
34. Inc E. Annual load profiles [Internet]. Electric Reliability Council of Texas Inc.; Available from: http://www.ercot.com/mktinfo/loadprofile
35. Joshi KA and Pindoriya NM. Impact investigation of rooftop Solar PV system: A case study in India. Innovative Smart Grid Technologies (ISGT Europe), 2012 3rd IEEE PES International Conference and Exhibition on. 2012. pp. 1–8.
36. Kersting WH. Radial distribution test feeders. IEEE Trans Power Syst [Internet]. 1991 [cited 2015 Apr 2];6(3): 975–85. Available from: http://www.scopus.com/inward/record.url?eid=2-s2.0–0026207084&partnerID=tZOtx3y1
37. Electric Reliability Council of Texas, Inc. – Annual Load Profiles [Internet]. ERCOT. 2014 [cited 2015 Apr 1]. Available from: http://www.ercot.com/mktinfo/loadprofile
38. American National Standard for Electric Power Systems and Equipment – Voltage Ratings (60 Hertz). ANSI C84.1–2011. National Electrical Manufacturers Association, Rosslyn, VA, 2011.
39. Suganthi L, Iniyan S, and Samuel AA. Applications of fuzzy logic in renewable energy systems – A review. Renew Sustain Energy Rev [Internet]. 2015 Aug [cited 2015 Apr 28];48: 585–607. Available from: http://www.sciencedirect.com/science/article/pii/S136403211500307X
40. Jordehi AR. Optimisation of electric distribution systems: A review. Renew Sustain Energy Rev [Internet]. 2015 Nov [cited 2015 Dec 29];51: 1088–100. Available from: http://www.sciencedirect.com/science/article/pii/S1364032115006516
41. EPRI. Assessment of transmission and distribution losses in New York. 2012;71178(Nyserda 15464). Available from: http://www.nyserda.ny.gov/-/media/Files/Publications/Research/Electic-Power-Delivery/epri-assessment-losses.pdf
42. Ponce-Cruz P and Ramírez-Figueroa FD. Intelligent control systems with LabVIEW™ [Internet]. Springer-Verlag London. London: Springer; 2010 [cited 2015 Dec 13]. XIII, 216. Available from: http://www.springer.com/us/book/9781848826830

Chapter 8

Case studies from selected countries – Romania and Italy

Morris Brenna[1], Virgil Dumbrava[2], Federica Foiadelli[1], George Cristian Lazaroiu[2], Michela Longo[1] and Catalina-Alexandra Sima[2]

8.1 Introduction

The development of electricity generation, transport, distribution and consumption was determined by switching from a power distribution structure to the smart grid supply system, in urban and newly developed areas, as well as where efficient use of renewable energy sources (RESs) for local customers at lower prices compared to power supply. The future power systems contain distributed energy resources (DER) and local loads, and can automatically connect or disconnect from the upstream network, as well as the electricity price feature of the smart grid compared to the upstream grid [1–3]. This new concept requires a high level of renewable energy and end-users, driven by the development of electronic power interfaces that respond quickly to intermittent new energy sources and changing load.

The future distribution networks will become smart grids [4,5], and the produced power must be delivered to the consumption points [6]. The future grids should be flexible, reconfigurable, resilient and self-healing. The widespread integration of active loads requires new systems of protection, new automation equipment and advanced management of generation and consumption [7–9].

There are still several open topics that need to be researched [10]. A research area can be dedicated to the use of storage systems to improve power quality and to better exploit RESs [11–13]. There are many advantages of storage capacity present within the networks where distributed generators are connected, but several issues should be considered.

Smart grids require specific energy storage systems and control systems to optimally exchange available power between sources and end-users [14–17].

[1] Department of Energy, Politecnico di Milano, Italy
[2] Department of Power Systems, University POLITEHNICA of Bucharest, Romania

In [14], a smart system for energy management was proposed to optimally coordinate power generation of distributed generators and storage systems so as to minimize operational costs. Within distribution grids, stochastic RESs such as photovoltaic (PV) systems and storage devices are optimally used to reduce operating costs and preserve storage utilization, considering the energy price for selling/purchasing energy [18]. Optimized smart grids operation should also consider local availability and local load state to ensure security of supply to local critical load in case of islanding appearance [19]. Stochastic energy management models have been proposed to cope with the uncertainty of generation/consumption and the impact on smart grids functioning with storage systems [20].

The introduction of electrochemical storage devices into electrical distribution grids through a dedicated section at DC, for better load management, is considered in [21]. The basic concept is to achieve the black start of a distribution grid section at MV without determining transient phenomena that is not accepted by the system. The results of a research collaboration with Italy's main electrical energy distributor, Enel Distribuzione, are reported and present a dynamic storage space model that exists in an experimental system within the Isernia project, funded by the Italian Electricity and Gas Regulator, with an objective to develop smart grid. The focus was on installing storage systems to control the line voltage, to achieve peak shaving and load variation. The storage system connected along the MV line is called DM 6040416 'Pesche'. This system that is supplied by Caprinone, a primary station (Centre of Italy) and it facilitates active and reactive energy flow for the medium voltage grid to determine load levels when distributed generators are connected to the system, lines along the line. In addition, this system controls also the grid voltage.

The modelling of the entire MV electrical system was carried on using the digital real-time simulator (RTDS) installed in the Enel Distribuzione testing centre in Milan. The RTDS simulator also allows real-life protection, relays and simulated network control [22].

8.2 Storage systems based on lithium-ion batteries

The use of the lithium as a material in rechargeable batteries is made in particular for the anode. This use is determined by the characteristics that it has, especially, small electronegativity, high specific capacity and low energy density. Lithium batteries are electrochemical storages, and are distinct from electrolytic technology and from materials that compose anode and cathode. Table 8.1 reports different characteristic parameters for different used materials. Table 8.2 reports the typical values of the performance characteristics of lithium-ion cells. The developed mathematical model allows evaluating the battery charge status (SoC) according to the supplied current for assessing the discharge time during the simulation. The adopted battery model consists in a controlled voltage sources in series with a constant resistor, as shown in Figure 8.1.

Table 8.1 Fundamental parameters of the different Li-ion cells

	NCA	**NMC**	**LMO**	**LFP**	**LTO**
	Nickel Cobalt Aluminium	**Nickel Manganese Cobalt**	**Lithium Manganese Oxide**	**Lithium Iron Phosphate**	**Lithium-Titanate**
Rated voltage [V]	3.7	3.7	3.7	3.3	2.2
Cathode	$LiNiCoAlO_2$	$LiNiCoMnO_2$	$LiMn_2O_4$	$LiFePO_4$	$Li_4Ti_5O_{12}$
Anode	C	C	C	C	
Power [W/kg]	H	G	A	A	A/L
Energy [Wh/kg]	H	H	G	A	H
Cycle life	G	G	A	A	H
Life	G	G	L	L	G
Development level	M	D/M	M	D	D
Cathode safety	L	L	A	G	A
Cell safety	P	L	L	A	G

H=high; G=good; A=average; L=low; D=development; M=mature

Table 8.2 Typical performance parameters of the Li-ion cells

Parameters	**Typical value**
Capacity of the cells (Ah)	0.1–10,000
Specific power (W/kg)	200–3,000
Specific energy (Wh/kg)	40–180
Energy efficiency (%)	80–95
Amperometric efficiency (%)	100
Typical discharge regime (*C* rate)	*C*/3–2*C*
Discharge maximum current (*C* rate)	1*C*–100*C*
Charge maximum current (*C* rate)	*C*/2–10*C*
Monthly self-discharge (%)	<1
Depth of discharge (DoD) 80% (cycles)	1,500–5,000
Operating temperature	−30–60 °C

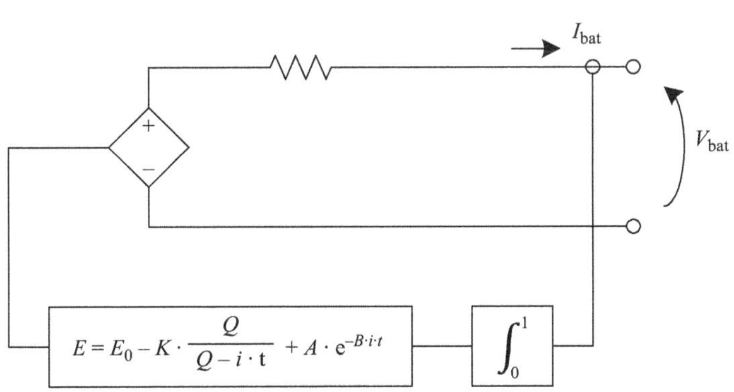

Figure 8.1 Battery nonlinear model

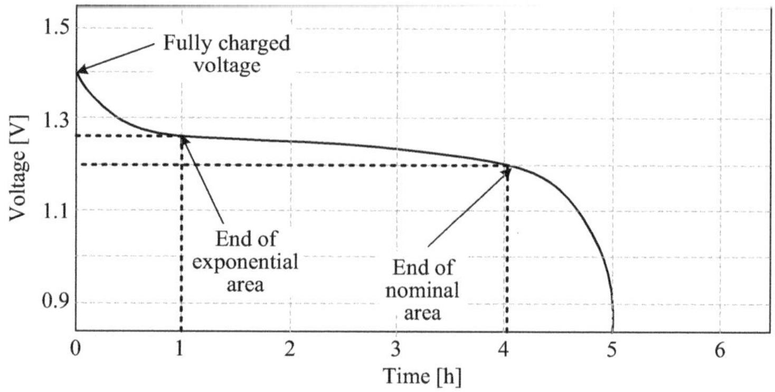

Figure 8.2 Typical discharge curve

This model has the same characteristics for charging and discharging the battery. The control of the voltage source is determined by a non-linear expression based on the SOC contemporary battery:

$$E = E_0 - K \cdot \frac{Q}{Q - \int idt} + A \cdot e^{-B. \int idt} \tag{8.1}$$

where E is the voltage of the battery (V), E_0 is the initial state of charge voltage that is constant when the system is fully charged (V), K is the bias voltage (V), Q is the capacity of the battery (Ah), $\int idt$ is the contemporary battery charge (Ah), A is the exponential amplitude (V), B is the inverse time constant (Ah^{-1}) and I is the current of the battery (A).

This considered model represents well different types of batteries, but its parameters are required to be accurately determined. The battery discharge characteristics can define the parameters. Figure 8.2 shows the characteristics that are typical for a battery (1.2 V) and a capacity of 6.5 Ah. The discharge characteristic is obtained for a direct current of 1.3 A. For determining the internal resistance, evaluating the variation of efficiency function of capacity and nominal voltage battery is important, as:

$$\eta = 1 - \frac{I_{nom} \cdot R \cdot I_{nom}}{V_{nom} \cdot I_{nom}} \text{ and using the } C \text{ rate expression} \tag{8.2}$$

$$I_{nom} = Q_{nom} \cdot (0.2/1hr) \rightarrow \eta = 1 - \frac{0.2 \cdot R \cdot Q_{nom}}{V_{nom}} \tag{8.3}$$

A good approximation for determining the internal resistance is an efficiency of 99.5 per cent, thus:

$$R = V_{nom} \cdot \frac{1 - \eta}{0.2 \cdot Q} \tag{8.4}$$

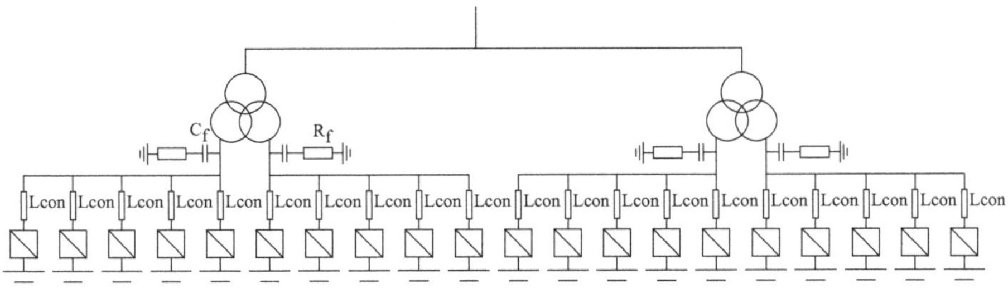

Figure 8.3 Electrical interconnection layout of the storage system

For determining the other parameters of the model, the points in the number of three are highlighted on the discharge characteristics shown in Figure 8.2 are considered: the full charge voltage, the exponential end and the end of the nominal area. The points are obtained from empirical tests and thus, a possibility for determining them is with the help of a regression function that contains numerous testing campaigns.

In particular, the storage considered here is made up of 20 stacks. Each stack has a series of 14 lithium batteries and has a dedicated inverter. For connecting to the MV distribution network, three winding transformers are used. Each secondary winding of the transformer has connected single-phase modules consisting of a battery rack and an inverter. Between the module and the secondary winding, there are placed RLC filters for mitigating the harmonics. Figure 8.3 illustrates the electrical layout.

For RTDS implementation, the modules attached to each secondary are viewed as a single module for decreasing the sum of the nodes. The electrochemical deposit has the capacity to provide respectively 0.8 MW and a 0.5 MWh.

8.2.1 System model for the Isernia project

For connecting the storage to the 50-Hz power system, a three-phase bridge converter with IGBT is used. The converter manages the active and reactive power flows between the storage and the distribution grid, ensuring a correct interface.

The control system allows power flow control independently on the active and reactive power. To obtain the desired waveform, it is necessary to use the pulse width modulation considering on space vector to achieve the right impulse sequence of the semiconductor [23]. This technique compared with the classical method is based on controlling the entire three-phase system and can be implemented on a microprocessor.

A necessary function for the converter of the interface is to energize a section of the MV distribution grid that passes into an area isolated from the mains. This technique is called 'blackout' and is used following an interruption to limit not-energizing the overall system.

The Carpinone substation is the main substation chosen for Isernia project in Italy. The substation is supplied from three high-voltage (HV) lines. Two 25 MVA HV/MV transformers are supplying 20 kV busbars, where 15 MV lines are connected.

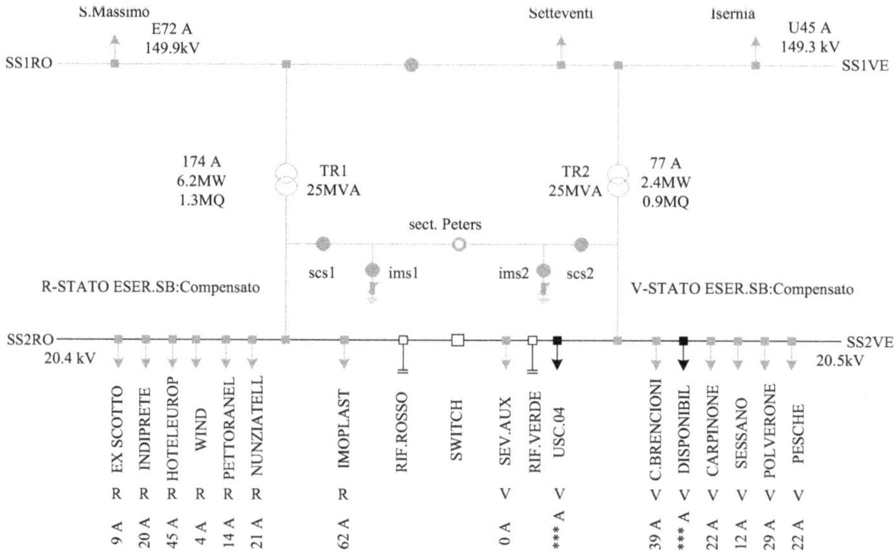

Figure 8.4 Layout of Carpinone substation [24]

Among these 15 lines, there is the DM 6040416 Pesche line. The layout of the substation is illustrated in Figure 8.4. The analysed distribution grid is DM 6040416 Pesche and it has connected all the loads and generators. The configuration is radial, with five generators connected for a total production of 3,686.1 kW. The total consumption of the DM 6040416 Pesche is represented by 44 loads. Among these, 36 loads are MV/LV secondary substations with in/out connection, and 8 are MV loads with four generators. The low voltage consumers are absorbing 1649.7 kW and 799.03 kVAr, and the other loads are requiring 1,567.4 kW and 759.13 kVAr.

The complexity of the test grid is high, and in RTDS a reduction of the network was considered by reducing the sum of nodes and keeping the voltage characteristic near the real value. An error of 0.25 per cent was considered acceptable. The loads are aggregated in 'clusters' (as shown in Figure 8.5(a)) or in 'sequence' (as illustrated in Figure 8.5(b)).

For reducing the number of nodes, a classification of the loads based on their typology is conducted. For the 'cluster' case, reducing the loads connected to nodes N_1–N_n to an equivalent load at N_p can be carried on by verifying that the voltage drop between N_p and the other nodes (N_1, N_2, ..., N_n) is below a considered threshold. For the 'sequence' case, reducing the loads connected to nodes N_1–N_n to an equivalent load at N_p is more difficult [24].

8.2.2 The black start of MV island

The necessity to black start a section of the MV network appears following a black out of the grid caused by a short circuit and occurrence of the island phenomena (shown in Figure 8.6).

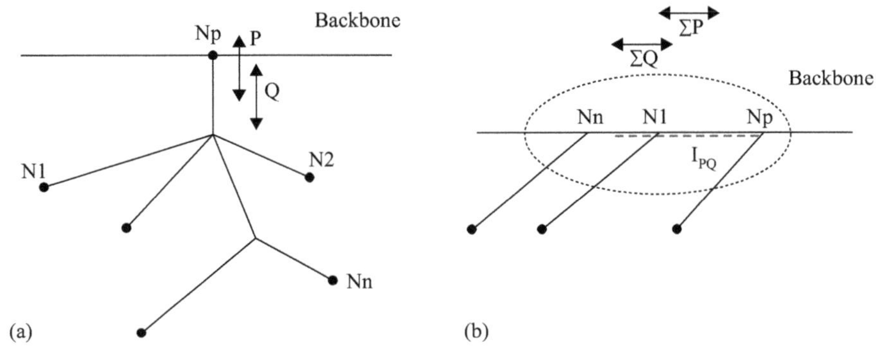

Figure 8.5 (a) Cluster and (b) sequence load typology on the DM 6040416 'Pesche' line

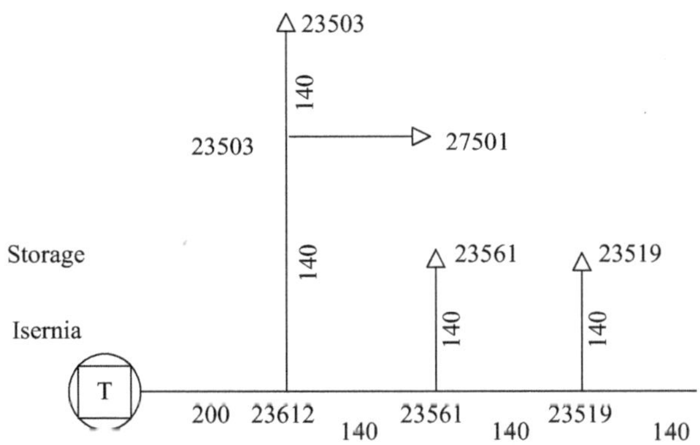

Figure 8.6 Island energized by the storage system

The black-start sequence after a network disconnection is:

- to turn on the DC/AC converter of the storage system in no-load condition (it is considered that the storage is initially fully charged);
- to connect the MV circuit breaker to supply the island.

In the first instants of turning on the storage system interface converter, an overvoltage occurs (Figure 8.7). To elude inrush currents that affects the inverter, a soft start-up can be considered.

After arrival at steady-state conditions, the MV breaker is connected and supplying the MV island. At the switch on instants, a voltage drop of 14 kV is occurring (time instant 3.7 s in Figure 8.7) determined by load supply and time-delay in the response of storage converter. After time instant of 5 s, the MV breaker is switched on and the DC/AC converter is maintaining the voltage at the threshold value of 20 kV. As can be seen, the control system of the storage system with converter brings the voltage to the established threshold in 1.3 s. Figure 8.8

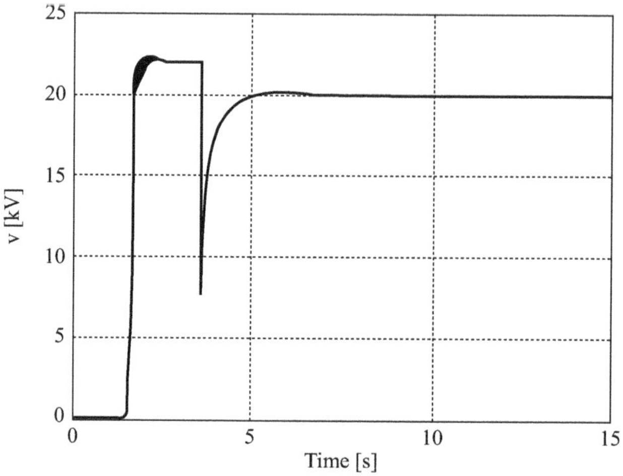

Figure 8.7 Variation of RMS voltage at the connection bus of storage system

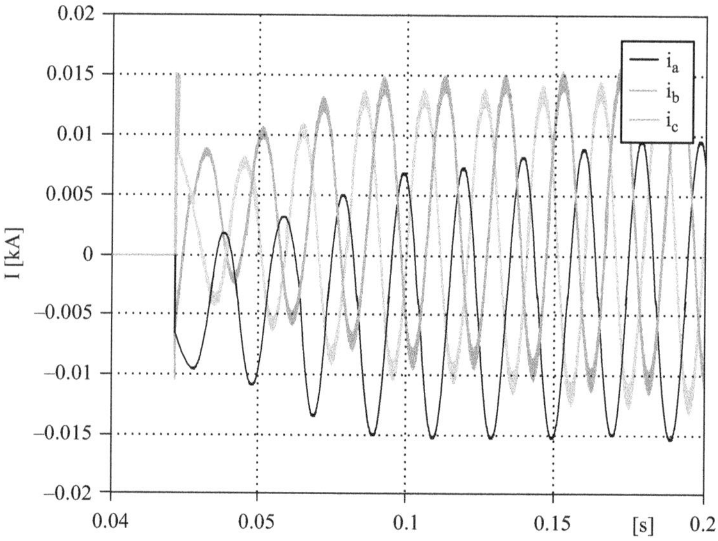

Figure 8.8 Three-phase currents injected by the storage system

illustrates the current generated by the converter of the storage system during the transition between black start and full load supply.

After full load supply, the MV island is energized by the storage system. In particular, the voltage and frequency are maintained by the DC/AC converter. The time period in this operating condition of the MV island is function of the capacity of the storage system. Within this time interval, in order to furtherly supply the

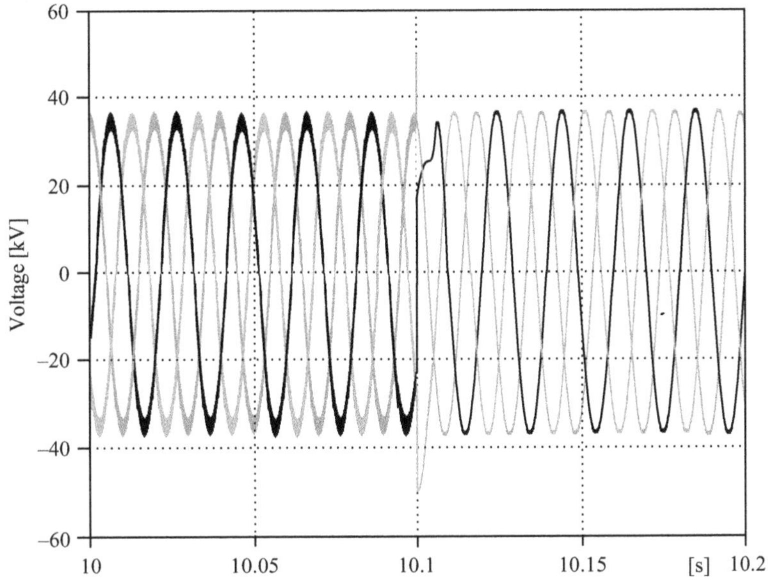

Figure 8.9 Switching process during the system restoration

loads, the mains supply interruptions should be cleared and the MV island should be connected to the upstream grid. The connection of the island with the mains requires the synchronization of frequency and voltage.

The connection with upstream grid was simulated for analysing the synchronization transicnts. Thc simulation was pcrformcd with an amplitudc diffcrcncc equal to five per cent of the nominal voltage and a phase shift of $1/6\ \pi$, being activated at time t = 10 s (Figure 8.7). The simulation result is illustrated in Figure 8.8, and the system does not produce important transients. Figure 8.9 represents the situation after system restoration.

8.3 Optimizing residential customer smart grid with storage devices

A smart grid implemented in Romania that uses local generation units (PV units and diesel generators), energy storage system based on batteries and non-interruptible and interruptible loads is analysed. In Figure 8.10, the considered smart grid is presented. The smart grid controller can exchange energy with the mains supply, so the energy surplus can be delivered (sold) to the main supply, and the energy consumption which cannot be covered by own generators are covered from the mains supply [25]. There are two categories of load: non-interruptible load (cannot be disconnected) and interruptible load (can be disconnected according to operating conditions and availability of local resources).

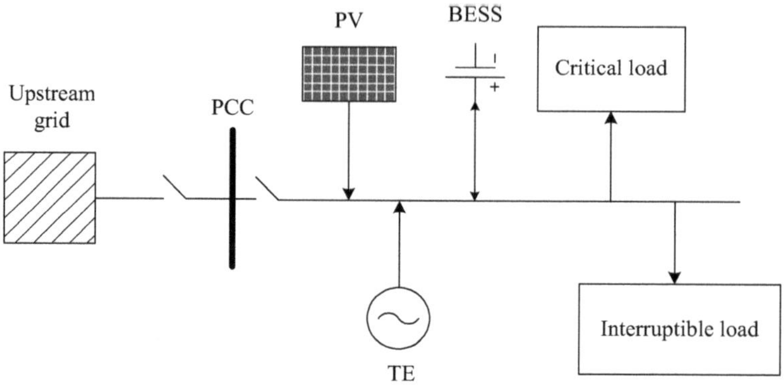

Figure 8.10 Schematic structure of the considered smart grid

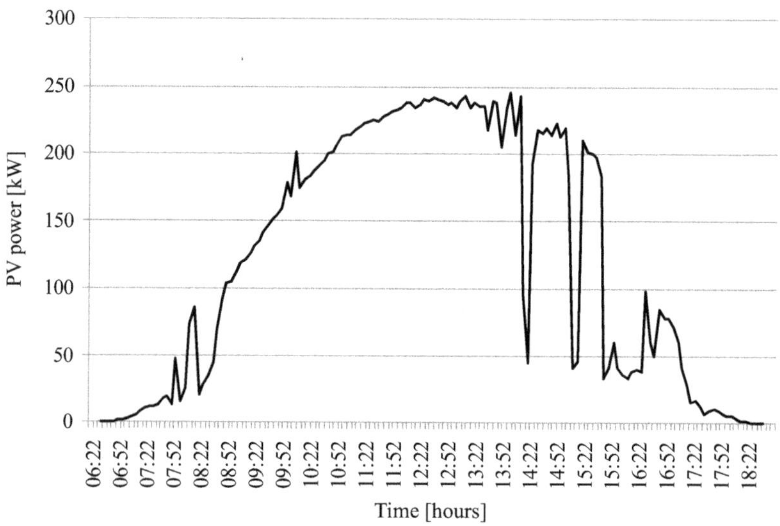

Figure 8.11 Recorded production curve of PV system

The smart grid is connected with the upstream grid. In this case, the entire load of the smart grid must be possible to be covered by the mains supply, so there are no interconnection limiting constraints between the smart grid and the upstream grid;

The PV systems are used on a large scale in the smart grids which are connected with the distribution grids from urban areas. The intermittent production of the PV systems, mainly in-time decay of their energy production, needs specific measures to be implemented. The production of these sources is uncertain and for this reason the forecasting methods are applied to better estimate their production. In Figure 8.11, is presented the recorded power production curve for a 300 kW PV plant, for a summer day [26].

A classical back-up source is used within the scheme as a dispatchable unit [27]. C_g is the quadratic cost function of the classical unit (CU) depending on the amount of energy $E_{CU}(t)$, at each time period t, expressed as:

$$C_g\left(E_{CU,g}(t)\right) = a_g \cdot E^2_{CU,g}(t) + b_g \cdot E_{CU,g}(t) + c_g, \quad \forall_g \in S_{CU}, \ \forall t \tag{8.5}$$

where a_g, b_g and c_g are the coefficients of the cost function for each unit within the set of CUs S_{CU}, expressed in Euro/MWh.

$u_g(t)$ (0 if unit is off, and 1 if the unit is on) is a binary variable which helps modelling the output power state of the CU:

$$u_g(t) \cdot P^{\min}_{CU,g} \leq P_{CU,g}(t) \leq u_g(t) \cdot P^{\max}_{CU,g}, \quad \forall_g \in S_{CU}, \ \forall t \tag{8.6}$$

where $P^{\min}_{CU,g}(t)$ and $P^{\max}_{CU,g}(t)$ are the minimum, respectively, maximum output powers of the CU g.

The start-up cost C^{SU}_g and shutdown cost C^{SD}_g associated with the classical fuelled unit are expressed with the constraints (8.7a) and (8.7b):

$$\begin{aligned} &C^{SU}_g(t) \geq S^U_g \cdot \left(u_g(t) - u_g(t-\tau)\right), \quad \forall_g \in S_{CU}, \ \forall t \\ &C^{SU}_g(t) \geq 0, \quad \forall_g \in S_{CU}, \ \forall t \end{aligned} \tag{8.7a}$$

$$\begin{aligned} &C^{SD}_g(t) \geq S^D_g \cdot \left(u_g(t-\tau) - u_g(t)\right), \quad \forall_g \in S_{CU}, \ \forall t \\ &C^{SD}_g(t) \geq 0, \quad \forall_g \in S_{CU}, \ \forall t \end{aligned} \tag{8.7b}$$

where S^U_g and S^D_g are the actual costs of turn on/off of the CU.

The flexibility and capability to level the produced power of RESs have made the storage battery energy systems to be more and more used in the existing distribution networks.

The storage system state-transition expression for the energy level $E_{st,i}(t)$ with finite capacity, for each unit i from the set of units S_{st} at any time t, is [28]:

$$\begin{aligned} &E_{st,i}(t) = E_{st,i}(t-\tau) + P^{chg}_{st,i}(t) \cdot \eta_i \cdot \tau - \frac{P^{dischg}_{st,i}(t)}{\eta_i} \cdot \tau \\ &E^{\min}_{st,i} \leq E_{st,i}(t) \leq E^{\max}_{st,i}, \quad \forall i \in S_{st}, \ \forall t \end{aligned} \tag{8.8}$$

where τ represents the time frame, and η_i are the conversion efficiency considering the energy losses that are smaller than one. $P^{chg}_{st,i}(t)$ and $P^{dischg}_{st,i}(t)$ are the charge and discharge powers of each storage unit i, under certain upper bounds (i.e. $P^{chg,max}_{st,i}$, respectively $P^{dischg,max}_{st,i}$) expressed as:

$$0 \leq P^{chg}_{st,i}(t) \leq P^{chg,\max}_{st,i}, \quad \forall i \in S_{st}, \ \forall t \tag{8.9a}$$

$$0 \leq P^{dischg}_{st,i}(t) \leq P^{dischg,\max}_{st,i}, \quad \forall i \in S_{st}, \ \forall t \tag{8.9b}$$

In (8.8) was considered that there is no transfer constraint between the upstream grid and the smart grid, which means that during grid-connected operation no load curtailment should occur. Otherwise, a constraint should be added to limit the power flow.

The goal is to minimize the smart grid operating costs [29]. Because the load curtailment has a high penalizing cost for the smart grid, the loads that can be interrupt will be disconnected only if it is absolutely mandatory during intervals that care characterized by large prices of electricity in the upstream grid or when the smart grid is islanded. The power demand $P_{crit\_l,j}(t)$ of the critical load j from the set of critical loads S_{crit_l} is constrained to be:

$$P^{\min}_{crit\_l,j} \leq P_{crit\_l,j}(t) \leq P^{\max}_{crit\_l,j}, \quad \forall j \in S_{crit\_l}, \quad \forall t \tag{8.10}$$

while the power demand $P_{inter\_l,k}(t)$ of the interruptible load k from the set of interruptible loads S_{inter_l} is bounded by:

$$0 \leq P_{inter\_l,k}(t) \leq P^{\max}_{inter\_l,k}, \quad \forall k \in S_{inter\_l}, \quad \forall t \tag{8.11}$$

By controlling the PV systems, CUs and storage devices with a supervisory control system, it is possible to supply of loads with minimum operating costs. This offers the possibility to use the internal resources as much as possible and also to determine the amount of transfer with the upstream grid at different periods functions of different electricity market prices and PV production forecasts [30,31]. Using an optimization model, it is possible to use the local generation whenever the electricity price is very high, and to command the storage charging whenever the price is low [32].

It is necessary that the smart grid to start to operate in an islanding operation mode in case of an upstream fault occurrence, which means that the smart grid will be separated from the faulted area. If the resources are available, the local loads will be supplied using resources that are available at that moment. In this situation, the supervisory controller will ensure the security and equilibrium of the production–consumption. The battery energy storage should be used such that to preserve its lifetime, reducing the number of charging/discharging cycles.

In the following, the forecasted PV production is considered perfectly accurate. The mathematical model for minimizing the operational costs (*Cost*) during the 24 hours operation mode can be expressed as:

$$[\text{MIN}]\text{Cost} = \sum_{t}^{T} \left[\begin{array}{l} \sum_{i \in S_{ThE}} [C_g(E_{CU}, g(t)) + C_g^{SU}(t) + c_g^{SD}(t)] + \\ \sum_{s \in S_{RES}} \pi_{PV,S}(t) \times P_{PV}^{curt}, s(t) \times \tau \\ + \sum_{j \in S_{crit\_l}} \pi_{crit\_l,j}(t) \times E_{crit\_l,j}(t) \times -\pi_{el}(t) \times P_{exch}(t) \times \tau \end{array} \right] \tag{8.12}$$

where π_{PV} is the penalty cost of PV curtailed power $P^{curt}_{PV,s}$, π_{crit_l} is the penalty value of lost load P_{crit_l}, π_{el} is the electricity market price, and $P_{exch}(t)$ is the quantity of power that is exchanged with the upper system in interval t. The power exchange is positive when the power is flowing towards the mains supply, or negative if the power is going towards the smart grid.

By using the energy balance constraint will ensure for each time frame τ, that the sum of total generated energy (the energy generated by the CU, the PV unit forecasted power and the storage system discharge) is equal to the total load (energy taken by the loads and for charging the storage system), plus the energy exchanged with the upstream grid:

$$\begin{aligned} &\sum_{g\in S_{CU}} E_{CU,g}(\tau) + \left(\sum_{s\in S_{RES}} P^{fcast}_{PV,s}(t) - \sum_{s\in S_{RES}} P^{curt}_{PV,S}(t) + \sum_{i\in S_{st}} P^{dischg}_{st,i}(t) \right) \cdot \tau = \\ &= \sum_{j\in S_{crit\_l}} E_{crit\_l,j}(\tau) + \sum_{k\in S_{inter\_l}} E_{inter\_l,k}(\tau) + \left(\sum_{i\in S_{st}} P^{chg}_{st,i}(t) + P_{exch}(t) \right) \cdot \tau \end{aligned} \tag{8.13}$$

The forecasted PV production value $P^{fcast}_{PV,i}(t)$ is accurately known, while the outputs at each time interval t are the variables $E_{CU,g}(t)$, $E_{crit\_l,j}(t)$, $E_{inter\_l,k}(t)$, $P^{cut}_{PV,S}(t)$, $P^{dischg}_{st,i}(t)$, $P^{chg}_{st,i}(t)$, $E_{st,i}(t)$ and $P_{exch}(t)$.

In Table 8.3, the capacities and other important characteristic data of all units from the simulated smart grid shown in Figure 8.10 are reported. In the studied cases, the following assumptions are made:

- The operation costs of RESs and battery storage are considered to be negligible;
- The thermal engines are supplied with gas, with fast turn on/off capabilities;
- The demand contains 60 per cent critical load, and 40 per cent loads that can be interrupt;
- Critical load curtailment costs are set to 50 Euro/kWh (this parameter can greatly influence the optimized operation of the smart grid;
- The electricity prices are typical average values on the day-ahead market, on a 24-hour interval.

The time horizon in which the smart grid is operating interconnected with the upstream grid is 24 hour. In Figure 8.12 are shown the results for the optimization

Table 8.3 Characteristic data of units

Unit	**Value**
PV	300 kW
Energy storage system (battery)	200 kWh
– Minimum level of the storage	40 kWh
– Efficiency of the conversion	0.9
Engine capacity	600 kW
– Coefficients of the cost function of the engine {a; b; c}	{0.03; 1; 1.3}
Maximum demand	800 kW

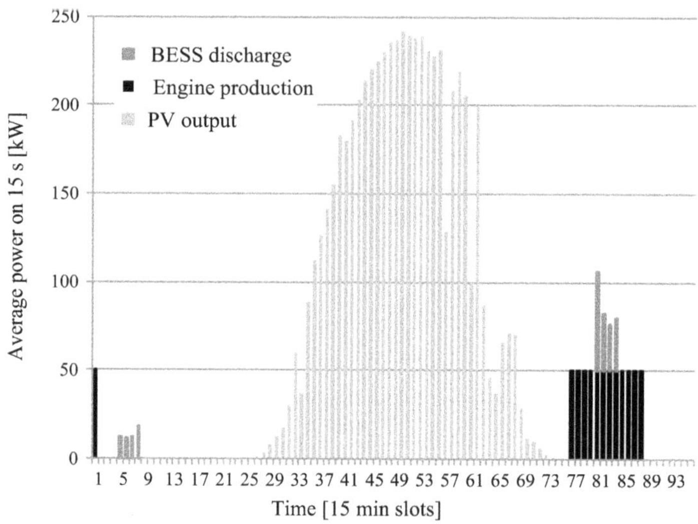

Figure 8.12 Resulting generation schedule within the smart grid

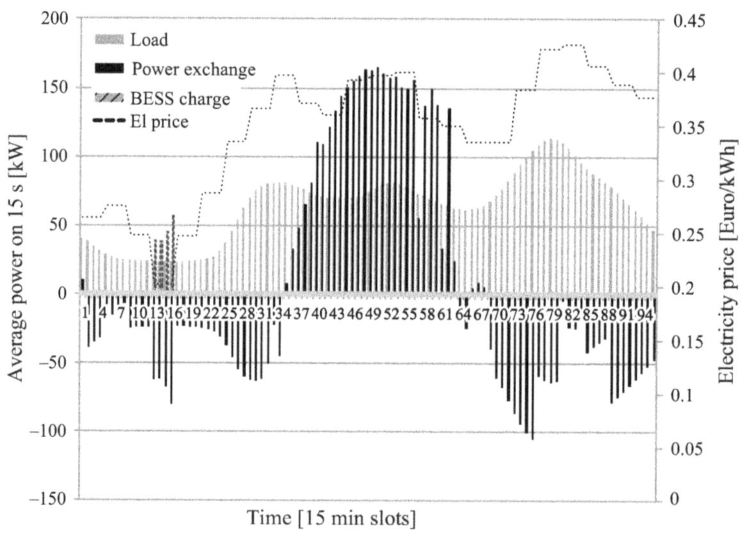

Figure 8.13 Resulting consumption schedule within the smart grid

model related to thermal engine operation and battery discharge, as well as the model input PV production. In Figure 8.13, it can be seen that the optimization model results related to consumption schedule within the smart grid. The electricity price follows the daily typical curve. The smart grid is buying power from the upper network to supply the demand during time periods when the electricity price

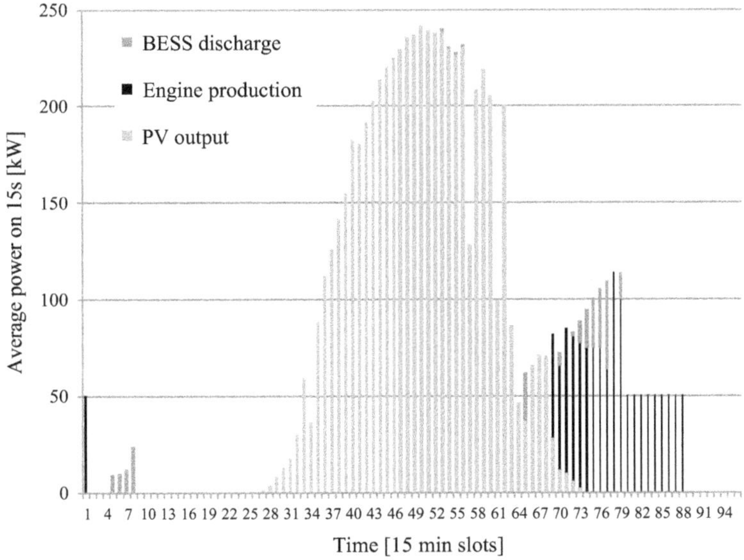

Figure 8.14 Resulting generation schedule within the smart grid in islanding state

is low (during the night). In addition, the storage energy system is charged in this period. In the time interval t = 20–34 (i.e. from hour 05:00 to hour 08:30), the electricity market price and the load are increasing. During this period, it is economical to purchase the required power from the upstream grid and not locally generating it with the thermal engine. In Figure 8.13, from time instant t = 34 (i.e. from hour 08:30), the PV installation starts producing the required power to balance the demand. When the electricity price is high (during midday), the surplus of local PV resources production is injected into the upstream grid and there are no constraints imposed to renewable sources that are connected to the network. After the 20th hour, the electricity price and demand are high and the PV production is zero. In this time period, the thermal engine is switched on and the battery is discharged to supply the smart grid demand.

In the upstream grid a fault occurs that determines the disconnection of the microgrid. The microgrid is operating in islanding state during the time interval t = 64–80 (i.e. between hours 16:00 and 20:00).

In Figure 8.14 are shown the results of the optimization model related to thermal engine operation and battery discharge, as well as the model input PV production. From Figure 8.15, the results of the optimization model related to consumption schedule within the microgrid can be seen.

From Figure 8.15, it can be seen that during the islanding operation, the upstream grid is no more offering support to balance the microgrid demand. Figure 8.15 illustrates that the thermal engine is switched on and it is maintained in operation between hours 16:00 and 20:00. For minimizing the operational costs and balancing the demand, the storage energy system is discharged. After hour 21, the

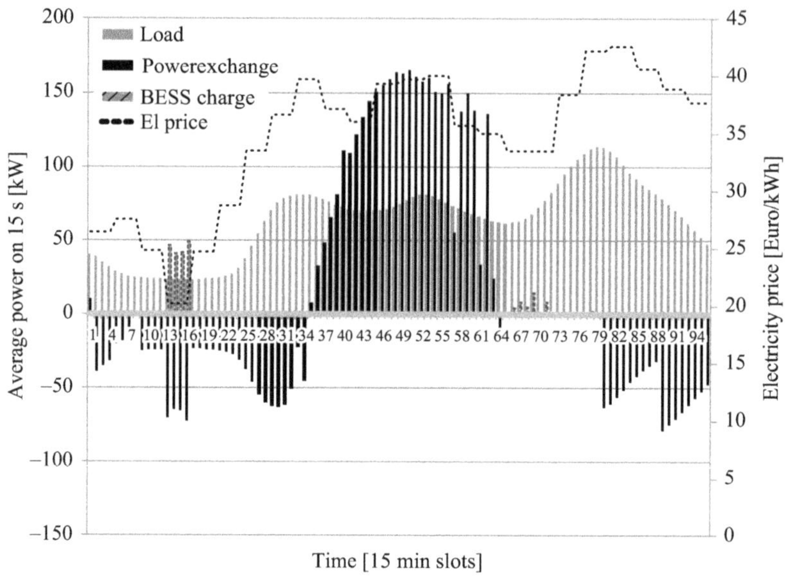

Figure 8.15 Resulting consumption schedule within the smart grid in islanding state

electricity price is decreasing and the thermal engine usage is lower. In the time interval $t = 66–70$ (the microgrid generation is exceeding the demand), the surplus is used for charging the storage energy system. After hour 21, the energy is purchased from the upstream grid instead of generating it locally using a thermal engine due to the fact that it is more economical to do so.

8.4 Conclusions

The analysis of various storage systems case studies in Italy and Romania was conducted. In Italy, the results from collaboration with Enel Distribution for black start function with a storage system with lithium batteries were presented. In this way, in case of a black-out, the energizing of an area of the medium-voltage power system can be achieved in an islanding operation way. A dynamic model for presenting the entire system was elaborated and through RTDS simulations considering the protection and control systems was achieved.

An optimization model for minimizing operational costs of a smart grid supplying cortical and interruptible loads in Romania was presented. The case study is based on a real facility equipped with renewable sources and storage systems for clean and reliable power supply. The analysis is based on perfectly know PV production data obtained from one year recorded measurements of the facility. The characteristic electrical data of the sources and loads within the smart grid are based on real data obtained from the producers.

Acknowledgement

This work was supported in part by a grant of the Romanian Ministry of Research and Innovation, CCCDI – UEFISCDI, project number PN-III-P1-1.2-PCCDI-2017-0404/31PCCD/2018, within PNCDI III.

References

[1] Tian, P., Xiao X., Wang K., and Ding, R. 'A hierarchical energy management system based on hierarchical optimization for smart grid community economic operation'. *IEEE Transactions Smart Grid*. 2016;**7**(1):2230–41.

[2] Lasseter, R., Abbas A., Marnay C., *et al., The CERTS Smart grid Concept*. CEC Consultant Report P500–03–089F. Sacramento, CA: California Energy Commission, 2011.

[3] Su, W. and Wang, J. 'Energy management systems in smart grid operations'. *Electricity Journal*. 2012;**25**(80): 45–60.

[4] Ardito, L., Procaccianti G., Menga G., and Morisio, M. 'Smart grid technologies in Europe: an overview'. *Energies*. 2013;**6**:251–81.

[5] Ipakchi, A. and Albuyeh, F. 'Grid of the future'. *IEEE Power and Energy Magazine*. 2009;**7**(2):52–62.

[6] Harris, A. 'Smart grid thinking—[power super grid]'. *Engineering & Technology*. 2009;**4**(9):46–49.

[7] Liu, B., Zhang, Y., Liu, C., He, W., and Bao, H. 'Reliability evaluation for distribution systems with distribution generation'. *European Transactions on Electrical Power*. 2010;**20**(7):915–26.

[8] Resende, F. O., Gil, N. J., and Peças Lopes, J. A. 'Service restoration on distribution systems using multi-smart grids'. *European Transactions on Electrical Power*. 2011;**21**(2):1327–42.

[9] Favre-Perrod, P., Kienzle, F., and Andersson, G. 'Modeling and design of future multi-energy generation and transmission systems'. *European Transactions on Electrical Power*. 2010;**20**(8):994–1008.

[10] Erol-Kantarci, M. and Mouftah, H. T. 'Smart grid forensic science: Applications, challenges, and open issues'. *IEEE Communications Magazine*. 2013;**51**(1):68–74.

[11] Brenna, M., Falvo, M. C., Foiadelli, F., *et al.* 'Challenges in energy systems for the smart-cities of the future'. Presented at 2012 IEEE International Energy Conference and Exhibition; Florence, Italy, 2012.

[12] Brenna, M., Foiadelli, F., Roscia, M., and Zaninelli, D. 'Synergy between renewable sources and electric vehicles for energy integration in distribution systems'. Presented at International Conference on Harmonics and Quality of Power; Hong Kong, 2012.

[13] Saber, A. Y. and Venayagamoorthy, G. K. 'Resource scheduling under uncertainty in a smart grid with renewables and plugin vehicles'. *IEEE Systems Journal*. 2012;**6**(1):103–09.

[14] Chen, C., Duan, S., Cai, T., Liu, B., and Hu, G. 'Smart energy management system for optimal smart grid economic operation'. *IET Renewable Power Generation.* 2011;**5**(3):258–67.

[15] Mitra, G. and Vallem, M. R. 'Determination of storage required to meet reliability guarantees on island-capable smart grids with intermittent sources'. *IEEE Transactions on Power Systems.* 2012;**27**(4):2360–67.

[16] Chaouachi, A., Kamel, R. M., Andoulsi, R., and Nagasaka, K. 'Multi-objective intelligent energy management for a smart grid', *IEEE Transactions on Industrial Electronics.* 2013;**60**(4):1688–99.

[17] Ross, M., Abbey, C., Bouffard, F., and Joós, G. 'Multiobjective optimization dispatch for smart grids with a high penetration of renewable generation'. *IEEE Transactions on Sustainable Energy.* 2015;**6**(4):1306–14.

[18] Lazaroiu, G. C., Dumbrava, V., Leva, S., and Roscia, M. 'Virtual power plant with energy storage optimized in an electricity market approach', Presented at International Conference Clean Electrical Power; Taormina, Italy, 2015.

[19] Lazaroiu, G. C., Dumbrava, V., Costoiu, M., Sima, C. A., and Leva, S. 'Optimizing DISCOs planning for networks with distributed energy resources', Presented at 2017 IEEE International Conference on Environment and Electrical Engineering and 2017 IEEE Industrial and Commercial Power Systems Europe; Milan, Italy, 2017.

[20] Cardoso, G., Stadler, M., Siddiqui A., *et al.* 'Smart grid reliability modeling and battery scheduling using stochastic linear programming'. *Electric Power System Research.* 2013;**103**:61–69.

[21] Brenna, M., Lazaroiu, G. C., and Tironi, E. 'High power quality and DG integrated low voltage DC distribution system'. Presented at 2006 IEEE Power Engineering Society General Meeting; Montreal, Canada, 2006.

[22] Brenna, M., De Berardinis, E., Delli Carpini, L., *et al.* 'Automatic distributed voltage control algorithm in smart grids applications'. *IEEE Transactions on Smart Grid.* 2013;**4**(20):877–85.

[23] Mirafzal, B., Saghaleini, M., and Kaviani, A. K. 'An SVPWM-based switching pattern for stand-alone and grid-connected three-phase single-stage boost inverters'. *IEEE Transactions on Power Electronics.* 2011; **26**(4):1102–11.

[24] Brenna, M., Foiadelli, F., Riva, S., Sapienza, G., Zaninelli, D. 'Dynamic model of a storage system with real-time simulation for power system reliability'. *International Transactions Electrical Energy Systems.* 2015; **25**(11):3109–21.

[25] Lazaroiu, G. C., Dumbrava, V., Balaban, G., Longo, M., and Zaninelli, D. 'Stochastic optimization of microgrids with renewable and storage energy systems', 2016 IEEE 16th International Conference on Environment and Electrical Engineering (EEEIC), Florence, Italy, 2016, pp. 1–5.

[26] Dumbrava, V., Lazaroiu, G. C., Leva, S., Balaban, G., Teliceanu, M., and Tirsu, M. 'Photovoltaic production management in stochastic optimized

microgrids', *UPB Scientific Bulletin, Series C: Electrical Engineering and Computer Science*. 2017;**79**(1): 225–244.

[27] Lazaroiu, G. C., Dumbrava, V., Leva, S., Roscia, M., and Teliceanu, M. 'Virtual power plant with energy storage optimized in an electricity market approach', 2015 International Conference on Clean Electrical Power (ICCEP), Taormina, Italy, 2015, pp. 333–338.

[28] Lazaroiu, G. C., Dumbrava, V., Roscia, M., and Zaninelli, D. 'Energy trading optimization of a virtual power plant on electricity market', 2015 9th International Symposium on Advanced Topics in Electrical Engineering (ATEE), Bucharest, 2015, pp. 911–916.

[29] Strbac, G., Aunedi, M., Pudjianto, D., *et al.*, 'Opportunities for energy storage: Assessing whole-system economic benefits of energy storage in future electricity systems'. *IEEE Power and Energy Magazine*. 2017;**15**(5):32–41.

[30] Wang, Z., Negash, A., and Kirschen, D. S. 'Optimal scheduling of energy storage under forecast uncertainties'. *IET Generation, Transmission & Distribution*. 2017;**11**(17):4220–26.

[31] Kanwar, N., Gupta, N., Niazi, K. R., and Swarnkar, A. 'Optimal distributed resource planning for microgrids under uncertain environment'. *IET Renewable Power Generation*. 2018;**12**(2):244–51.

[32] Akram, U., Khalid, M., and Shafiq, S. 'Optimal sizing of a wind/solar/battery hybrid grid-connected microgrid system'. *IET Renewable Power Generation*. 2018;**12**(1):72–80.

Chapter 9

Case studies from selected countries – the USA

Musa Yilmaz[1] and Heybet Kilic[2]

9.1 Introduction

To establish a sustainable energy system and overcome energy and environmental crisis caused by the use of fossil fuels, a new energy revolution is taking shape with electricity energy. It is characterized by the development and large-scale use of renewable energy. With the energy prospect, government investment and policy support, development of microgrids is gradually emerging storage applications [1–5]. The improvement and mercantile of energy storage systems will have a prominent effect in smart grid applications in terms of future power system models which is illustrated in Figure 9.1 [6].

By procurement of energy whenever required, energy storage will

- improve the reliability of electricity supply;
- reduce the variability and uncertainty, increase the stability;
- better integrate renewables into the system;
- increase the efficiency (economic and utilization) of the existing generation and transmission facilities;
- decrease the requirement for extra transmission entities; and
- be the preferential supplier for ancillary services.

To compare challenges and benefits of energy storage system (ESS), the role of energy storage in smart grid is explained as shown in Table 9.1.

Over the past years, ESSs have achieved merchant or exhibition level all over the world; the following is a list of the advanced technologies for storage systems [7–14]:

- Lead ACID (flowed, VRLA, Advanced);
- Lithium ion (LFP, NMC, LTO);
- NAS, Na-$NiCl_2$ batteries; and
- Flow batteries (ZBB, VRB).

[1]Batman University Faculty of Engineering and Architect, Electric and Electronics Engineering, Turkey
[2]Electrical and Energy Department, Diyarbakır Technical Sciences Vocational School, Dicle University, Turkey

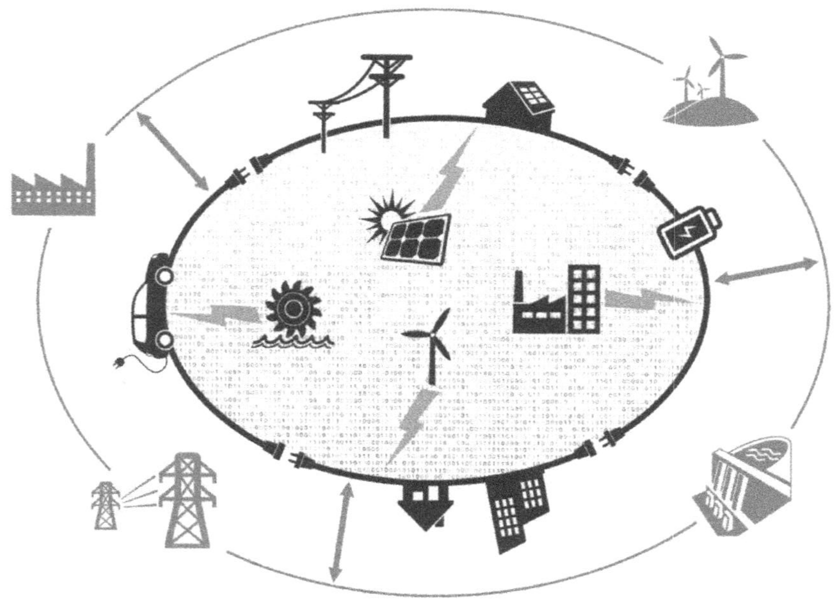

Figure 9.1 Future energy system

Table 9.1 The role of energy storage in smart grid

Parameter	Fuel	Generation	Transmission	Distribution	Services
Challenges	Volatility	Low utilization	Congestion	Disruptions	Dirty power
ESS Benefits	Hedge risk	Baseload arbitrage	Higher utilization	Stable power	Quality power

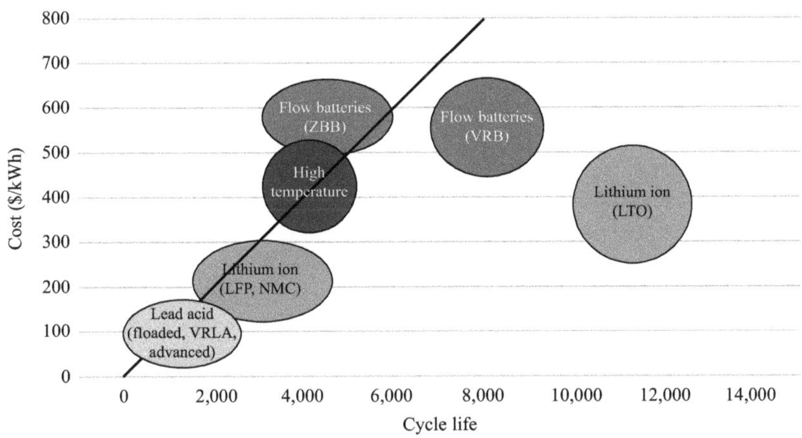

Figure 9.2 Show cost and the life cycle of the different type storage system technologies

From Figure 9.2, cost and the life cycle of the different types of storage system technologies and the following comments can be deduced:

- Different battery types have different cycle life depending on which life cycle cost is considered ($/kWh/cycle). Lead acid batteries have a very low cost but comparatively low cycle life. Advanced lead acid batteries have a higher cycle life at a higher cost.
- NAS batteries have an intermediate cycle life between lithium ion batteries and flow batteries.
- Flow batteries have a high cycle life and a moderately high cost.
- Lithium ion battery systems may be designed for high-power or long-duration applications.
- LTO batteries can offer lowest marginal cost due to longer cycle life and high charge/discharge rates.

The cost per storage cycle can be defined as:

$$\text{Cost per storage cycle}(\$/\text{kWh}) = \frac{\text{Energy storage cost}(\$/\text{kWh})}{\text{Cycle}(\#) * \text{Round trip Efficiency}(\%)} \tag{9.1}$$

9.2 Market segmentation of energy storage market

The energy storage systems are important innovations for electric production and consumption, and at the same time, it is the feature of traditional electrical power technology. Moreover, it is sufficient for different application areas since it includes renewable energy grid concretion, power transmission and distribution, distributed generation, microgrid, etc. [7–18]. As a flexible power source, energy storage systems can be broadly classified and applied to power generation, transmission, and distribution. The applications of energy storage can be classified as shown in Figure 9.3.

9.3 Case studies

The U.S. Department of Energy (DOE) has provided about $ 4.5 billion to modernize the electricity grid. A life storage project is accepted for ten financial statements for the year lifecycle assessments. This is a special case standard value for a variety of reasons [19–22]. Due to the very limited number of available resources for this study in the United States, we aimed to distinguish conditions that

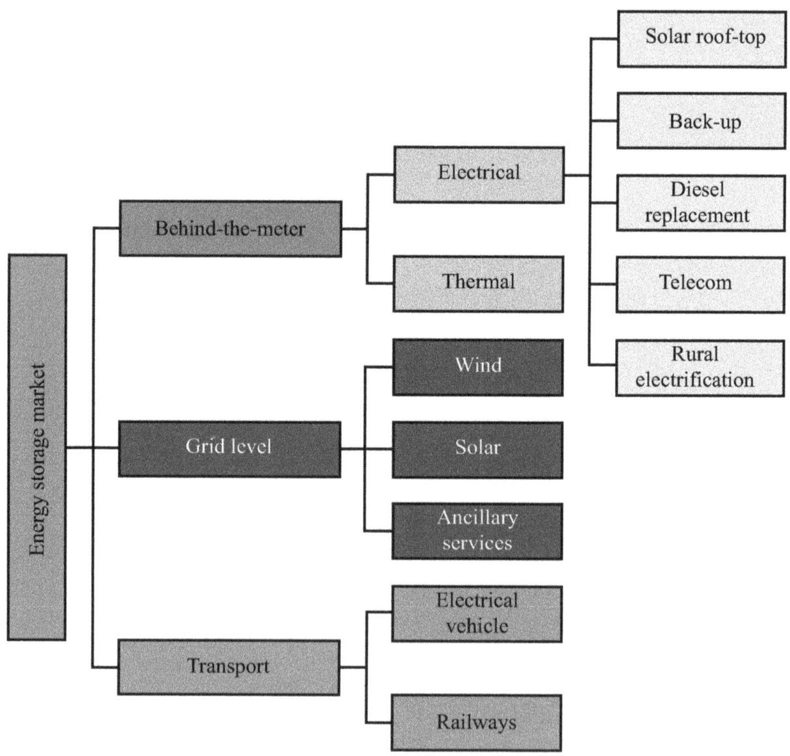

Figure 9.3 Division for applications of energy storage

are demonstrated by several perimeters (dissimilar cases and framework types in different climatic conditions). But we have tried to perceive the situations that represent diverse perimeters (several states and kinds of systems in different climatic conditions), because of the limited number of facilities available for this study within the United States [23,24].

Pecuniary advantages and maximal market back demand forecasts for the United States are depicted in Figure 9.4.

Primary results and key assumptions from the guide are provided in Table 9.2 and Figure 9.4. Table 9.2 includes five criteria for the 17 principal profits characterized in the study. Capacity indicates the range of power ratings of the storage systems that are applied for a given benefit. The benefit indicates the present value of the respective profit type for ten years (inflation of 2.4%, discount rate of 10%) [1–7].

Served total electric load by the grid is an important index that highlights the maximum probable market potential. The peak load is market size. The US values are based on data from NERC. Table 9.3 indicates the values project peak load in the United States [2,3,24].

The American Recovery and Reinvestment Act (ARRA) supported 16 projects for energy storage demonstration (Figure 9.5). The projects ranged from facility

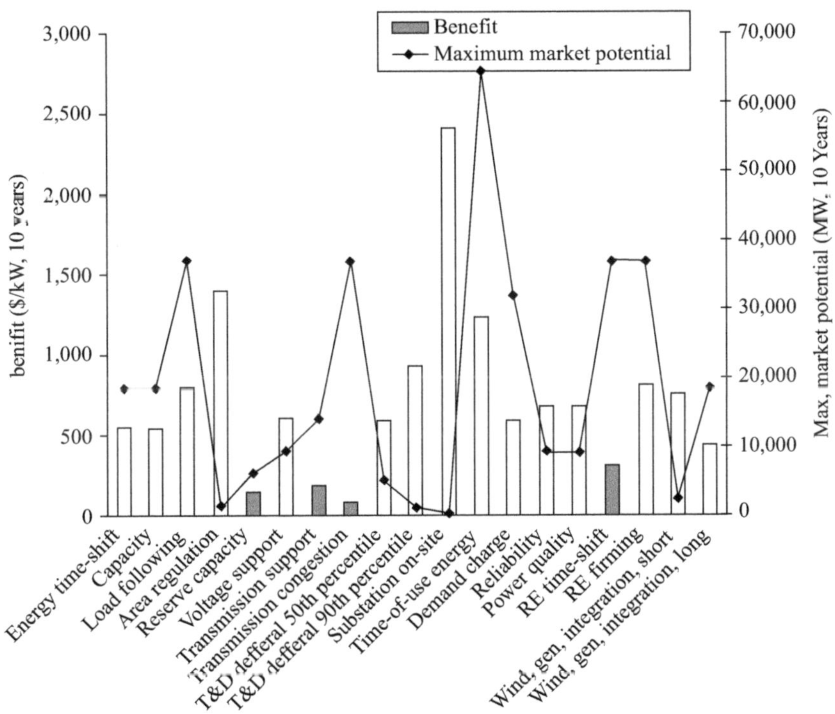

Figure 9.4 Pecuniary advantages and maximal market back demand forecasts for the United States

Table 9.2 Presumptions and results

		Discharge state		Capacity (Power: W, kW, MW, GW)		Profit ($/kW)		Potency (MW,10 years)	Austerity ($Million)
	Profit	**Low**	**High**	**Low**	**High**	**Low**	**High**	**U.S.**	**U.S.**
1	Power supply capacity	4	6	1,000 kW	0.5 GW	360	711	18,417	9,838
2	Energy time-delay	2	8	1,000 kW	0.5 GW	400	700	18,417	10,129
3	Area regulation	15 min	30 min	1,000 kW	40 MW	786	2,011	1,012	1,415
4	Load following	2	4	1,000 kW	0.5 GW	600	1,000	36,834	29,467
5	Reserve capacity (Power Supply)	1	2	1,000 kW	0.5 GW	58	226	5,986	844
6	Transmission support	2 s	5 s	0.01 GW	0.1 GW	193		13,813	2,646
7	Voltage support	15 min	1	1,000 kW	0.01 GW	400		9,209	5,525
8	Transmission congestion relief	3	6	1,000 kW	100 MW	32	141	36,834	3,168
9.1	T&D Upgrade deferral 90th percentile	3	6	0.25 GW	2 MW	758	1,079	997	916
9.2	T&D Upgrade deferral 50th percentile	3	6	0.25 GW	5 MW	487	687	4,986	2,912
10	Substation on-site power	8	16	1,500 W	5 kW	1,800	3,000	250	600
11	Demand charge management	5	11	0.05 MW	0.01 GW	582		32,111	18,695
12	Time-of-use energy cost management	4	6	1,000 W	0.01 GW	1,228		64,228	78,743
13	Electric service reliability	5 min	1	200 W	0.01 GW	358	979	9,208	6,154
14	Electric service power Quality	10 s	1 min	200 W	0.01 GW	358	978	9,209	6,154
15	Renewables capacity firming	2	4	0.01 MW	0.5 GW	708	915	36,834	29,909
16	Renewables energy time-shift	3	5	0.01 MW	0.5 GW	234	389	36,834	11,455
17.1	Wind generation grid integration, short time	10 s	15 min	200 W	0.5 GW	500	1,000	2,302	1,727
17.2	Wind generation grid integration, long time	1	6	200 W	0.5 GW	100	782	18,417	8,122

Table 9.3 U.S. peak load and peak load growth

Peak load, 2008 (MW)	796,479
Generation capacity, 2008 (MW)	925,916
Reserve margin (%)	16.2
Peak load growth rate expected (%/year)	1.80%
Load estimate, 2017 (MW)	928,850
Load growth forecast, 2009 to 2017 (MW)	124,371

and technology demonstrations to large-scale energy storage systems. This investment had a prominent affirmative effect on the energy storage industry. The ARRA project's objective is to summarize the lessons learned and provide recommendations for future investments of the DOE. The developers focused on mitigating the ramp and changing the time display a reference plant. This pilot study would expand the state of current demonstration systems and would direct resources to the deployment of 50 MW plants [23–26].

9.3.1 East Penn Manufacturing Project

Lyon Station, PA., October 12, 2011—East Penn Manufacturing Co., Inc., and its auxiliary Ecoult announced the start of Public Labor-intensive Co. of Prosperity Energy Storage Project of New Mexico (PNM) in Albuquerque, NM (Figure 9.6). This project associates a sophisticated energy storage solution valve-regulated lead-acid (VRLA) and UltraBattery with a 500 kW solar plant installed separately. Its aim is to ensure synchronous voltage flattening for constant power levels and peak changes to yield power whenever it is most required [23–25].

The system is 90% available during the programmed operating times. From January 1, 2015, the total return of the start-up was 6,140 MWh. In terms of performance, the system learns the tantamount of 2 cycles at 100% depth of discharge (DOD) every day, if it works at 2 MW. When managing 3 MW, the systems learns the tantamount of three 100% DOD cycles per day [16,23].

The PJM project changed its dynamic frequency control-signal and sets up to compensate for its payment structure while the integration of rapid-response resources in the regulatory market.

9.3.2 New Mexico Project

The New Mexico Public Service (PNM) addresses this matter as moiety of projects; a smart grid exhibition project being carried out with the Electric Power Research Institute (EPRI) and the energy storage program of the Smart Grid, DOE [23–26]. First, PNM shows that disposals energy stored in batteries to reduce voltage ripple by smoothing the battery and managing peak demand through battery change at the same time. Second, PNM is cooperating in an exhibition of a public micronetwork for commercial buildings contrived and built by the Japanese Government's Industrial Technology and New Energy Development Organization (NEDO).

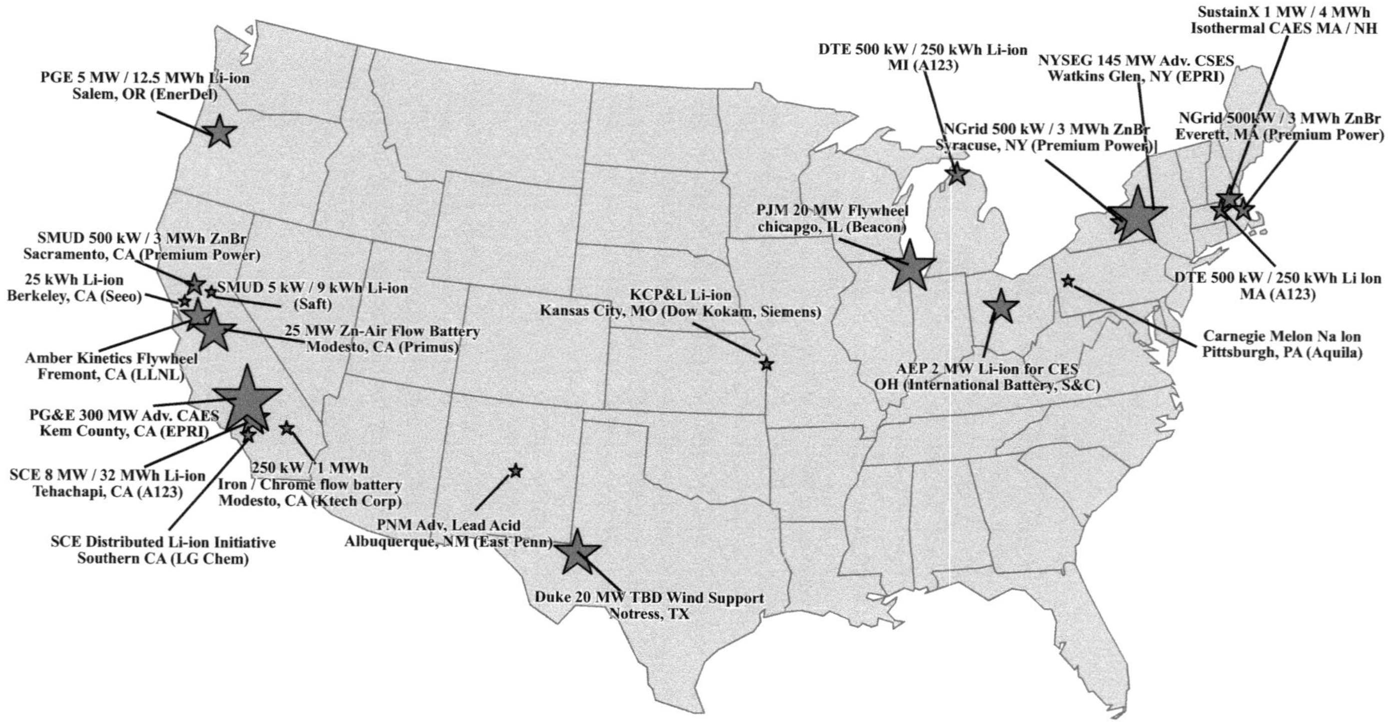

Figure 9.5 ARRA-funded utility energy storage projects

Figure 9.6 East Penn UltraBattery installation

The PNM project was the first of 16 ARRA energy storage demonstration projects to be brought online, and the process had started in September 2011. These objectives were achieved by the installation of two battery systems (one force and one power) in a 0.5 MW solar PV system to compose a distributable DG. The energy battery uses the Ecoult/East Penn-system—Advanced lead-acid battery and has an output of 250 kW for 4 h (1,000 kWh total power). The power battery (the Ecoult/East Penn) is used to soften UltraBattery system and has a power rating of 500 kW for 41 min (334.34 kWh). The place is entrenched in a ramification southeast of Albuquerque (New Mexico) [23].

The main challenges in the design of the system are concerned with the concretion of the battery system in the PNM network. First, utility operators and management groups were not familiar with the functionality and profits of energy storage. Second, it was hard to acquire or improve the necessary management software to control an energy storage resource on the distribution level. Therefore, the regarding communication is active potential applications for the storage of energy at the distribution level and to divide, making it more effective to optimize and control of energy storage capacity [23].

9.3.3 Auwahi project site in Maui, Hawaii

NEC Energy Solutions provides a lithium iron phosphate battery (nano phosphates) in Maui, Hawaii, to smoothen ramp rates in a 21 MW wind park. The battery has a capacity of 11 MW/4,300 kWh. This was established to meet heating rates to operate a wind farm to the local allegiance necessities [25].

The system was moved by truck from Massachusetts to California, and then shipped directly to Hawaii, where a truck transferred it to the project area.

The project consists of nine battery containers, three inverters, and three transformers, together with the thermal management and switching units [25].

The electronic components of the power in the installation were available on the market, but nonetheless Dynapower Corporation of Vermont has developed them in-house for the project. The actual utilization of the signboard capability for this specific project was not revealed, however the developed battery system is able to achieve/provide 100% DoD download. Lithium-ion battery technology nano phosphate has been restricted to the use of Department of Defense or expanded to low state of charge, unlike the technologies of lead-acid batteries [25].

The total cost of the project was driven by the cost of the equipment. This is achieved in the conversion of current and storage by the inverter and the battery cells. Thermal management equipment (cooler) consumes the greatest amount of electricity as part of the auxiliary power consumption.

9.3.4 SC Tehachapi energy storage

The storage system of 8 MW/32 MWh of lithium-ion batteries in the Techapi wind farm in California seeks several objectives: to improve the performance of the network, connect and assist in the integration of renewable energy sources in a connection zone where up to 4.5 GW wind generation is expected. Having been completed in 2014, it is already in operation. The batteries are accompanied by two inverters, each of 4.5 MVA of nominal power. The set, which is connected to 66 kV, provides numerous functions to the system due to its design and the flexibility provided by the storage of high capacity in power and energy. At the time it is the largest storage project (excluding pumping) in the United States.

The permitting and construction process was much simpler due to the far-flung area and a large single upright block transformer station that was already in possession of SCE. This allowed to take advantage of the SCE switchgear standards, eliminating the need for the typical approval process within the boundaries of large cities and other thickly populated locations. The dimensions of the location imposed restrictions on the form of the building, leading to an L-shaped structure [23]. To simplify the approval process and the available circumstances within the substation, the connection was arranged as temporal. The requirement of a persistent link would be an extension of the switchgear, which would have implied an extended ISO certification, the dilation of the switchgear footmark, and a peripheral effect valuation.

9.3.5 Duke Energy Notrees wind storage

Duke Energy, a large power company in the United States, is using Younicos' control method to operate the processes of a 36 MW energy storage system on the 153 MW Notrees wind farm site in West Texas. The Notrees Battery Park was invested in portion by a Smart Grid award from the US DOE. It was built by Duke Energy in 2012 and is the storage facility together with the largest wind energy in North America. The associated plant at Notrees provides flexible and environmental-friendly resource to the Texas Electrical Reliability Council (ERCOT) that manages the network and operates approximately 75% of the deregulated market in the state. The Battery Park is placed in a 20,000 square foot building adjoining the wind farm ramification. It is linked to the distribution level of the 34.5 kV system of the wind farm and is connected to the ERCOT network from end to end with an identical link to the wind farm [1–26].

The main objectives of the project are to obtain the following:

- Combine with discontinuous production of renewable energy;
- Demonstrate offering the advantages of using rapid-response energy storage system services for network control;
- Improve the utilization of assets that flow through the storage of energy during nonpeak generation periods; and
- Approve that a battery solution can be triggered by market rate or the set times with the ramp control.

Procurement of rights to land and the location close to the collector substation, the O & M Construction of the wind farm and collaborated area, and the high-voltage (HV) radial transmission line of wind farms were early challenges. The wind system served has no provisions for the installation, so that the process of change in the lease serfdom has long.

While it is very suitable for the affirmation of the wind level, the application of frequency regulation challenges the current battery. Simultaneous participation in the scheme and regulation of markets requires the operation of the system in a state of partial load. First operations highlighted quickly the adverse effect on the BESS. The goal moves to the contribution in the upstream regulatory rate. Also upstream of among 1 PM and 5 AM daily fees [23].

The configuration of the modules is with a number of sensors in each column. If a column has poor voltage, the poor cells will be manually specified and relocated. In a future dilation or implementation, it would be preferable to implement that at the cellular level.

9.3.6 Aquion

Finally, several different types of new batteries capable of storing the energy production of a wind or solar power plant in an efficient and cost-effective way are being hooked up to the electricity network. These network batteries could lower the cost of renewable energy by satisfying the difficulties of intermittence, which arises when the sun does not shine or the wind does not blow. The prototype project 10 kWh (5 h discharge) includes a 1 kV electrical charge system, which the project ARRA supplied together with the current M-Line battery module (estimate of 25 kWh) [23].

There are similar network battery projects taking off in California (USA), driven by new legislation contrived to entrench changes in renewable energies (see Old materials allow new large capacity batteries). Last month, Ambri, a Cambridge-based start-up that makes molten metal batteries, broadcasted that it would link its batteries to the network in Massachusetts, New York, Hawaii, and Alaska (see giant batteries for a new type of electrical network).

The new battery technologies are not cheaper than the natural gas plants which are used now, preferably to support renewable energies. But it has to do mainly with the fact that the batteries are still being produced on a small scale.

Aquion and Ambri target appropriate markets. The price of electricity is especially high in Hawaii, so the value of batteries to store wind and solar energy increases. Hawaii's microgrid will support a 176 kW solar panel plant large enough for 40 medium-sized homes, and store 1,000 kWh of energy, sufficient to supply an ordinary American home for more than a month. The Aquion factory could install 200 such projects each year.

Ambri's pilot projects are aimed at a slightly dissimilar situation in which the important ascpect is to reduce the peak demand of energy in the network. If customers can store electricity in their own homes, they will reduce the demand on the network at key moments and therefore cut back on their electric bill.

One of Ambri's pilot projects could also directly serve to a main issue in New York City (USA), where the power grid that imports energy into the downtown is occasionally so forced that it is about to break down. The company's batteries could serve to reduce demand and the requirement to build more power grid. They could also supply as reinforcements in the case of blackouts, says Ambri executive director Philip Giudice. The latter is a major concern of the city that has been buffeted by severe weather events in recent years, such as Superstorm Sandy [23–26].

9.3.7 Seeo

Originally developed in partnership with United Technologies Corporation (UTC), Vionx's exclusively licensed vanadium redox flow battery (VRFB) offers a safe and flexible solution built to exacting utility standards so that it will continue to perform for long years ago and beyond. At the heart of Vionx's VRB design is the battery stack flow field design from the UTC. The stack-level tests are underway. In 14 months, it has advanced from the pivot point to having accumulated 3,000 h run time in the first batch. Recently, they tested 3 kW, 10 kW-cells and 6 cells, 20 cells. The real battery cell support at two times the power per unit field is compared to competitor cells. The UTC predicts a lifetime of more than 2,000 h for the first batteries and a goal for a shelf life of 20 years for the generation of batches [2–5,23–26].

Vionx has a contract for the construction of 160 kW, plan extended in Fort Devens. system supply to for 4 h. The current project end time for 2018 (for 24 months). Vionx installed a plant with a battery progression 500 kW. The following portion supplies a digest of the lessons from the Vionx project.

Vionx expected that the greatest strength its technology provides is the long execution time at low cost. In particular, applications in which lithium batteries discharge between 4 to 10 hours are of interest. Its core competence is the battery technology: batteries, modules, system scale and software. Vionx preferred working with a shareholder for the concretion, installation, and commissioning. The first frameworks are principally geared to storage at the distribution level. The most important applications are the shift of T & D and renewable and firming integration [25].

9.3.8 Xtreme Power on Kodiak Island, Alaska

This storage system is an isolated electrical system has been in operation since 1997 and has been updated in 2008 (after more than 11 years of operation, over its initial useful life set at 8 years); it consists of a 1 MW lead-acid battery, with power and capacity to supply power for 1 h and 24 min. It connects to the voltage level of 12.47 kV in a power system where the main source of energy comes from two 4.9 MW hydro units and a 3.3 MW diesel group. It is installed in a small town in Alaska with 1,000 inhabitants and two large power consumers: a cold-storage factory and a large sawmill (the latter currently does not exist).

The fundamental objective of the storage system was to find a source of rapid generation when industrial consumption soars and to absorb excess production from hydraulic generation, and to allow it to work in a more stable operating regime. Thus, it is capable of supporting 800 kVA continuous loads and charging tips up to 1,200 kVA. At that time, it represented an investment of 1.6 million euros.

Still in operation, there are numerous system data that support the global technical and economic benefit resulting from the installation of the storage system. For example, until 2008 there was a total saving of more than six and a half million dollars since its installation. From the environmental point of view, it meant avoiding the consumption of more than 5.3 million gallons of diesel, as well as the pollution associated with its combustion. Highlighting continuous operation throughout the year, as well as an operating regime in which it maintains its SOC between 50% and 80%, ensures a long life of the system [23]. The expanded solution of lead-acid batteries has to be sufficient for that enforcement, due to the fact that the system stays in a status of high charge and can be unloaded rapidly for very curt periods. Since the advantages of the lead-acid batteries result from a better economy than the lithium ion batteries, this type is also profitable. The corporation says that the temperature is the most significant operational circumstance. It should remain within 20–30 °C inside the container. The fitment costs depend on travel expenditures to the remote location. Those visitations will be carried out quarterly or less frequently, and monitoring is carried out during the year. The other important variable costs are ancillary loss of parasitical loads. Advanced lead acid batteries are recyclable and almost 100% recyclable at the end of its useful life [25].

9.3.9 California off-grid PV systems

A 55 kWh aqueous hybrid ion (AHI) battery and 11 kW photovoltaic (PV) array was installed on a California ranch by Aquion Energy. This was a constant accession to credible performance without the consumption of diesel. The area is outside the supply range of public utilities. There were some options for energy. Expansion of the network service on the area was so pricey. Therefore, the holder of the ranch invested in a microgrid installation with the goal of minimizing or eliminating the generation of diesel for daily energy supply [22,26]. There were no incentives in place, however the rebate on hedge was sufficiently appealing for the area to establish the system without public funds. It was found that the performance, environmental impact, and cost profile of the battery were better than lead and alternative choices. The Aquion cost decomposition is listed as follows.

The whole grid consists of the following elements:

- 3 Aquion M100-HV battery cells (55 kWh) [25];
- 40 SolarWorld 270 W solar modules;
- Dynapower Corporation Energy Conditioning System;
- Unirac SolarMount;
- FWPV-8 Outback ceiling-mounted combo box;
- Wall-mounted Solar Wall Combination Box MNPV4HV-Disco-3R wall mounted by MidNite;
- Crouse-Hinds combinable box CCB6SL F15; and
- 30 kW reserve diesel generator.

9.3.10 Maryland PV and solar system

It was installed for client and ancillary services. Energy storage LLC ensured a 500 kW AC energy storage system and the system was combined with Konterra's inverter in Maryland. The ESS has 300 kWh lithium-ion batteries of AllCell technologies. The ESS and inverters were linked to PV solar cells of 400 kW [25].

9.3.11 NaS battery storage system in Presidio

It is a 4 MW/32 MWh NaS battery system installed in 2010 to support the city of Presidio in Texas, powered by a 69 kV, 60-mile-long T-line. This installation, in addition to guaranteeing the continuity in the supply during 8 h in case of loss of the supply transport line, or even during maintenance work of the same, has allowed to postpone investment in new lines of transport, in addition to providing benefits in relation to the quality and reliability of the electrical signal supplied thanks to the rapid performance of the battery. It must be borne in mind that the characteristics of the feeding to this city, with a single line of more than 50 years in service, caused numerous interruptions in the supply, which was worsened with the high frequency of electrical storms that occurred in the area (near the border with Mexico). As an example, between 2001 and 2006, there were 247 power cuts, with nine of them as an interruption in service of more than 6.8 h on average. In addition, between July 8 and September 8, 2007, 81 events were recorded with a very poor voltage signal. In its day it was the first and largest battery in the United States of this technology.

9.3.12 DTE Energy

The aim of the ARRA project was to prove the value of the supported charge of energy, frequently called Community Energy Storage (CES). The project also contains the provision of batteries for electric vehicles (EV) for a secondary use case, as a different system of batteries of 500 kW for PV integration. Eighteen CES units were installed, with the below features. Seventeen of the units' distribution circuits are supplied from a selected substation called Trinity. A system is with the battery in an underground repository. The transformer consists of the inverter, the arched ceilings and a distribution pad [23].

The following functions of the network are evaluated by the tea project [9]:

- Voltage support/VAR;
- Renewable energy time change;
- Smoothing of renewable generation;
- Shaved peak circuit;
- Arbitration of energy, for example, discharge during a high price of LMP; and
- Circuit model commands.

When the project was built up, the CES system was recognized in national laboratory for IEEE-1547 certification. This makes it possible for the electric-service-system project engineers to accept new plants into the microgrid without elaborated studies for each installation. The first project was successfully tested with the test-installation of the DNV-GL (Kema) battery. It has been tested for basic functionality, including use cases listed below.

- Efficiency return;
- Maximum test of the shaving profile;
- Frequency regulation profile test;
- Island test; and
- Harmonics analysis [23].

In more than ten studies, the round-trip-efficiency was typically assorted in an interval of 87% and 90%. The system's capability to trace an agent frequency control-signal was also checked. Moreover, DNV GL carried out feasibility research on the rate of regulatory services and the maximum utilization of the BESS in the Trinity circuit.

The primer CES was established in a practice lab, to check documents of IEEE standards, work instructions and practice on the U-teams, which would install and manage the CES in the Trinity circuit. As the engineers and teams in a new technology project are crucial for approval, it also authorizes collection of feedback that can be integrated in the design and operating developments in the future.

The commitment of the community was the main point to the accomplishment of first project. The municipality was linked and information about the project was made available, despite the fact that the CES has been installed in the public services easement. The owners of nearby houses were also linked directly in order to respond to interrogatories about the installation of the battery. Circuit modeling was conducted to define the load of the distribution transformer. Potential sites have been identified such that customers can be served for at least 2 h in case of interruption.

9.3.13 Emergency program shelters, Florida

In this program, more than 115 units of 10 kW PV plants have been installed together with lead-acid batteries of 25 kWh in schools in Florida, basically seeking to supply power in an uninterrupted manner in case of failure of the external network in an area with high frequency of hurricanes, and to reduce the electric cost of the school thanks to solar production. The high capacity of these shelters in the case of hurricanes, which can accommodate up to a total of 50,000 people, makes these centers an ideal location for these mixed systems of renewable energies and storage. It contributes to increase the so-called resilience of the electrical system and is an example of distributed storage.

9.4 Conclusions

Large network-linked demonstration projects were essential for enhancement of the effectiveness of the technical and test-yield samples for dissimilar systems. Some notable findings and recommendations are listed below:

- The transport of energy storage system supplementary may take longer than anticipated when the Department of Homeland Security Scales consumption to over demand. This limits the travel to sunlight and delayed surrender.
- Codes for stationary energy storage in a reduction of the complication of the availability and structures operation.
- The use of small-bushel test-system can decrease progress and concretion time. Benefits include:
  - test of performance;
  - environmental tests;
  - security tests;
  - power conversion control tests; and
  - and public service network interface tests.
- For the storage of energy in the distribution level, the first installation in the public services training center can provide examination of engineering document standards; verification of operational processes; and training for the U-teams [23], the installation and maintenance of CES.
- Simplifying the necessities for the DOE debt warranty schedule and the reduction of the minimum size of the loans would be advantageous for next energy storage demonstration projects, particularly against the background of the deficiency of the presence of attempt stock funds for a long-term demonstration.
- Supplementary DOE assistance for future energy storage demonstration projects will aid to rise the espousal by simplifying the verification of the efficiency of the energy storage technologies.
- Different participators in the industry focus on the shift of T & D and markets that change with time. The participators have shown plants <1 MW, with the aim of finally advancing installations on a scale of 20 MW to 50 MW. The participators offerings to the requirement for an ensuing exhibition that generally as a sample project, on a scale of 1,000 kW to 5 MW as a requisite reduction level hereinabove with installations> 20 MW.
- Business administration participators agree that the incidence of arrangement auxiliary service provides a commercially viable market for the storage of energy connected to the network. It is a public concurrence about the financial

value of a number of markets on the customer side. Nevertheless, there are a number of opinions about the financial viability of the markets on which the display utility side on the frequency regulation. The shift of T & D and the change of the schedule were cited more often than the next likely options.

- A competitive bid to further support the successful connection with large-scale ARRA energy demonstration projects would be beneficial. The development of operations is necessary in order to maximize revenue or benefits for the network and the outcomes. In case of the high preliminary investment, a relatively small investment would ensure that the results of the projects will be connected during the anticipated life of the system.
- Service companies, in general, and vertically integrated service companies, in particular, have in the past established that it is difficult to define the optimum application for energy storage. The participators (companies, organizations, universities) are notice that the experimentation from the ARRA program, future installations more specific and accurate.

References

[1] Divya, K.C. and Østergaard, J. (2009). "Battery energy storage technology for power systems—An overview." Electric Power Systems Research 79(4): 511–520.

[2] Tant, J., Geth, F., Six, D., Tant, P., and Driesen, J. (2013). "Multiobjective battery storage to improve PV integration in residential distribution grids". IEEE Transactions on Sustainable Energy, 4(1), 182–191.

[3] Yang, Y., Li, H., Aichhorn, A., Zheng, J., and Greenleaf, M. (2014). "Sizing strategy of distributed battery storage system with high penetration of photovoltaic for voltage regulation and peak load shaving." IEEE Transactions on Smart Grid, 5(2): 982–991.

[4] Kentli, F. and Yilmaz, M. (2017). "Improving tracking efficiency of two-axis sun tracking systems." Energy Harvesting and Energy Efficiency. Springer, Cham, 179–203.

[5] Adhikari, S. and Fangxing, L. (2014). "Coordinated Vf and PQ control of solar photovoltaic generators with MPPT and battery storage in microgrids." IEEE Transactions on Smart Grid, 5(3): 1270–1281.

[6] Zobaa, A.F. (2014). "Optimal multiobjective design of hybrid active power filters considering a distorted environment." IEEE Transactions on Industrial Electronics, 61(1): 107–114.

[7] Yilmaz, M. and Kentli, F. (2015). "Increasing of electrical energy with solar tracking system at the region which has Turkey's most solar energy potential." Journal of Clean Energy Technology, 3, 287–291.

[8] Neves, D., Silva, C.A., and Connors, S. (2014). "Design and implementation of hybrid renewable energy systems on micro-communities: a review on case studies." Renewable and Sustainable Energy Reviews, 31: 935–946.

[9] Yilmaz, M. (2017, March). The prediction of electrical vehicles' growth rate and management of electrical energy demand in Turkey. In Green Technologies Conference (GreenTech), 2017 Ninth Annual IEEE (pp. 118–123). IEEE.
[10] Bizon, N., Tabatabaei, N.M, Blaabjerg, F., and Kurt, E. (2017). Energy harvesting and energy efficiency. Springer International Publishing.
[11] Sisternes, D., Fernando, J., Jenkins, J.D., and Botterud, A. (2016). "The value of energy storage in decarbonizing the electricity sector." Applied Energy, 175: 368–379.
[12] Marion, G., Mathias, U., and Christina, D. (2017). The use case and smart grid architecture model approach, Springer, DOI 10.1007/978-3-319-49229-2.
[13] Mariam, L., Basu, M., and Conlon, M.F. (2016). "Microgrid: architecture, policy and future trends." Renewable & Sustainable Energy Reviews, 64: 477–489.
[14] Khan, A.R., Mahmood, A., Safdar, A., Khan, Z.A., and Khan, N.A. (2016). Load forecasting, dynamic pricing and DSM in smart grid: a review. Renewable & Sustainable Energy Reviews, 54: 1311–1322.
[15] Wan, C., Zhao, J., Song, Y., Xu, Z., Lin, J., and Hu, Z. (2015). "Photovoltaic and solar power forecasting for smart grid energy management." CSEE Journal of Power and Energy Systems, 1: 38–46.
[16] The Climate Group (2016). Current situation and prospect of the development of smart grid in China; Beijing, China.
[17] Zobaa, A.F., Vaccaro, A., and Lai, L.L. (2016). "Guest editorial enabling technologies and methodologies for knowledge discovery and data mining in smart grids." IEEE Transactions on Industrial Informatics, 12(2): 820–823.
[18] Rohjans, S., Uslar, M., Bleiker, R., *et al.* (2010). Survey of smart grid standardization studies and recommendations. In: Proceedings of IEEE Conference on Smart Grid Communications, Germany.
[19] Kanabar, M.G., Voloh, I., and McGinn, D. (2012). Reviewing smart grid standards for protection, control, and monitoring applications. In: Proceedings of IEEE Conference on Innovative Smart Grid Technologies, USA.
[20] SMB Smart Grid Strategic Group (SG3) (2016). IEC Smart Grid Standardization Roadmap.
[21] Lin, G.X. (2014). "An overview on the development of smart grid and distributed energy", Beijing, China: China High-tech Enterpises;1: 141–142.
[22] Zobaa, A.F. and Abdel Aleem, S.H.E. (2014). "A new approach for harmonic distortion minimization in power systems supplying nonlinear loads." IEEE Transactions on Industrial Informatics, 10(2): 1401–1412.
[23] Bender, D., Byrne, R., and Borneo, D. (2015). "ARRA energy storage demonstration projects: Lessons learned and recommendations, A study for the DOE energy storage systems program" Sandia Report, Sandia National Laboratories, SAND2015-5242.
[24] Eyer, J.M., Iannucci, J.J., and Corey, G.P. (2010). "Energy storage for the electricity grid: Benefits and market potential assessment guide, A study for

the DOE energy storage systems program." Sandia Report, Sandia National Laboratories, SAND2010-0815.

[25] BYD Energy (2015). "Case studies: battery storage" The International Renewable Energy Agency (IRENA).

[26] Colak, I., Bayindir, R., Fulli, G., Tekin, I., Demirtas, K., and Covrig, C.F. (2014). "Smart grid opportunities and applications in Turkey." Renewable and Sustainable Energy Reviews, 33: 344–352.

Chapter 10

The use and role of flywheel energy storage systems*

Trevor J. Bihl[1]

10.1 Introduction

Flywheels, in short, are machines that store kinetic energy in a rotating mass. The flywheel operates via a "flywheel effect" whereby its rotation is maintained via its own inertia [1]. Although ancient science did not understand the mechanics of flywheels, ancient engineers were able to develop spindle whorls (by 6000 BC) and potter's wheels (by 3000 BC) which both used their own inertia to maintain motion [1]. Work on flywheels gradually expanded until the industrial revolution, whereby they were incorporated into engine design to smooth rotations and damp vibrations [1–3]. Early uses of flywheels in electrical systems included regulation of power generation systems [4] and damping vibrations [2, 3]; roles they continue in today [5]. However, flywheels have seen additional use as a form of battery and interest is increasing in this use as material developments are made.

Flywheels have a long history as both a load-equalizing and energy storage mechanism, cf. [6, 7]. Additionally, in conjunction with batteries, flywheels found early use at power plants for power quality control [6]. However, until advancements were made in materials and design in the 1970s, see [8], flywheels found few uses beyond small and individual applications. Since these advances, flywheels have found uses for space-based batteries, uninterruptible power supplies (UPSs), and power quality improvement [7, 9, 10].

Flywheel technology up to the 1970s employed steel flywheels and bearings which required frequent or annual replacement [11, 12]. Thus, the application of flywheels for energy storage was initially limited. Proposals for advanced flywheels for energy storage began in the 1970s and pushed toward advancements in materials and efficiencies [13, 8]. Advances made in flywheel energy storage (FES) systems have yielded systems viable for some situations, e.g., UPS, and proposals and research for large-scale implementations [12]. Additionally, as discussed in

*The views expressed are those of the author and do not reflect the official policy or position of the US Air Force, Department of Defense or the US Government.

[1]Air Force Institute of Technology, Wright Patterson AFB, OH, USA

[14], future work with FES systems could also involve the integration of multiple energy storage approaches.

Various efforts, cf. [7, 15–20], have reviewed FES systems; this work incorporates these reviews, extends when possible, and reviews recent efforts and flywheel research in other disciplines. After the introduction, this chapter first briefly presents a background on flywheels and design considerations. Then applications and research efforts are discussed. A conclusion then follows.

10.2 Background

FES systems generally operate with the components seen in Figure 10.1, which illustrates an early 1900s flywheel load equalizer. In essence, the system seen in Figure 10.1 was a precursor to FES systems in development today. FES systems operate with a combined motor/generator that first rotates the flywheel to store energy and then, when energy is needed, the flywheel turns the motor/generator to convert its rotational energy to electrical energy [7].

While, conceptually, FES systems today operate similarly to the system in Figure 10.1, differences are apparent in their look, a conceptualization of a modern FES system is presented in Figure 10.2. These differences are due to advancements which include lower friction through magnetic suspension of the flywheel, advanced components, vacuum or low-pressure containers to reduce friction from air resistance, and a containment system in case of failure.

Kinetic energy, in joules (J) or watt-seconds (Ws), stored in flywheels can be expressed as

$$W_{KE} = \frac{1}{2} J \omega^2$$

where the rotational speed of the flywheel is defined as ω, in radians per second, and J is the inertia of the flywheel in kilogram, meters squared [22, 23]. The

Figure 10.1 Flywheel load equalizer from the early 1900s, adapted from [6]

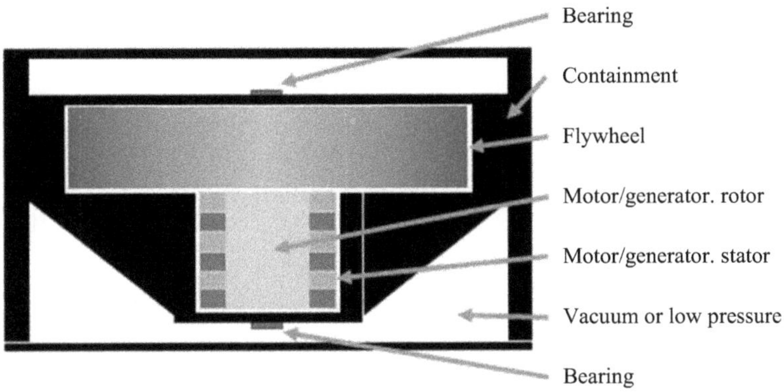

Figure 10.2 Generic representation of a flywheel energy storage (FES) system, adapted from [21, 7]

computation for inertia then involves the shape, mass, and physical properties of the flywheel and its material [7].

10.2.1 Design considerations

Design considerations of FES systems include all aspects, such as materials, flywheel shape, containment, and bearings. Thus, developing long and long-lasting FES systems has been a consistent research theme with much work in advanced materials, cf. [24, 25, 7], and advanced bearings to limit mechanical contacts and points of friction, e.g., by magnetic levitation of the flywheel [8].

10.2.1.1 Material considerations

Given the general kinetic energy relationship for flywheels, to increase the energy stored in a given flywheel, one can either increase the rotational speed or increase the mass. However, structural properties of materials and the flywheel shape should be considered, and one is thus interested in the maximum energy density with respect to volume, in J/m^3,

$$W_v = K\sigma$$

and maximum energy density with respect to mass, in J/kg,

$$W_m = \frac{K\sigma}{\rho}$$

where K is a shapefactor, ρ is the mass density, and σ is the strength (maximum stress) of the material [7]. Thus, two characteristics are critical in determining energy density of a flywheel: its shape of and the material properties [16].

A variety of materials have been considered or proposed for flywheel construction in FES systems. Table 10.1 presents various materials of possible interest and was compiled from [24, 25, 7]. Energy densities were computed in Table 10.1 by assuming a maximum possible shape factor of 1.0; thus, the results in Table 10.1

Table 10.1 Theoretical flywheel material comparisons, ordered by strength, data from[a–f]. Maximum energy density computed with K = 1.0

Material	Density (kg/m^3), ρ	Strength (MPa), σ	Maximum energy density (MJ/kg)
Birchwood[f]	650	137	0.21
Mild steel[f]	7,800	340	0.04
Standard electrical steel[f]	8,000	400	0.05
Aluminum alloy (AlMnMg)[a]	2,700	600	0.22
Fiber-reinforced plastic (E-Glass/EP 60%)[f]	2,200	960	0.44
Alloy special steel ($42CrMo_4$)[f]	7,800	1,100	0.14
Titanium ($TiAl_6Zr_5$)[a]	4,500	1,200	0.27
Metal matrix composite[f]	3,300	1,450	0.44
Carbon-fiber composite (S2)[a]	1,920	1,470	0.766
Carbon-fiber composite (AS4C)	1,510	1,650	1.093
Steel (AICI 4340)[a]	7,800	1,800	0.231
Carbon-fiber composite (M60J)[f]	1,500	2,010	1.34
Carbon-fiber composite (M30S)[a]	1,553	2,760	1.777
Vapor-grown carbon nanofibers[d,e]	2,000	2,920	1.46
Spectra 1000[b]	968	3,000	3.099
E-Glass[b]	2,546	3,102	1.219
Carbon-fiber composite (T1000G)[a]	1,664	3,620	2.175
Kevlar 49[b]	2,546	3,620	1.422
Technora T220[b]	1,384	4,123	2.979
T-700 Graphite[b]	1,799	4,826	2.682
T-1000 Graphite[b]	1,799	6,895	3.833
Single-wall carbon nanotubes (low end)[d,e]	1,300	50,000	38.462
Single-wall carbon nanotubes (high end)[d,e]	1,300	500,000	384.615
Multiwalled carbon nanotubes (low end)[d,e]	1,750	10,000	5.714
Multiwalled carbon nanotubes (high end)[d,e]	1,750	60,000	34.286

[a]from [24]
[b]from [25]
[c]from [7]
[d]from [18]
[e]from [26]
[f]from [27]

are optimistic. Additionally, Table 10.1 includes materials under development, e.g., carbon nanotubes, with potential or proposed applications to FES systems. Despite the theoretical advantage many materials might have, risks exist for flywheel failures in composite materials, and some large-scale FES systems have worked to reduce friction and losses in steel flywheels FES systems, e.g., Amber Kinetics.

A variety of shape factors have been considered as seen in [7], ranging from 0 to 1.0. Shape factors depend on the homogeneity of the material and assumptions about height to diameter of the flywheel [7]. For example, for homogenous materials where the height of the flywheel is small compared with the diameter, general shape factors would be 1.0 for a disc and 0.305 for a flat bar with a hole in the middle [7]. Additionally, three-dimensional interactions should be considered in flywheel shape design [7]. Notably, three-dimensional interaction issues extend to spoke design, as seen in the simulation work of [28].

10.2.1.2 Bearing considerations

Both mechanical and magnetic bearings have been developed for FES systems. One limitation in the development of FES systems was the reliance on mechanical bearing until research into viable magnetic bearings was conducted in the 1970s and 1980s [8]. While mechanical bearings have advantages in being known technologies, they require lubrication/service, have losses due to friction, and might require active cooling systems [18, 29]. Magnetic bearings have advantages over mechanical bearings in that they (1) reduce losses by having few mechanical contacts and points of friction and (2) allow high-speed flywheel rotations. Additionally, magnetic bearings function better in vacuum containment system than mechanical bearings [18]. However, magnetic bearings have less mature standards and are not always ideal and require (1) active control and (2) backup systems in case of failure [18, 29].

10.2.1.3 Motor/generator considerations

FES systems employ an electric machine, a combined motor/generator which has two purposes: (1) acting as a motor to spin the flywheel by drawing power from the supply and (2) acting as a generator to extract energy from the flywheel by using its rotation to generate electricity [15]. Some options and considerations exist in selecting the type of electric machine for FES systems: asynchronous, permanent magnet synchronous machines (PMSM), and variable reluctance machines (VRMs) [15]. Asynchronous machines might be preferred for high-power applications, but these are not appropriate for FES systems operating in a vacuum due to cooling requirements because of rotor losses [15]. PMSMs have no rotor losses and can operate in a vacuum, but these bring a risk of accidental demagnetization [15]. VRMs are preferable for many applications since they can operate in a vacuum (low rotor losses) and they do not have demagnetization concerns; however, they are slightly more expensive than asynchronous machines [15].

10.2.1.4 Power electronics

Power electronics concerns are dictated by motor/generator considerations and specifications [16]. Many different topologies are possible, e.g., DC–AC, AC–AC, AC–DC–AC, etc., and depend on tradeoffs between the end use, efficiencies desired, and the energy source [16]. For example, a DC–AC system might be desirable for a FES system connected between a photovoltaic source or wind source and the grid [30].

10.2.1.5 Containment

The highest risk in a flywheel system is due to rotor failure [10]. Since the flywheel is likely rotating at high speeds, failures can produce dangerous projectiles and containment is a serious concern for FES systems. Due to the amount of energy involved, failures can result in damages, injuries, and deaths, as seen at quantum energy storage in 2015 [31]. Containment is needed to: (1) reduce risks from flywheel failures, but also for (2) provide vacuum, or appropriate low-pressure conditions frequently needed for high-speed flywheels, and (3) general housing and mounting of the FES system components [24].

Failure modes differ in rotors depending on the material, e.g., metal or composite, with metal rotors being more predictable due to metals being homogeneous isotropic materials [10]. Composite materials are by nature anisotropic and failures can be unpredictable [10]. However, some general behaviors are known: metal rotors tend to fail at machined or welded areas while composite rotors tend to fail due to delamination or disintegration (bursts) [10]. Composite failures can also produce highly flammable carbon dust [10]. Containment systems have been developed to provide traps, e.g., honeycomb layers, to collect dust and particles, combinations of layers to absorb energy in the event of a failure, and multiple chambers to provide for vacuums and failure protection [10].

10.2.1.6 High-speed versus low-speed FES system considerations

Although various advantages exist with an FES system using the most advanced materials and components, these increase costs and complexity. Due to tradeoffs in design relative to cost and complexity, high-speed (>10,000 rpm) operations are not viable for all systems and FES systems can be roughly divided into low-speed and high-speed devices [32]. Low-speed FES systems generally use conventional materials, e.g., steel or alloy, and are used for terrestrial FES systems in power grids and employ large-sized flywheels [32]. High-speed FES systems have generally been used in aerospace applications, are smaller in size than low-speed FES systems, and employ advanced materials and vacuum containers due to the speeds employed [32]. Due to the speeds involved, high-speed FES systems tend to operate in absolute vacuums to reduce friction due to air; however, low-speed FES systems can operate in partial vacuums or use alternative gases to reduce, but do not eliminate friction due to air/gas [32]. Although high-speed FES systems are generally used for aerospace or transportation applications, as technologies developed for high-speed FES systems become more mature, these can be expected to be adopted for grid applications, e.g., the UPS system developed in [33]. General characteristics of both high- and low-speed FES systems are presented in Table 10.2 as per discussions in [15, 32, 17].

10.2.2 Comparisons with alternative energy storage system

A variety of measures can be used to compare energy storage systems, price, storage duration, power/energy ratings, etc. One basic approach to comparing competing energy storage systems is via a “Ragone plot” which considers energy density and power density [35, 36]. Figure 10.3 provides a Ragone plot for flywheels in comparison with batteries and capacitors, adapted from [37]. An ideal energy storage system would operate with high energy density abilities and high power density abilities, i.e., the upper right of Figure 10.3. In general, some comparison between storage methods can be made with this plot; for example, batteries are seen to provide high energy density, but relatively low power density when compared to other methods [37]. Thus, batteries are good for storing energy but not quickly discharging it [37]. Comparatively, flywheels can transfer energy quicker than batteries, but they cannot store as much energy [37].

Table 10.2 General differences between low-speed FES systems and high-speed FES systems, adapted from [15]

Attribute	Low-speed FES system	High-speed FES system
Rotation speed[a,b]	<6,000 rpm	10,000–100,000 rpm
Materials[a–c]	Traditional (Steel, alloys, etc.)	Advanced (composites, etc.)
Applications[a–c]	Terrestrial, power grids	Aerospace, transportation
Bearings[a–c]	Mechanical, magnetic, or hybrid	Magnetic
Generator/motor[a–c]	Asynchronous, permanent magnet synchronous, and reluctance	Permanent magnet synchronous, and reluctance
Confinement[a,b]	Vacuum, partial vacuum or light gas	Absolute vacuum
Integration of flywheel with generator/motor[a,b]	None or partial	Full or partial
Containment weight[a,d]	2 × flywheel weight	½ × flywheel weight
Relative cost[b]	1	5 times the cost of comparable low-speed FES

[a]from [15]
[b]from [32]
[c]from [17]
[d]from [34]

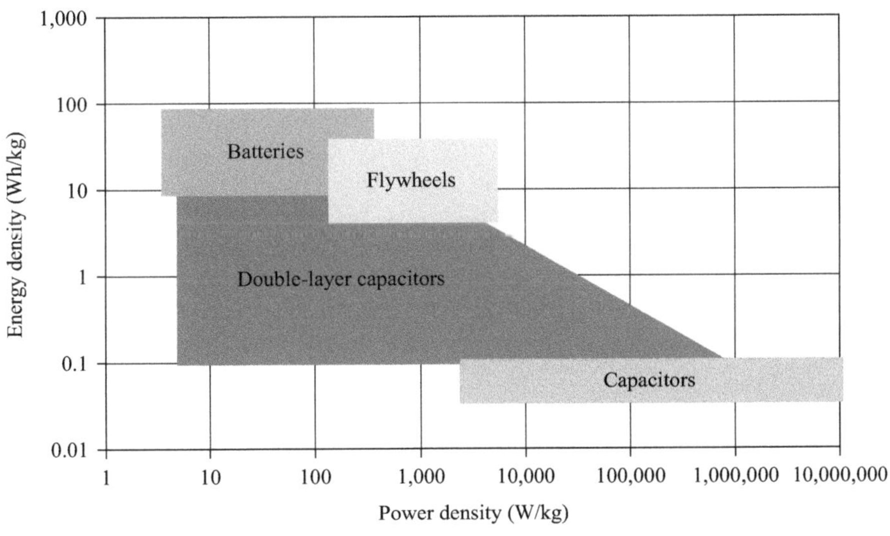

Figure 10.3 Comparison of energy storage systems by energy and power density, adapted from [37]

Table 10.3 Qualitative comparison of energy storage methods, adapted from [27, 41–43]

Column heads	Li-Ion battery	Pb battery	Ni-Cd battery	Capacitors	FES
Cost	**	**	**	*	*
Energy density	***	**	***	*	*
Power density	**	**	***	***	***
Standby losses					*
Lifetime/aging	**	**	*	**	***
Efficiency	***	***	***	***	**
Charging time	*	*	**	***	***
Environmental concerns	*	*	*	***	***
Manufacturing effort	*	**	*	*	**

*is fair,
**is good,
***is excellent

Beyond, energy density and power density, considerations of additional factors are necessary in comparing energy storage systems. Such factors include reliability, safety, temperature and humidity limitations, and self-discharge [38]. Table 10.3 presents a further qualitative comparison of FES systems versus competing energy storage systems, adapted and extended from [27]. Details seen in Figure 10.3 are also reflected in Table 10.3 whereby flywheels and capacitors are seen to have higher power densities, but lower energy densities, than batteries. As presented in [27], advantages of FES systems are seen where aging effects, whereby repeated charging/discharging cycles reduce the abilities of an energy storage system, need to be minimized, a low charging time is desirable and/or high power density is needed. Additionally, battery technologies have known issues with environmental protection and disposal concerns, which are minimal in FES and capacitor systems [39].

However, FES systems have deficiencies with respect to standby losses, thus the motor/generator is needed to consistently keep the flywheel at speed [40]. Additionally, FES systems generally cost more than competing energy storage technologies; however, this is a notable changing as FES systems become more mature, cf. \$1,000–\$5,000 per kWh and \$250–\$350 per kW in 2009 [41] versus \$200–\$150,000 per kWh and \$30–\$700 per kW in 2015 [42].

10.2.3 Sizing considerations

When integrating FES systems into a grid, appropriately sizing the system is important. In general, sizing considerations for energy storage include considerations for both the generation methods and the storage methods [44, 45]. To accurately size FES systems, accurately modeling the discharge rates in conjunction with costs can determine the appropriate amount of FES systems to incorporate [46]. Optimizing sizing methods have been developed for sizing [47]; and specifically, for FES systems, optimal control law methods have been proposed for sizing [48]. Additionally, total lifecycle costs should be considered for an effective sizing and storage decision [49].

10.2.4 Available FES systems

FES systems are produced by a variety of manufacturers for different purposes, as detailed in Table 10.4. Table 10.4 extends upon the work of [18] with additional companies and condensed FES system applications. In general, companies produce FES systems for specific purposes (transportation, UPS, and microgrids) and focus on applications suitable for these domains. Additional details could be gathered by examining individual companies and comparing for a specific application. The list in Table 10.4 also excludes entities (research laboratoties, companies and universities) engaged in FES system research since such entities might produce prototypes, but not commercially available units. Additionally, companies that are "crowdfuned" and companies that appeared to be discussion groups were not included, e.g., Velkess and Vox Solaris, since it was not possible to determine if they have commercially produced or plan to produce any units. Notably, Table 10.4 shows 39 manufacturers active in 2017, a significant increase over the ten mentioned in 2002 [50].

10.2.5 FES system research

It is naturally hard to determine what groups conduct research in FES systems. Authors in [18] provided a short list of universities known to conduct research into flywheels; however, this list was subjective and only involved examining affiliations in recent journals and thesis. To gain a larger view of the universities, one could examine a repository of known publication types, e.g., ProQuest's database of dissertations, or one could text mine or laboriously examine affiliations based on journal/conference papers, theses/dissertations, and technical reports. For brevity, and to generate some understanding of the universities where research, ProQuest was searched to determine the quantity of FES related dissertations. Although this approach excludes both research organizations, e.g., US Air Force Research Laboratory [51], and corporate research, this approach is more systematic and objective than the approach used by [18].

A ProQuest search on "flywheel," yielded 3,908 PhD dissertations granted by 455 institutions between 1924 and 2017 (2,365 by 398 universities between 2000 and 2017 alone). To narrow the scope, this search was constrained to "flywheel energy storage," which yielded 277 PhD dissertations granted by 122 institutions between 1954 and 2017. Figure 10.4 presents a Pareto chart of universities based on the number of FES-related dissertations and Figure 10.5 presents a Pareto chart of countries based on the number of FES-related dissertations produced at their degree-granting institutions.

In Figure 10.4, institutions with 1–4 dissertations on FES-related research between 1954 and 2017 were grouped into subgroups based on the quantity, i.e., 68 universities had only 1 dissertation on FES-related research in this time period. These subgroups are at the end (right side of the Pareto chart); to compute the

Table 10.4 Flywheel energy storage (FES) system manufacturers, adapted and extended from [18]

Company	FES system applications	Country
ABB	Microgrids	Switzerland
Active Power & Piller Group GmbH	UPS, active power acquired by Piller in 2016	Germany (Piller), US (Active Power)
Acumentrics	UPS	The US
Adelwitz Technologiezentrum (ATZ)	Transportation	Germany
AFS Trinity	Transportation	The US
Alstom	Transportation	France
Amber Kinetics Inc.	Microgrids	The US
American Flywheel Systems Inc.	Aerospace	The US
Applied Materials Technology Inc.	Aerospace	The US
Atrex Energy, Inc.	UPS	The US
Beacon Power, LLC	Large grids	The US
Boeing	Aerospace and mobile power	The US
Calnetix Technologies, LLC	UPS, grids	The US
Centre for Concepts in Mechatronics (CCM)	Transportation	The Netherlands
Caterpillar	UPS	The US
Dow-UT, Inc.	Aerospace	The US
Elytt	Transportation	Spain
Flybrid Systems	Transportation	The UK
Flybus	Transportation	The UK
Flywheel Energy Systems	Multiple	Canada
Foster-Miller Inc.	Aerospace	The US
General Electric	UPS	The US
GKN Hybrid Power	Transportation	The US
Kinetic Traction Systems	Power quality, train stations	The US
LaunchPoint Technologies	UPS, train stations, military	The US
Oerlikon	GYROBUS in 1950s	Switzerland
ORKESS	Grid	Germany
Pentadyne Power Corporation	UPS	The US
PowerThru	UPS	The US
Power Tree	Aerospace and grids	The US
Quantum Energy Storage	Microgrids	The US
Ricardo	Excavators	The UK
Rotonix USA, Inc.	UPS, power quality, grids	The US
Satcon Technology Corporation	Transportation and space applications	The US
Seakeeper	Ship stabilization	The US
Socomec	UPS	France
Stephentown Spindle LLC	Grid/integration	The US
Stornetic GmbH	Grid	Germany
Temporal Power Ltd	Grids	Canada
Texas Bus	Power buffers	The US
Traxler, mecos	UPS	Switzerland
U.S. Flywheel Systems	Transportation	The US
Vycon & Calnetix	Train stations, cranes	The US
Williams Advanced Engineering	Power quality	The UK

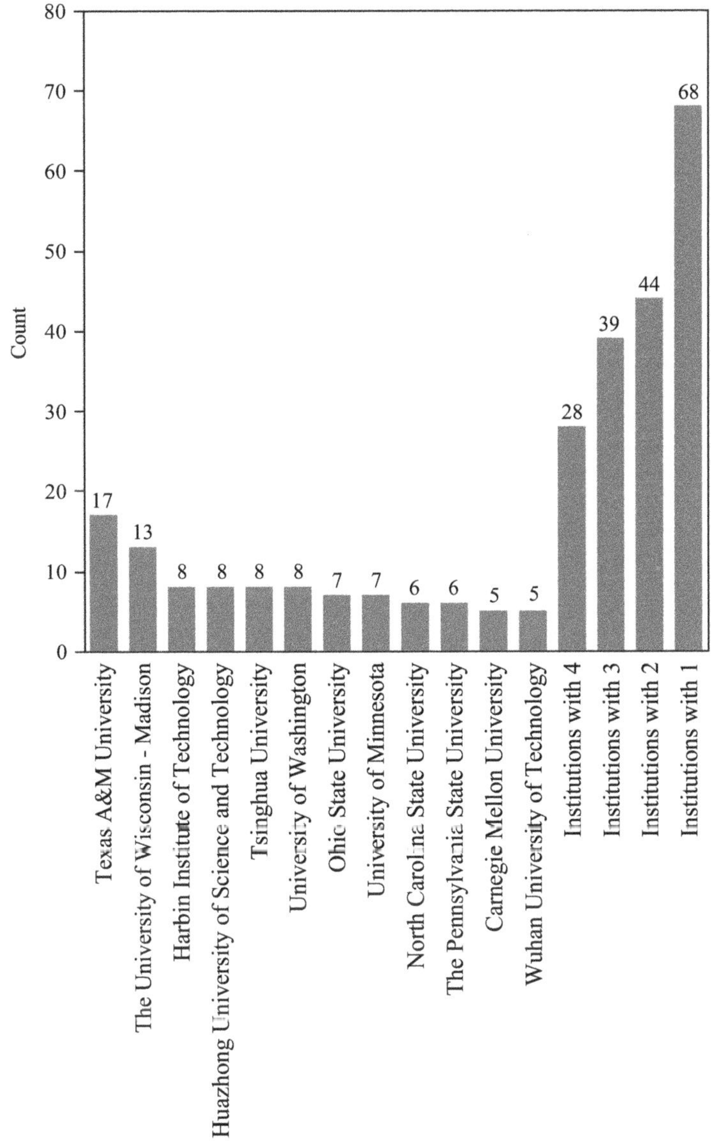

Figure 10.4 Pareto chart of universities associated with flywheel-related research

number of universities in these subgroups, divide the quantity by the subgroup size (for example, 39/3 equals 13 universities producing 3 PhD dissertations on FES research). As seen in Figure 10.4, while the vast majority of FES research appears to be individual efforts or small research groups, based on small quantities, but a few universities have produced a large number of FES research, e.g., the University

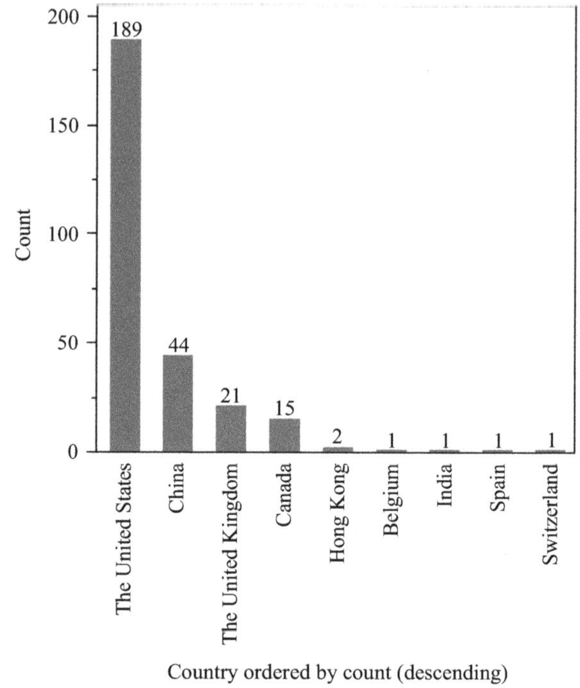

Figure 10.5 Pareto chart of countries based on the quantity of flywheel energy storage-related research (based on the number of PhD dissertations)

of Texas at Austin and the University of Wisconsin at Madison. From here, one could further look at years of activity and types of research conducted, as seen in [18]. Figure 10.5 presents a similar Pareto chart of the countries where PhDs were granted for "FES"-related topics. Notably, PhDs were mostly granted by institutions in the United States, followed by China, the United Kingdom (including Northern Ireland and Wales) and Canada.

10.3 Applications and research of flywheels energy storage systems

Flywheel applications and considerations exist from generation to the end-user. While flywheel considerations are integral to engine design, cf. [52–54], of interest herein are applications in the transmission and distribution (T&D) system and those by end-users. Additionally, tangential applications and research areas should also be considered since developments in these areas can possibly be leveraged for FES systems.

10.3.1 Uninterruptible power supply

UPSs are systems which provide emergency, backup, power for outages and prevent power quality issues [55, 56]. Generally, UPSs sit between the power grid and the devices, such as computers for small UPSs, or facilities, for large UPSs, they are protecting and wait unit needed [55, 56]. Many UPSs systems employ batteries; however, aging of batteries is a concern for UPSs since aged batteries last for shorter amounts of time after multiple charging/discharging cycles [57]. Confidence in the availability and reliability of UPS systems can be critically important since some are deployed in critical infrastructure applications, e.g., hospital emergency rooms [58]. While flywheel UPS cannot hold a charge for months or years, as a battery can, flywheel UPS systems have a significant advantage over batteries since aging effects are minimal [50]. Additionally, since most power disturbances in the United States last for less than 1 s, the short storage time of flywheels might not be a concern since many would not be expected to perform constantly [50].

10.3.2 Power quality

FES UPS systems have found a use in improving power quality since at the early 1900s, cf. [6, 59]. Since power quality is related to generator rotor speeds, adding virtual or physical machines to a grid has been proposed for more control over power quality [60, 61]. In this role, FES systems, e.g., flywheel UPSs, have been proposed for integration in grids for power quality improvement [61, 62, 59].

One example implementation of FES systems for power quality management is the 20 MW Stephentown Advanced Energy Storage plant in Stephentown, NY, USA, conceptualized in Figure 10.6. The Stephentown Advanced Energy Storage plant used 200 Beacon Power's Gen4 FES systems for storage and was the first (project began in March 2012) commercial FES plant solely dedicated to power

Figure 10.6 Conceptualization of the Stephentown Advanced Energy Storage Plant, from [63]

quality [63]. Lessons learned from the Stephentown Advanced Energy Storage plant were used in developing the Hazle FES plant in Hazle, PA, USA, which came online at 20 MW in 2014 [64]. Additional FES plants are also in development.

10.3.3 Integration with renewable energy systems

FES systems have been specifically proposed for load leveling with renewables. Since renewable energies, e.g., wind and tide, are intermittently available, effective energy storage is a consideration in expanding the application of renewable energy sources [65–67, 17]. FES systems have thus been proposed for integration with renewable energy systems to provide for energy when demanded and with short notice [68]. Additionally, FES systems could provide an extra role in renewable integration by adjusting for the variable power generation of wind energy [40] and solar [27].

10.3.4 Energy harvesting

Energy harvesting considers extracting ambient energy from the environment [69]. In general, energy harvesting includes a variety of methods, such as solar, piezoelectric, vibrations, thermoelectric, electromagnetic (magnetic field and radio frequency), and wind [69–75]. Largely, energy harvesting is a concern for remote applications whereby collecting energy is advantageous rather than relying on batteries, which must be replaced.

Recent research has examined flywheels mounted in ships to serve as energy harvesters, whereby flywheels are used to stabilize a ship and extract and store energy from waves using gyroscopic precession [76].

Gyroscopic precession, where the motion of a platform results in an input torque and variation in the spin of the flywheel, has been used for flywheel-based attitude control [76]. Thus, the wave energy harvesting approach of [76] essentially reverses the concepts used in spacecraft control moment gyros whereby the flywheels in the gyros are used to rotate and control the attitude of a satellite [77].

10.3.5 Aerospace flywheel systems

Space-based attitude control has employed flywheel systems since the 1980s for attitude control [10]. FES systems are desirable for space systems due to their long-term stability [10]; this is logical since charge/discharge cycles of energy storage systems can occur frequently, as seen in [78]. Additional efforts have also focused on vibration and attitude control in conjunction with FES and release [79].

Flywheels are also employed for attitude control, an important consideration in space systems to ensure that the satellite is directed precisely for operation, with control moment gyros being one flywheel-based system for attitude control [10, 80]. To reduce the space and weight used in a satellite, combined energy storage and attitude control systems are also of considerable interest [51, 10]. With further developments in this area, various proposed large-scale space power systems, see [81–84], could feasibly incorporate combined flywheel attitude control and energy storage.

10.3.6 Tangential applications

10.3.6.1 Building motion and vibration control

Motion control, i.e., from wind and vibrations, is a major concern in tall buildings with passive, active and hybrid approaches in use [85]. Although not a heavily used method, flywheels (active gyroscopic stabilizer systems) have been proposed for vibration and motion control in buildings [86] with experimental demonstration on the HRHE Tower in Tokyo by [87] pioneering the use of flywheels for building gyro-stabilization methods. Complications in flywheels for stabilization include controller development [88], and that the angular velocity of the flywheel has a feedback effect on the structure it is stabilizing [89]. However, properly designed flywheel systems for building control could incorporate the turning moment of the flywheels to create systems that are more compact than competing building motion control systems [90, 85].

10.3.6.2 Transportation applications

Beyond storing energy and damping vibrations from the engine [54, 91], FES systems have found various uses in automotive applications. The first commercial use of a flywheel powered vehicle was the 70 passenger GYROBUS was an early (1950s) electric vehicle developed by the Oerlikon company in the 1950s [92, 93] and in use for 17 years in Switzerland and South Africa [94]. The GYROBUS operated by employing a flywheel to store energy to power the bus between stops, whereby at each bus stop a connection was made to the electric grid to spin the flywheel up to speed [92]. However, the GYROBUS was limited by a 2-min charge time at each stop and the technology [94].

Limitations in batteries and recent interest in electric vehicles have pushed toward renewed interest in flywheels in vehicles [95]. Additionally, FES systems have been proposed for high-speed trains to provide for acceleration, speed control on grades, and energy harvesting from braking [96].

Limitations of flywheels for energy storage in vehicles include gyroscopic aspects of a large spinning mass which change the driving dynamics [7]. However, this can be mitigated by placing the flywheel in a gimbal system or operating flywheels in pairs to cancel any net torque [7, 50].

10.3.6.3 Catapult launching systems

The Electromagnetic Aircraft Launch System (EMALS) was developed to provide more control and smoother operation than conventional steam system in launching aircraft from aircraft carriers [97, 98]. FES systems were included in EMALS to replace steam accumulators on the US Navy aircraft carriers to take advantage of a comparatively reduced charge time, more control over energy used to launch aircraft, and reduced space [50, 97–99]. However, the EMALS system is still under development and study since the version used on the USS Gerald R. Ford has an ability to overstress the frames of some aircraft [100].

10.4 Conclusions

This chapter reviews FES systems, their design consideration, implementations, and research areas. Attention was paid to research in areas outside of energy storage to point researchers to other domains, e.g., aerospace and vibration control, which might have research efforts in flywheels systems that would be leveraged for FES systems.

References

[1] Genta, G. (2014). *Kinetic energy storage: Theory and practice of advanced flywheel systems*. London: Butterworth-Heinemann.

[2] Gasche, F. G. (1910). Interaction of flywheels and motors when driving roll trains by induction motors. *Proceedings of the American Institute of Electrical Engineers*, *29*(6), 913–930.

[3] Specht, H. C. (1909). Function of fly-wheels in connection with electrically operated rolling mills. *Proceedings of the American Institute of Electrical Engineers*, *28*(6), 633–642.

[4] Fernald, R. H. and Orrok, G. A. (1916). *Engineering of power plants*. New York, NY: McGraw-Hill Book Company.

[5] Xin, Q. (2011). Overview of diesel engine applications for engine system design-Part 3: Operating and design characteristics of different applications. SAE Technical Paper, No. 2011–01–2180.

[6] Peck, J. (1909). Flywheel load equalisers. *Journal of the Institution of Electrical Engineers*, *43*(196), 174–191.

[7] Bolund, B., Bernhoff, H., and Leijon, M. (2007). Flywheel energy and power storage systems. *Renewable and Sustainable Energy Reviews*, *11*(2), 235–358.

[8] Kirk, J. A. and Studer, P. A. (1977). Flywheel energy storage—II: Magnetically suspended superflywheel. *International Journal of Mechanical Sciences*, *19*(4), 233–245.

[9] Dunn, M. (2011). Start it up: Flywheel energy storage efficiency. *National Consortium for Specialized Secondary Schools of Mathematics, Science and Technology (NCSSSMST) Journal*, *16*(2), 30–33.

[10] Babuska, V., Beatty, S. M., and Fausz, J. L. (2004). A review of technology developments in flywheel attitude control and energy transmission systems. *Aerospace Conference*, 2784–2800.

[11] Crisp, J. N., Pinson, J. D., Anderson, L. A., and Bishop, W. S. (1979). *Analysis of remote site energy storage and generation systems*. Dayton, OH: University of Dayton.

[12] Fiske, O. J. and Ricci, M. R. (2008). Third generation flywheels for high power electricity storage. *Proceedings of the 19th International Conference on Magnetically Levitated System and Linear Drives*, 553–559.

[13] Post, R. and Post, S. (1973). Flywheels. *Scientific American*, *229*(6), 17–23.

[14] Carrasco, J., Franquelo, L., *et al.* (2006). Power-electronic systems for the grid integration of renewable energy sources: A survey. *IEEE Transactions on Industrial Electronics*, *53*(4), 1002–1016.

[15] Pena-Alzola, R., Sebastian, R., Quesada, J., and Colmenar, A. (2011). Review of flywheel based energy storage systems. *International Conference on Power Engineering, Energy and Electrical Drives (POWERENG)*, 1–6.

[16] Amiryar, M. E. and Pullen, K. R. (2017). A review of flywheel energy storage system technologies and their applications. *Applied Sciences*, *7*(3), 1–21.

[17] Khodadoost Arani, A., Karami, H., Gharehpetian, G. B., and Hejazi, M. S. (2017). Review of flywheel energy storage systems structures and applications in power systems and microgrids. *Renewable and Sustainable Energy Reviews*, *69*, 9–18.

[18] Hedlund, M., Lundin, J., Santiago, J. D., Abrahamsson, J., and Bernhoff, H. (2015). Flywheel energy storage for automotive applications. *Energies*, *8*(10), 10636–10663.

[19] Wicki, S. and Hansen, E. (2017). Clean energy storage technology in the making: An innovation systems perspective on flywheel energy storage. *Journal of Cleaner Production*, *162*, 1118–1134.

[20] Arani, A., Karami, H., Gharehpetian, G., and Hejazi, M. (2017). Review of flywheel energy storage systems structures and applications in power systems and microgrids. *Renewable and Sustainable Energy Reviews*, *69*, 9–18.

[21] Pjrensburg (2012, April 29). *A rendering from a solid-works model, edited to include labels, in png format.* Retrieved from https://commons.wikimedia.org/w/index.php?curid=19258681

[22] Diaz-Gonzalez, F., Sumper, A., and Gomis-Bellmunt, O. (2016). *Energy storage in power systems.* West Sussex: John Wiley & Sons.

[23] Shepherd, W. and Shepherd, D. W. (2014). *Energy studies.* London: Imperial College Press.

[24] Long, Z. and Zhiping, Q. (2007). Review of flywheel energy storage system. *Proceedings of ISES World Congress*, 2815–2819.

[25] Bitterly, J. G. (1998). Flywheel technology: Past, present, and 21st century projections. *IEEE Aerospace and Electronic Systems Magazine*, *13*(8), 13–16.

[26] Al-Saleh, M. H. and Sundararaj, U. (2011). Review of the mechanical properties of carbon nanofiber/polymer composites. *Composites Part A: Applied Science and Manufacturing*, *42*(12), 2126–2142.

[27] Buchroithner, A., Haan, A., Preßmair, *et al.* (2016). Decentralized low-cost flywheel energy storage for photovoltaic systems. *International Conference on Sustainable Energy Engineering and Application (ICSEEA)*, 41–49.

[28] Dasseh, M., Samer, Y., and Webb, J. F. (2015). A study on maximizing the energy density of a system by choosing a suitable flywheel. *Global Journal of Technology and Optimization*, *6*(1), 1–6.

[29] McMullen, P. and Hawkins, L. (2012). Long term backup bearing testing results. *International Symposium on Magnetic Bearings*, 1–11.

[30] Boukettaya, G. and Krichen, L. (2014). A dynamic power management strategy of a grid connected hybrid generation system using wind,

photovoltaic and flywheel energy storage system in residential applications. *Energy*, *71*, 148–159.

[31] California Department of Industrial Relations, Cal/OSHA (2015, Dec. 9). *Cal/OSHA Cites Poway Energy Technology Company $58,025 for Industrial Flywheel Explosion that Injured Four Workers*. Retrieved from PR Newswire: http://www.prnewswire.com/news-releases/calosha-cites-poway-energy-technology-company-58025-for-industrial-flywheel-explosion-that-injured-four-workers-300190913.html

[32] Emadi, A., Nasiri, A., and Bekiarov, S. B. (2004). *Uninterruptible power supplies and active filters*. Boca Raton, FL: CRC Press.

[33] Werfel, F. N., Floegel-Delor, U., Riedel, T., *et al.* (2010). Towards high-capacity HTS flywheel systems. *IEEE Transactions on Applied Superconductivity*, *20*(4), 2272–2275.

[34] Sapowith, A. and Handy, W. (1982). *A composite flywheel burst containment study*. Report number: No. UCRL-15452; AVSD-0350-81-RR. Lawrence Livermore National Laboratory.

[35] Christen, T. and Carlen, M. W. (2000). Theory of Ragone plots. *Journal of Power Sources*, *91*(2), 210–216.

[36] Ragone, D. (1968). Review of battery systems for electrically powered vehicles. SAE Technical Paper, No. 680453.

[37] Kuseian, J. (2013). *Naval power systems technology development roadmap*. United States Navy, PMS320, Electric Ships Office.

[38] Scherson, D. A. and Palencsár, A. (2006). Batteries and electrochemical capacitors. *Interface*, *15*(1), 17–22.

[39] Ma, Y., Yang, P., Zhou, X., and Gao, Z. (2016). Research review on energy storage technology. *International Conference on Mechatronics and Automation*, 159–164.

[40] Cimuca, G. O., Saudemont, C., Robyns, B., and Radulescu, M. M. (2006). Control and performance evaluation of a flywheel energy-storage system associated to a variable-speed wind generator. *IEEE Transactions on Industrial Electronics*, *53*(4), 1074–1085.

[41] Chen, H., Cong, T. N., Yang, W., Tan, C., Li, Y., and Ding, Y. (2009). Progress in electrical energy storage system: A critical review. *Progress in Natural Science*, *19*, 291–312.

[42] Sabihuddin, S., Kiprakis, A. E., and Mueller, M. (2015). A numerical and graphical review of energy storage technologies. *Energies*, *8*, 172–216.

[43] Masaud, T., Lee, K., and Sen, P. (2010). An overview of energy storage technologies in electric power systems: What is the future? *North American Power Symposium (NAPS)*, 1–6.

[44] Masaud, T., Oyebanjo, O., and Sen, P. (2017). Sizing of large-scale battery storage for off-grid wind power plant considering a flexible wind supply–demand balance. *IET Renewable Power Generation*, *11*(13), 1625–1632.

[45] Berrada, A. and Loudiyi, K. (2016). Operation, sizing, and economic evaluation of storage for solar and wind power plants. *Renewable and Sustainable Energy Reviews*, *59*, 1117–1129.

[46] Walawalkar, R., Apt, J., and Mancini, R. (2007). Economics of electric energy storage for energy arbitrage and regulation in New York. *Energy Policy*, *35*(4), 2558–2568.

[47] Okinda, V. and Odero, N. (2015). A review of techniques in optimal sizing of hybrid renewable energy systems. *International Journal of Research in Engineering and Technology*, *4*(11), 153–163.

[48] Hearn, C., Lewis, M., Pratap, S., *et al.* (2013). Utilization of optimal control law to size grid-level flywheel energy storage. *IEEE Transactions on Sustainable Energy*, *4*(3), 611–618.

[49] Zakeri, B. & Syri, S. (2015). Electrical energy storage systems: A comparative life cycle cost analysis. *Renewable and Sustainable Energy Reviews*, *42*, 569–596.

[50] Hebner, R., Beno, J., and Walls, A. (2002). Flywheel batteries come around again. *IEEE Spectrum*, *39*(4), 46–51.

[51] Lappas, V., Richie, D., Hall, C., Fausz, J., and Wilson, B. (2009). Survey of technology developments in flywheel attitude control and energy storage systems. *Journal of Guidance, Control, and Dynamics*, *32*(2), 354–365.

[52] Andrews, L. and Porter, R. (1909). The use of large gas engines for generating electric power. *Journal of the Institution of Electrical Engineers*, *43*(196), 3–40.

[53] Cummins, C. L. (1976). Early IC and automotive engines. SAE Technical Paper, No. 760604.

[54] Fonte, M., Anes, V., Duarte, P., Reis, L., and Freitas, M. (2015). Crankshaft failure analysis of a boxer diesel motor. *Engineering Failure Analysis*, *56*, 109–115.

[55] Racine, M. S., Parham, J. D., and Rashid, M. H. (2005). An overview of uninterruptible power supplies. *Proceedings of the 37th Annual North American Power Symposium*, 159–164.

[56] Gurrero, J. M., García De Vicuña, L., and Uceda, J. (2007). Uninterruptible power supply systems provide protection. *IEEE Industrial Electronics Magazine*, *1*(1), 28–38.

[57] Sauer, D. U. and Wenzl, H. (2008). Comparison of different approaches for lifetime prediction of electrochemical systems—Using lead-acid batteries as example. *Journal of Power Sources*, *176*(2), 534–546.

[58] Agbamu, P. O., Menkiti, I. D., Ohuoba, E. I., and Desalu, I. (2017). Critical incidents and near misses during anesthesia: A prospective audit. *Journal of Clinical Sciences*, *14*(1), 18–24.

[59] Richard, T., Belhomme, R., Buchheit, N., and Gorgette, F. (2001). Power quality improvement case study of the connection of four 1.6 MVA flywheel dynamic UPS systems to a medium voltage distribution network. *Transmission and Distribution Conference and Exposition*, 253–258.

[60] Chen, Y., Hesse, R., Turschner, D., and Beck, H. P. (2011). Improving the grid power quality using virtual synchronous machines. *International Conference on Power Engineering, Energy and Electrical Drives (POWERENG)*, 1–6.

[61] Lawrence, R. G., Craven, K. L., and Nichols, G. D. (2003). Flywheel ups. *IEEE Industry Applications Magazine*, *9*(3), 44–50.
[62] Jiancheng, Z., Lipei, H., Zhiye, C., and Su, W. (2002). Research on flywheel energy storage system for power quality. *International Conference on Power System Technology (PowerCon)*, 496–499.
[63] US Department of Energy (2017, Jun. 1). *Stephentown Spindel*. Retrieved from Energy.gov: https://energy.gov/lpo/stephentown-spindle
[64] Arseneaux, J. (2015). *20 MW Flywheel frequency regulation plant (No. DEOE0000200)*. Hazleton, PA: Hazle Spindle LLC.
[65] Hart, E. K., Stoutenburg, E. D., and Jacobson, M. Z. (2012). The potential of intermittent renewables to meet electric power demand: Current methods and emerging analytical techniques. *Proceedings of the IEEE*, *100*(2), 322–334.
[66] Barton, J. P. and Infield, D. G. (2004). Energy storage and its use with intermittent renewable energy. *IEEE Transactions on Energy Conversion*, *19*(2), 441–448.
[67] Angelis-Dimakis, A., Biberacher, M., Dominguez, J., et al. (2011). Methods and tools to evaluate the availability of renewable energy sources. *Renewable and Sustainable Energy Review*, *15*(2), 1182–1200.
[68] Bryden, I. G. and Macfarlane, D. M. (2000). The utilisation of short term energy storage with tidal current generation systems. *Energy*, *25*, 893–907.
[69] Chalasani, S. and Conrad, J. M. (2008). A survey of energy harvesting sources for embedded systems. *In Proceedings of the IEEE Southeastcon*, 442–447.
[70] Lu, X., Wang, P., Niyato, D., Kim, D. I., and Han, Z. (2015). Wireless networks with RF energy harvesting: A contemporary survey. *IEEE Communications Surveys & Tutorials*, *17*(2), 757–789.
[71] Weimer, M. A., Paing, T. S., and Zane, R. A. (2006). Remote area wind energy harvesting for low-power autonomous sensors. *IEEE Power Electronics Specialists Conference*, 1–5.
[72] Moghe, R., Yang, Y., Lambert, F., and Divan, D. (2009). A scoping study of electric and magnetic field energy harvesting for wireless sensor networks in power system applications. *In Proceedings of the IEEE Energy Conversion Congress and Exposition*, 3550–3557.
[73] Blach, N. T., Lake, R. A., and Coutu, R. A. (2016). Design of ferroelectric MEMS energy harvesting devices. *National Aerospace and Electronics Conference (NAECON) and Ohio Innovation Summit (OIS)*, 16–22.
[74] Kumar, R., Subramanyam, G., and Chodavarapu, V. (2016). Energy harvesting flexible regenerative power source for wearable devices. *National Aerospace and Electronics Conference (NAECON) and Ohio Innovation Summit (OIS)*, 129–132.
[75] Bhagavatula, V., Wesson, W., and Rudell, J. C. (2011). Integrated CMOS transceivers applied to defense applications in a wide area radio network for sensor (WARNS) communication. *National Aerospace and Electronics Conference (NAECON)*, 344–347.

[76] Townsend, N. and Shenoi, A. (2012). A gyroscopic wave energy recovery system for marine vessels. *IEEE Journal of Oceanic Engineering, 37*(2), 271–280.

[77] Leve, F. A. and Fitz-Coy, N. G. (2010). Hybrid steering logic for single-gimbal control moment gyroscopes. *Journal of Guidance, Control, and Dynamics, 33*(4), 1202–1212.

[78] Bihl, T., Heidenreich, J., Allen, D., and Hunt, K. (2009). SPECTTRA: A space power system modeling and simulation tool. *7th International Energy Conversion Engineering Conference*, 1–4.

[79] Park, J. (2008). *MIMO active vibration control of magnetically suspended flywheels for satellite IPAC service.* Texas A&M University: PhD Dissertation.

[80] Danielson, C. R., Frank, N. W., and Wilson, B. (2006). *Simulation of flywheel energy storage system.* Albuquerque, NM: Sequoia Technologies.

[81] Drexler, K. and Heppenheimer, T. (1977). On the feasibility of small power satellites. *Journal of Energy, 1*(3), 200.

[82] Mankins, J. C. (1997). A fresh look at space solar power: New architectures, concepts and technologies. *Acta Astronautica, 41*(4–10), 347–359.

[83] Oda, M. (2005). Building a space solar power system. *Proceedings of 2nd International Conference on Recent Advances in Space Technologies (RAST)*, 24–27.

[84] Macauley, M. K. and Shih, J.-S. (2007). Satellite solar power: Renewed interest in an age of climate change? *Space Policy, 23*(2), 108–120.

[85] Morales-Beltran, M. and Teuffel, P. (2013). Towards smart building structures: Adaptive structures in earthquake and wind loading control response–a review. *Intelligent Buildings International, 5*(2), 83–100.

[86] Kareem, A., Kijewski, T. and Tamura, Y. (1999). Mitigation of motions of tall buildings with specific examples of recent applications. *Wind and Structures, 2*(3), 201–251.

[87] Kazao, Y., Takahara, K., Yamada, M., and Sakamoto, S. (1992). Active vibration control of a structure using gyrostabilizers. *First International Conference on Motion and Vibration Control*, 158–163.

[88] Townsend, N. and Shenoi, R. (2014). Control strategies for marine gyrostabilizers. *IEEE Journal of Oceanic Engineering, 39*(2), 243–255.

[89] Ünker, F. and Çuvalci, O. (2015). Vibration control of a column using a gyroscope. *Procedia – Social and Behavioral Sciences, 195*, 2306–2315.

[90] Higashiyama, H., Yamada, M., Kazao, Y., and Namiki, M. (1998). Characteristics of active vibration control system using gyro-stabilizer. *Engineering Structures, 20*(3), 176–183.

[91] Wilson, D. E. (1986). The design of a low specific fuel consumption turbo compound engine. SAE Technical Paper, No. 860072.

[92] The Oerlikon Electrogyro (1955). *Automobile Engineer*, 559–566.

[93] Clerk, R. C. (1964). The utilization of flywheel energy. *SAE Technical Paper, No. 640047.*

[94] Lampe, D. (1974). Superflywheel: The 'battery' that spins. *Popular Mechanics, 142*(5), 124–127, 169.

[95] Olivier, J.-C., Bernard, N., Mendoza, L., and Bourguet, S. (2014). Techno-economic optimization of flywheel storage system in transportation. *Symposium de Génie Electrique*, 1–5.
[96] Thelen, R. F., Herbst, J. D., and Caprio, M. T. (2003). A 2 MW flywheel for hybrid locomotive power. *IEEE Vehicular Technology Conference*, 3231–3235.
[97] Doyle, M., Sulich, G., and Lebron, L. (2000). The benefits of electromagnetically launching aircraft. *Naval Engineers Journal*, *112*(3), 77–82.
[98] Doyle, M. R., Samuel, D. J., Conway, T., and Klimowski, R. R. (1995). Electromagnetic aircraft launch system–EMALS. *IEEE Transactions on Magnetics*, *31*(1), 528–533.
[99] Swett, D. W. and Blanche, J. G. (2005). Flywheel charging module for energy storage used in electromagnetic aircraft launch system. *IEEE Transactions on Magnetics*, *41*(1), 525–528.
[100] Capaccio, T. (2015, Mar. 27). Some Navy jets can't launch from the carrier Ford. *The Virginian-Pilot*. Retrieved from http://hamptonroads.com/2015/03/navy-jets-too-much-fuel-cant-launch-new-aircraft-carrier

Chapter 11

Forecasting and optimisation for multi-purpose application of energy storage systems to deliver grid services: case study of the smarter network storage project

Neal S Wade[1], David M Greenwood[1], Panagiotis Papadopoulos[2], Christos Keramisanos[2] and Adriana Laguna[2]

11.1 Introduction

The recent advent of grid-scale energy storage systems (ESS) has been the result of both technology and market advancements. To reach the current levels of confidence in the coming wide-scale adoption of energy storage, the real-world operation of large-scale ESS has been needed to prove that effective operation can be achieved. This was the purpose of the Smarter Network Storage (SNS) project delivered by UK Power Networks from 2013 to 2016. This chapter draws heavily on the development, implementation and operation of the systems required to support the technical and commercial management of the SNS project.

Section 11.2 gives an overview of the SNS project, explaining the motivation for the project and the objectives that were tackled. To be a financially viable project, it was necessary to engage in the provision of grid services to third parties in addition to managing a network capacity constraint. Section 11.3 describes the types and characteristics of these grid services. In order to fully engage with grid services, a number of systems needed to be put in place, these are described in the next sections. Section 11.4 outlines the reasons that forecasting and optimisation systems are required and their objectives in general, Section 11.5 details the design of these systems in SNS and Section 11.6 explains how these are used to schedule operations of the ESS. The operational experiences are detailed in Section 11.7, in which the successful delivery of single and multiple services is shown. A system always has to operate within a number of practical considerations. Some steps beyond the SNS project are outlined in Section 11.8, before a brief conclusion summary in Section 11.9.

[1]School of Engineering, Newcastle University, Newcastle-upon-Tyne, UK
[2]Innovation Team and UK Power Networks Services, UK Power Networks, London, UK

11.2 SNS project overview

Building on the earlier, small-scale UK Power Networks led battery project at Hemsby in Norfolk, UK [1], the SNS project is a pioneering demonstration project, testing the deployment of grid-scale battery storage on the distribution system. Specifically, SNS was developed to trial primarily how energy storage could be used as an alternative to traditional network reinforcement whilst also being utilised for a range of other system-wide services in a variety of business models.

To do this, SNS sought to consider the suitability of storage for providing multiple services to the system, the potential revenues available from different uses, the viability of different business models and the barriers within the regulatory frameworks to deployment of storage in such business models. Therefore, SNS was designed to explore the technical, commercial and regulatory feasibility of the application of grid-scale storage.

11.2.1 System characteristics

The SNS project involved the installation of a grid-scale 6 MW/7.5 MVA/10 MWh advanced Li-ion storage device on UK Power Networks' network. The facility is connected to the 11 kV side of UK Power Networks' Leighton Buzzard 33/11kV substation.

Experience from the SNS trials provides unique understanding of the commercial deployment of such a facility. SNS is designed to provide multiple services to the electricity system as shown in Figure 11.1, with the primary applications being deferral of distribution network reinforcement (i.e. peak-shaving (PS)) and secondary applications being reactive power support, frequency control, reserve, Triad avoidance and tolling.

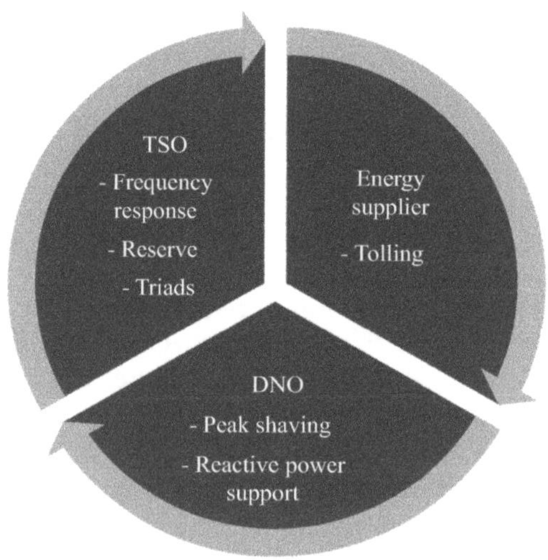

Figure 11.1 Services and beneficiaries provided by the ESS of the SNS project

11.2.2 SNS software solution

The operation of SNS relies upon a Smart Optimisation and Control System (SOCS) software solution to support the day-to-day operation of the energy storage device. The complete solution performs a variety of tasks including scheduling the services on the storage device and controlling in real time the storage device. There are three main different subsystems that comprise the SOCS solution:

1. Forecasting, Optimisation, Scheduling System (FOSS):
   This system is responsible for forecasting the demand at Leighton Buzzard and producing an optimised schedule for the operation of the ESS. The optimised schedule ensures that the storage system is available to support the local network by providing PS when required while providing for the remaining time the most profitable services to maximise the value from its operation.
2. Battery Energy Storage System Management (BESSM):
   This system is responsible for the real-time monitoring and control of the battery system. It executes in real time the optimised schedule that it receives from Forecasting, Optimisation, Scheduling System (FOSS).
3. Ancillary Services Provider (ASP):
   This system/agent is the intermediary between the battery owner and the Transmission System Operator (TSO). It is responsible for declaring the availability of the storage system in ancillary services markets, triggering dispatch of storage in real time when required and collecting data on the delivery of the services declared and delivered to the TSO.

These subsystems produce outputs that require intercommunication to facilitate the ultimate action of dispatching the energy storage device. The communication between these subsystems is facilitated via the SOCS messaging system.

11.2.3 SNS learning

Experience from SNS provides learning in respect of the following:

- planning, designing and constructing a grid-scale battery project in Great Britain (GB);
- developing a new optimisation and control system and trialling the commercial arrangements for shared use of energy storage;
- demonstrating how 6 MW/10 MWh of Li-ion storage can be deployed on the distribution network to support security of supply;
- trialling the multi-purpose application of storage for a range of different system benefits to help maximise value e.g. investment deferral and ancillary services and associated commercial arrangements;
- understanding of the opportunities and challenges for different business models; and
- identifying barriers within the regulatory and legal framework to the deployment of distribution-connected storage and proposing suggested measures to address these barriers.

Information on all of these aspects can be found on the UK Power Networks (UKPN) innovation website [2]. This chapter goes into substantial detail on the forecasting and optimisation components of the project.

11.3 Grid services

ESS are increasingly being installed within transmission and distribution networks; these systems can gain revenue by providing different services to transmission and distribution system operators. The services available have different requirements, duty cycles and market structures; therefore, specific applications may be better suited to particular energy storage technologies. This section describes the available services and provides analysis of operational and market data from the GB electricity system. However, equivalent services exist in other systems [3], and consequently this review is broadly applicable internationally. Each service description is accompanied by a table, quantifying some of the key requirements for delivery.

11.3.1 Distribution network services

Distribution network services take advantage of the topological location of the ESS, and can therefore only be offered by a limited number of ESS. These services can sometimes be fulfilled using reactive power as well as real power. However, there is currently no formal market structure or framework for DNOs to procure these services from ESS operators (although there are limited trial schemes [4]).

11.3.1.1 PS/Power flow management

Power flow management uses the production or consumption of real power to alleviate the thermal constraints within power networks. A specific example of this is PS, which refers to the reduction of electricity demand at times of peak consumption.

Electricity demand varies throughout the day; in the United Kingdom, this peak typically occurs in the early evening, driven by residential electrical appliances use. In the majority of cases, the peak demand only occurs for a small fraction of the time [5], but the generation, transmission and distribution systems are constrained by it. A PS scheme attempts to reduce the demand peak either literally, through demand side response [6] or by offsetting the demand by supplying power through distributed generation or energy storage local to the demand [7].

Electricity demand is highly variable, depending on time of day, day of week, month and season, as well as environmental variables such as temperature and whether it is light or dark. Consequently, the PS time, power and energy will all vary from day to day.

Figure 11.2 shows the demand over 24 h at a UK substation during a winter day. Between 17:30 and 20:00, the demand exceeds the circuit's loading limit, so the EES would need to offer PS for those 3.5 h in the event of a circuit or transformer outage. The highest peak exceeds the demand by around 3 MVA, so the

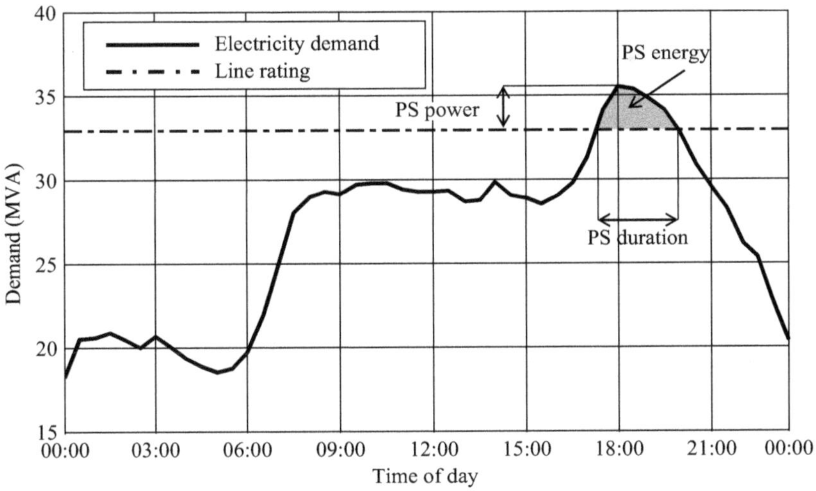

Figure 11.2 PS needs to take place when the existing infrastructure cannot support the demand peak. A storage system must meet both power and energy requirements to successfully offset the peak

Table 11.1 Power flow management service parameters

Real/reactive power	Both, primarily real
Availability fee	n/a
Utilisation fee	n/a
Available hours	Site-specific, but typically weekdays 16:00–20:00 in the UK residential load-dominated networks
Power:energy ratio (MW:MWh)	Site-specific, but typically 2-3:1
Minimum power	Site-specific

EES would need a converter rated at 3 MVA to successfully offset the peak. Finally, the total area between the demand curve and the line rating is around 5 MWh; this is the energy capacity that would be required for the EES to meet the PS demand. Service parameters for PS in GB are shown in Table 11.1. Other systems, for example, those with air-conditioning-driven peaks, may have significantly different service requirements.

11.3.1.2 Voltage control

In electrical networks, the voltage must be kept within statutory limits to ensure safe operation, and minimise the chances of a fault leading to customer disconnection. In the United Kingdom, the voltage limits are between ±10% and ±6%, depending on the voltage level. On LV feeders, the tolerance is +10%/−6%, which corresponds to a range of 216–253 V. Conventionally, the voltage of a feeder is highest at the point of connection to the wider network – a secondary transformer, in the case of a radial LV

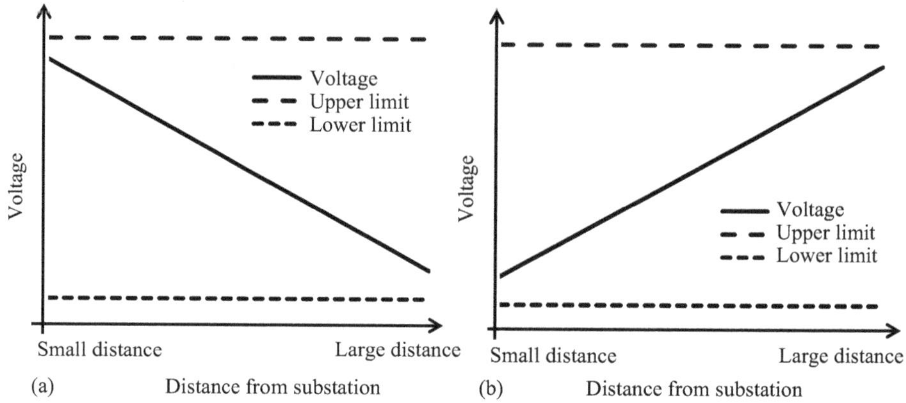

Figure 11.3 Illustrations of voltage (a) fall and (b) rise on a distribution network feeder

Table 11.2 Voltage control service parameters

Real/reactive power	Both
Availability fee	n/a
Utilisation fee	n/a
Available hours	Site-specific; peak demand, peak PV generation or both
Power:energy ratio (MW:MWh)	Site specific, but typically little energy at higher voltage levels, where reactive power can be used
Minimum power	Site-specific

feeder. The voltage then falls as the feeder supplies each customer, leading to a voltage profile like the one shown in Figure 11.3(a). In some cases, in particular, where there are high levels of distributed generation such as domestic PV, the voltage can *rise* along the length of the feeder, as illustrated in Figure 11.3(b). In some cases, both can occur at different times of a day, at different positions on the feeder or both. The network operator is required to keep the voltage within pre-defined limits, however in conventional LV networks, there are limited technical solutions available to achieve this.

ESS has the capability to affect the voltage in its local area by exchanging real or reactive power with the network. Unlike conventional methods, such as transformer tap changers, ESS can react almost instantly, in fine steps, and can be sited down the feeder, rather than at the substation. The characteristics of the voltage control service are shown in Table 11.2.

11.3.2 Transmission system services

Transmission system services are procured by the system operator (SO) to aid with system balancing and control, and ensure a reliable and high-quality service. These services are location independent (with exceptions in the event of transmission constraints), and are therefore highly competitive. The market is dominated by conventional peaking plant, such as open-cycle gas turbines and pumped hydro storage; the services are not designed with distributed ESS in mind, which represents a barrier to entry in many cases, although the SO is adapting the markets to incorporate them. The specific services described here are from the GB market, operated by the National Grid (NG) [8], and the data presented are publicly available.

11.3.2.1 Short-term operating reserve

The SO needs to access sources of additional power during operation, which can take the form of either demand reduction or generation, to deal with unforeseen shortfalls in supply caused by increased demand or unavailable generation. Short-term operating reserve (STOR) providers are required to be on standby at specific times of day, and to provide a pre-specified change in power output in response to a signal from the SO. STOR operates through an open tender, with providers bidding to supply the service in each of six STOR seasons during the year. These operational slots consist of a few hours on each day, and differ between weekdays and weekends, and providers are typically remunerated for both availability and delivery. To participate in the service, a provider must meet several criteria:

- Offer a minimum of 3 MW demand reduction or generation (this can be aggregated);
- Maximum response time for delivery of 4 h following instruction from the SO, although typical contract durations are 20 min or less;
- The ability to deliver the contracted MW for a continuous period of not less than 2 h;
- Have a recovery period after provision of reserve of not more than 20 h;
- Be able to deliver at least three times per week.

Data covering the number of calls, duration of calls and payments for availability and utilisation are available from NG. Analysis of these data has been carried out to quantify typical STOR usage patterns. Figure 11.4 shows the number of calls per day and the average remuneration in each of the six seasons, while Figure 11.5 shows cumulative probability distributions of service delivery.

Table 11.3 shows the probability of a given number of calls on a typical day in each season, and Table 11.4 gives an overview of the service parameters.

11.3.2.2 Frequency response

Frequency response (FR) services are used to either maintain the frequency within regulation (50 $\pm$ 0.5 Hz) or to bring the frequency back to 50 Hz, following a period of deviation (such as particularly high or low demand, loss of supply or interconnector). As with STOR, FR is operated using a market framework, and fees are paid for both availability and delivery. Some FR services can be offered continuously for entire days or weeks. This is not always an option for EES, because following any delivery, the state of charge (SoC) must be adjusted before the EES is available to perform the service again. FR services can be power-to-grid (PtG),

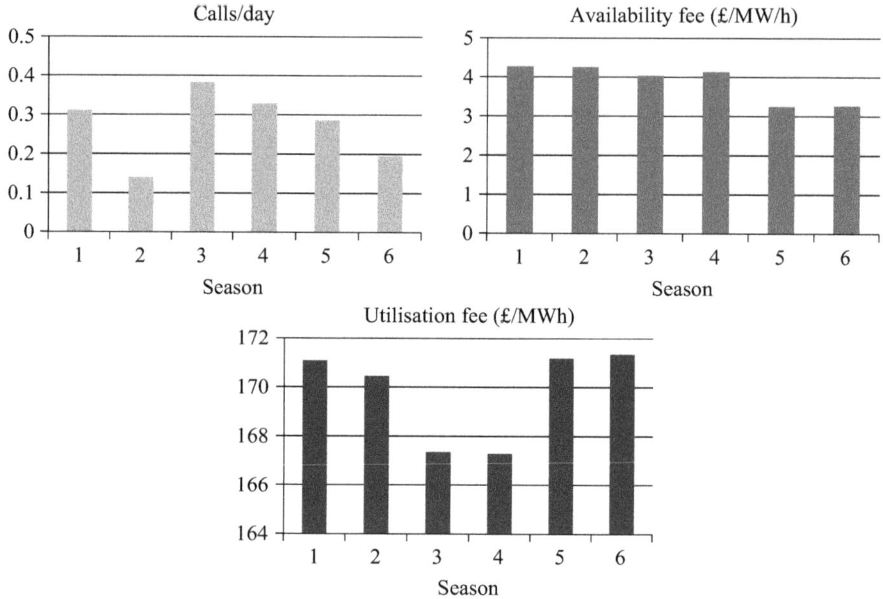

Figure 11.4 Bar charts of the expected calls per day, average availability fee and average utilisation fee for STOR services in the UK

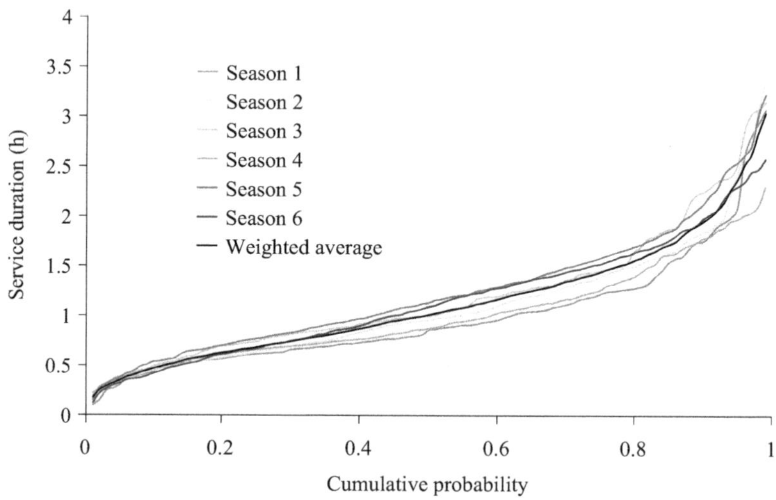

Figure 11.5 Cumulative probability of STOR call duration

Table 11.3 The probability of 0, 1 or 2+ STOR calls on a given day in each season

Calls/day	**Season 1 (%)**	**Season 2 (%)**	**Season 3 (%)**	**Season 4 (%)**	**Season 5 (%)**	**Season 6 (%)**	**Weighted average (%)**
0	73.3	86.9	68.2	72.0	75.1	82.3	79.1
1	22.8	12.2	26.1	23.7	21.5	16.0	18.3
2 or more	3.9	0.9	5.7	4.4	3.4	1.7	2.6

Table 11.4 STOR service parameters

Real/reactive power	Real
Availability fee	3–5 £/MW/h
Utilisation fee	165–175 £/MWh
Available hours	Season-specific
Power:energy ratio (MW:MWh)	1:2 (1 call)
Minimum power	3 MW

power-from-grid (PfG) or bidirectional. There are currently three types of commercial FR procured by the TSO:

- Dynamic FR: The amount of power delivered is directly governed by the system frequency.
- Static FR: The service power is pre-determined, and is delivered in full following a frequency deviation of a certain size.
- Frequency control by demand management (FCDM): It provides a response through interruption of supply to demand customers. It is mainly required to manage large deviations, and although unlikely to be provided by ESS, it is included here for completeness.

The difference between the dynamic and static services is illustrated in Figure 11.6.

NGs have made two years of frequency data available at a resolution of 1 s. These data were used in [9] to simulate two years operation of ESS with power to energy ratios of 1:1 and 2:1, responding to frequency deviations of ±50.05 Hz. The response was carried out using a droop control curve which required rated power for deviations greater than 0.5 Hz. The efficiency of the ESS was ignored. Every time the SoC reached the limits (100% or 0%), it was artificially reset to a nominal operating point of 50%. The 2:1 ESS hit its operating limit 283 times and the 1:1 ESS hit its operating limits 155 times [9].

Frequency services are also broken down into primary response, which is required as quickly as possible, and attempts to contain the deviation in frequency, and secondary response, which is used to help the frequency recover to its nominal value.

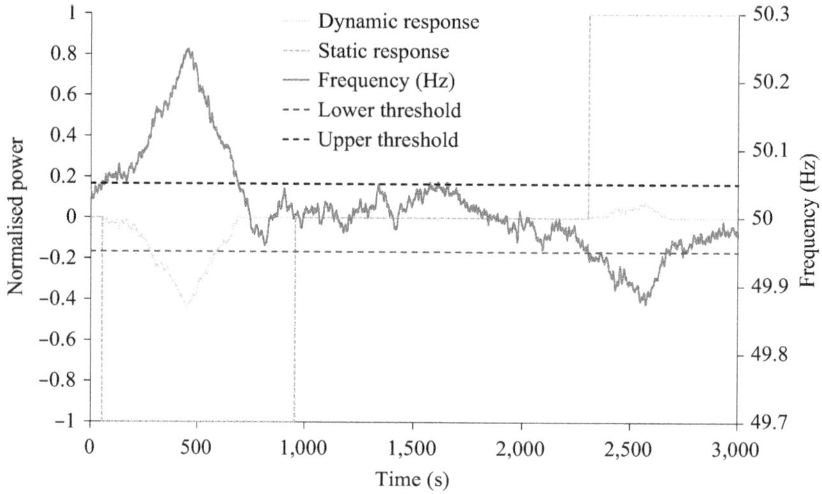

Figure 11.6 Static and dynamic responses to two frequency events

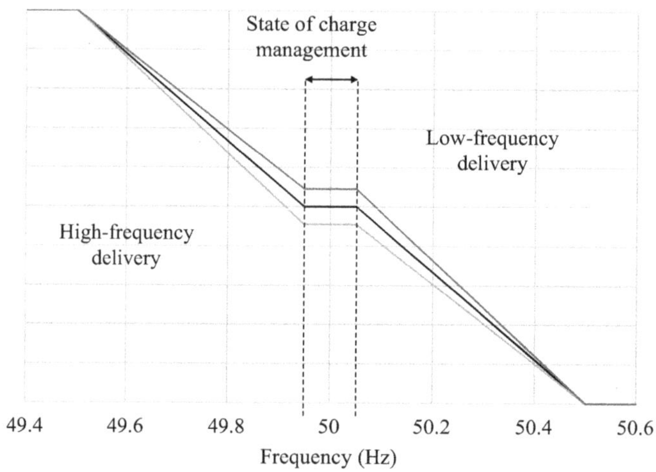

Figure 11.7 The response curve for EFR. The upper and lower bounds show the required power output, as a proportion of the maximum tendered power, for a given frequency. The deadband around 50 Hz allows an ESS to manage its SoC using a proportion of its tendered power [9]

11.3.2.3 Enhanced frequency response

Enhanced frequency response (EFR) is a new frequency regulation service being procured by NG [10]. It is explicitly designed to enable delivery via ESS. Figure 11.7 shows the response curve for EFR under the wide band service (which requires a response at ±50.05 Hz). There is also a narrow band service, which requires a response at ±50.015 Hz. When the frequency is within the deadband, the

Table 11.5 Frequency response service parameters

	SFR	DFR	EFR
Real/Reactive power	Real	Real	Real
Availability fee	3–5 £/MW/h	8–20 £/MW/h	7–12 £/MW/h
Utilisation fee	0	0	0
Available hours	Continuous	Continuous	Continuous
Power:energy ratio (MW:MWh)	2:1 (1 call)	1:1–3	2:1
Minimum power	1 MW	1 MW	1 MW

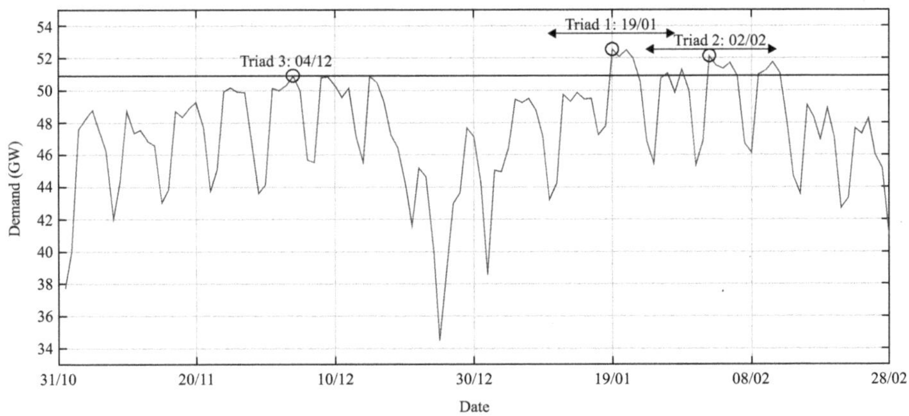

Figure 11.8 The daily maximum demand for winter 2014–15 – data courtesy of the National Grid

provider is permitted to vary their power exchange with the grid within ±9% of the tendered power; this can be used to manage the SoC of an ESS [8].

EFR has demanding requirements which are well suited to being fulfilled by an ESS [10]:

- Full response must take place within 1 s of a frequency deviation occurring.
- The provider should deliver the service in either direction (PtG or PfG).
- The provider must be able to provide 100% capacity for a minimum of 15 min.
- The provider should maintain an operational availability of 95% to qualify for the full payment.

The requirements for each frequency service are shown in Table 11.5.

11.3.2.4 Triad

Triad refers to three half-hour (HH) settlement periods between November and February that are used to determine charges for demand customers with HH metering [11]. These periods are defined as the three periods with the highest system demand separated by at least ten days. To illustrate this concept, the maximum demand on each day for winter 2014–15 is shown in Figure 11.8. The three Triad periods have been identified on this figure; the greatest demand peak was on

January 19 so this is identified as Triad 1. The second highest peak, on January 21, is excluded for occurring within ten days of Triad 1, so the next highest point on February 2 is identified as Triad 2. Following this, many of the other high demand peaks occur within ten days of either Triad 1 or Triad 2; Triad 3 is consequently identified as December 4, even though at 50.899 GW, it is actually the 52nd highest demand value observed during the year. This highlights a difficulty in predicting the Triad periods – due to the ten-day separation rule, the demand during the smaller Triad periods may be quite close to the typical demand peaks.

A customer charge through the transmission use of system charge (TUoS) is based on their demand during the Triad periods. The average demand during these settlement periods is multiplied by a value (23–46 £/kW); the value is dependent on the geographic region, with higher costs in regions more likely to suffer from transmission constraints. Triad revenue is a saving from these payments; to maximise the saving, the ESS needs to deliver maximum power during all three Triad periods. Table 11.6 gives an overview of the main parameters for the Triad services.

11.3.2.5 Capacity market

The capacity market (CM) is a mechanism to ensure that the system is able to meet its peak demand. CM contracts provide a steady revenue stream in exchange for mandatory delivery of either generation or demand curtailment during system stress events (times where the availability of generation is deemed to offer inadequate security of supply). In the auction to provide the service for 2020/21 [12], the clearing price was 22.50 £/kW/year; 3.2 GW of these contracts were awarded to storage. The service parameters for the CM are shown in Table 11.7. The CM will not begin operation until 2018, so it is difficult to anticipate the frequency and duration of system stress events. These events are expected to last no more than 4 h,

Table 11.6 Triad service parameters

Real/reactive power	Real
Availability fee	0
Utilisation fee	20–50 £/kW (average over three Triad periods)
Available hours	A HH period on winter weekdays
Power:energy ratio (MW:MWh)	2:1
Minimum power	n/a

Table 11.7 CM service parameters

Real/reactive power	Real
Availability fee	22.50 £/kW/year
Utilisation fee	0
Available hours	System stress events (1–4 h)
Power:energy ratio (MW:MWh)	1:1–4
Minimum power	2 MW

but with the majority lasting 1–2 h, this leads to difficult decision-making for ESS developers, as the required energy capacity may be redundant in the majority of cases. Table 11.7 gives an overview of the key parameters of the CM service.

11.3.3 Energy trading services

While the services offered to the network operators provide revenue in exchange for supporting the network, the services in this section are concerned with energy transactions.

11.3.3.1 Trading agreements

The different agreements that are offered by an energy supplier are typically the following:

Fixed-price power-purchase agreements – These are designed to provide a fixed price for energy supply across a long-term period (1–3 years). While this provides price certainty over the period, the product is designed for highly predictable load patterns and has limited flexibility for changing operational profiles.

Index-linked power-purchase agreements – This is similar to the fixed-price power-purchase agreement (PPA) above, but the purchase and sale price are linked to a specific market index providing some additional flexibility in load pattern, in return for higher risk around the pricing at a particular time throughout the contracted period.

Flexible power-purchase agreements – This is a framework agreement for a long-term period. There are various means of locking in prices, and as such is more suited to highly predictable operating patterns where monthly volume forecasts can be provided.

Energy-trading service agreements – This allows full access to the UK power market from HH periods up to long-term power exchanges. The asset operator can buy and sell volumes of power according to the assets operational requirements under a single contract. Under this arrangement, the risk of energy imbalance is carried by the asset operator. However, it allows short notice changes to positions.

Tolling agreement – Under such an agreement, the supplier takes commercial control of the storage device in return for a £/MWh capacity fee and an utilisation fee when the plant was called to run. The supplier would optimise the operation of the assets within the wholesale markets. The asset operator would continue to retain responsibility for physical operation.

11.3.3.2 Arbitrage

Arbitrage describes the practice of taking advantage of a price difference between two or more markets; in the case of ESS, it refers to taking advantage of the varying price of energy as supply and demand fluctuate. An ESS can buy energy during low price periods (such as during the night), store the energy, then sell it during high price periods (such as during the evening peak). An example 24-h price period in GB is shown in Figure 11.9. The profitability of this activity depends on the price volatility, but is significantly influenced by the round trip efficiency and energy capacity of the ESS [13], and less significantly by the impact on the ESS lifetime through degradation mechanisms. Table 11.8 gives an overview of the energy trading service parameters.

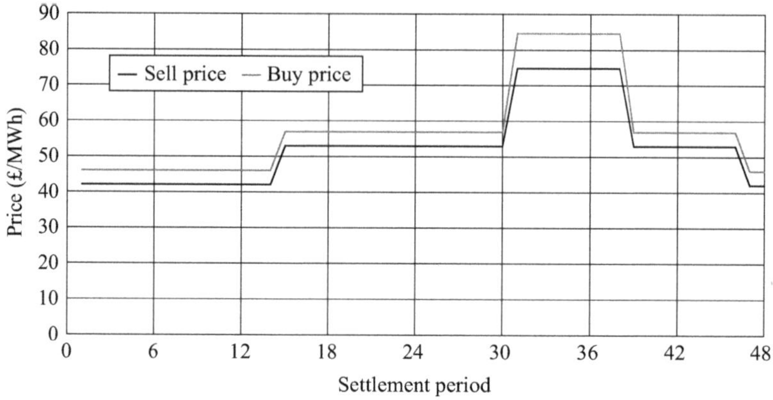

Figure 11.9 An example price profile for a 24-h period (comprising 48 HH periods)

Table 11.8 Energy trading service parameters

Real/reactive power	Real
Availability fee	0
Utilisation fee	Price-dependent
Available hours	Continuous, but price-dependent
Power:energy ratio (MW:MWh)	n/a
Minimum power	n/a

11.4 Objectives of forecasting and optimisation

The SNS project sought to prove that an ESS can effectively provide multiple services to a range of system participants. All the service offerings that are available need to be recorded and codified in a way that allows them to be compared with each other, this listing of services is referred to as a service portfolio. Information included in this should be sufficient to define the returns and operating profile that are associated with each kind of service delivery.

The value that can be obtained from providing individual grid services varies on a range of timescales. Within each day and throughout the weeks and seasons, the magnitude of power needed for ancillary services changes. As the number of service providers changes in response to market opportunities, the rewards associated with availability for and delivery of ancillary service provision fluctuate between tender rounds.

The cost of energy changes in every 30-min settlement period in response to the availability of particular generating units and the actions required to balance supply and demand in real time. Decisions made on when to operate the energy store in charging or discharging mode have direct implications on the time of day at which energy is supplied to or received from the system. As such, the time of use will have an associated energy cost which influences the return that can be expected from providing the given service.

To maximise the financial returns that an energy storage owner is able to capture, a running schedule must be formulated that evaluates both the revenue that participation in service delivery will provide and the costs that will be incurred. This evaluation is dependent on several sources of information, each of which is subject to differing levels of uncertainty. Some information is obtained from third parties and operational experience, while some has to be calculated on a routine basis.

In the case that a demand-PS service is within the service portfolio or there is a power flow constraint on the network to which an ESS is connected, it is critical to know the customer demand in the relevant area of the distribution network. This calls for a demand forecasting function from which decisions can be made on the need to reserve power and energy to manage the demand peak or constraint.

The demand forecast, codified service portfolio and energy prices provide the complete information needed for optimisation. This process is formulated to maximise the financial return to the ESS operator. There are many considerations in designing an optimisation process that depend upon the nature of the services that need to be included and the interactions that have to be undertaken with third parties.

11.5 SNS system design

The starting point for the SNS project was the need to defer network reinforcement by providing energy into a primary substation during the peak demand periods. Only providing peak network capacity through energy storage is not competitive in comparison to network reinforcement, but if other services are provided in the considerable proportion of time that PS is not required, the business case can stack up. This fundamental requirement for PS necessitates a demand forecast which is described next. Following that, the process of optimisation is described, in which services are selected from the portfolio based on the maximum return that can be obtained for the storage operator. The final output of forecasting and optimisation is an operating schedule that details the periods in which services are delivered or adjustments to the storage system's SoC are made.

The design principles for the forecasting and optimisation functions in the SNS project were reported in the logical architecture design document which is available in full online [14]. Key features from this document are combined with additional information in the following sections.

11.5.1 Principles of forecasting and applications in SNS

A demand forecast of whole substation average power in each HH period in the forward schedule is required. To align with the timescales on which scheduling takes place, the demand forecast is made from 1 to 8,760 h in advance. Separate forecast models were developed for short-term (1-week) and long-term (12-month) windows.

The data used in the demand forecasting algorithm are historical demand, historical meteorological data, forecast meteorological data (5-day ahead at 3-h resolution) and an annual load growth (or decrease) trend. The only meteorological data used is temperature, although wind speed and solar insolation could be important factors if renewable generation is a significant presence in the network area under consideration.

11.5.1.1 Approaches to demand forecasting

Two branches of approach to demand forecasting were considered: time-series methods and end-use load modelling. Time-series methods are considerably more mature and the available data are appropriate for using this approach. Estimates of the success that could be achieved through an end-use load modelling approach are not well defined and as such, it is not suitable for this important aspect of SNS. These alternative approaches and subsequent variants are illustrated in Figure 11.10.

Time-series-based techniques fall into two broad categories: statistical methods or artificial intelligence techniques. Statistical methods fit parameters to an analytic expression that uses the endogenous past demand profile and possibly exogenous data, such as temperature and economic output [15]. A range of artificial intelligence techniques have been used in demand forecasting, including artificial neural networks [16] which typically make use of plentiful historical data to uncover underlying patterns of variation and carry this forward in time, without seeking to identify the underlying causes of demand variation.

The literature suggests that artificial intelligence techniques do not generally offer significant improvements over more established statistical methods. In some cases they perform considerably less well and since they are based on data rather

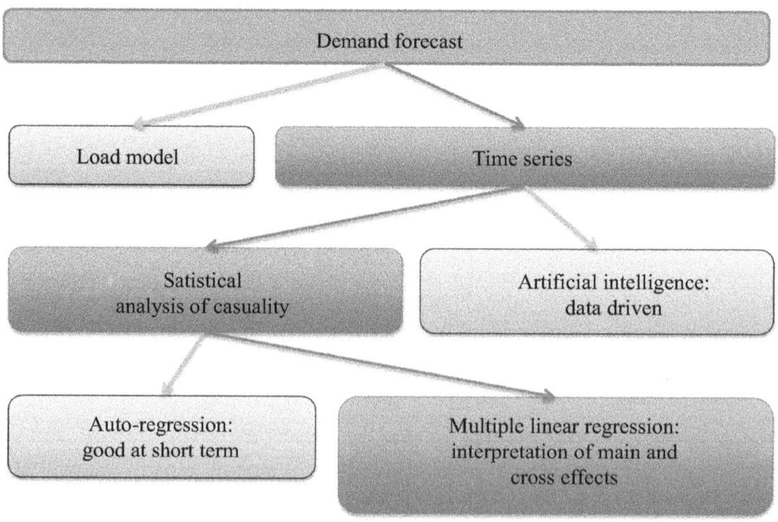

Figure 11.10 Demand forecasting solution hierarchy

than explanatory variables, it can be difficult to discern why a technique is or is not performing to a satisfactory standard.

Statistical models can again be divided into a number of alternatives including semi-parametric, auto-regressive or multiple linear regressive. Auto-regressive models use the endogenous historical demand data alone and can work well for short-term load forecasting. A multiple linear regression (MLR) takes in any number of endogenous and exogenous data sources.

Regardless of the method used, over-fitting of data has to be avoided. An over-fitted model will perform very well within the data that was used to construct the model (in-sample), but out-of-sample performance is likely to be compromised by variations that we not seen in-sample. Over-fitting can be worse than under-fitting.

Demand forecasting horizons are typically split into three timescales, short [15], medium [17] and long term [18]. Different techniques and variations within techniques have been applied at these alternative timescales as the requirements and explanatory variables can be different for each timescale. Within-day (typically hourly) meteorological forecast data have been shown to improve short-term and medium-term demand forecasts [17]. Weather forecast data ceases to improve demand forecasting accuracy once a 31-day timescale is reached. The error in the weather forecast can be sufficiently large so as to create a worse error in the demand forecast than would have been present without any forecast weather data.

When considering the timescales on which demand forecasting can be improved, refinement of a demand forecast is only warranted if there is an improvement in the accuracy of an explanatory variable.

It is possible to provide a confidence interval for upcoming demand, using a process called density forecasting [18]. A procedure called bootstrapping can be used to accomplish this using the statistical characteristics of the explanatory variables in the forecasting model to give a resulting spread in the demand forecast. Since it is required to account for the worst-case scenario, a density forecast was not used in the SNS solution.

11.5.1.2 Demand forecasting technique

The chosen method for forecasting is an MLR as this provides a transparent, easily calculable online solution that clearly links the causal data to the resulting demand [19]. The accuracy was expected to be acceptable and the model can be updated in an off-line process if performance falls below the necessary level.

An MLR is expressed as a summation of pre-calculated β_i terms (with $i = 0, 1, \ldots, n$), explanatory variables X_i (with $i = 1, \ldots, n$) and an error term, e representing the forecasting error. When combined, this gives

$$Y = \beta_0 + \beta_1 X_1 + \beta_2 X_2 + \ldots + \beta_n X_n + e. \tag{11.1}$$

The β_i terms are calculated offline in a study that examines the correlation between all of the explanatory variables that are available and considered to have a causal effect on the demand.

Table 11.9 List of explanatory variables for demand forecasting and their range

Variable	Range
Time of day	1–48
Days of week	1–7
Month	1–12
Season	1–4
Daylight savings	0–1
Type of day	1–3
Special day	0–1
Daylight	0–1
Temperature	n/a
Previous demand	n/a

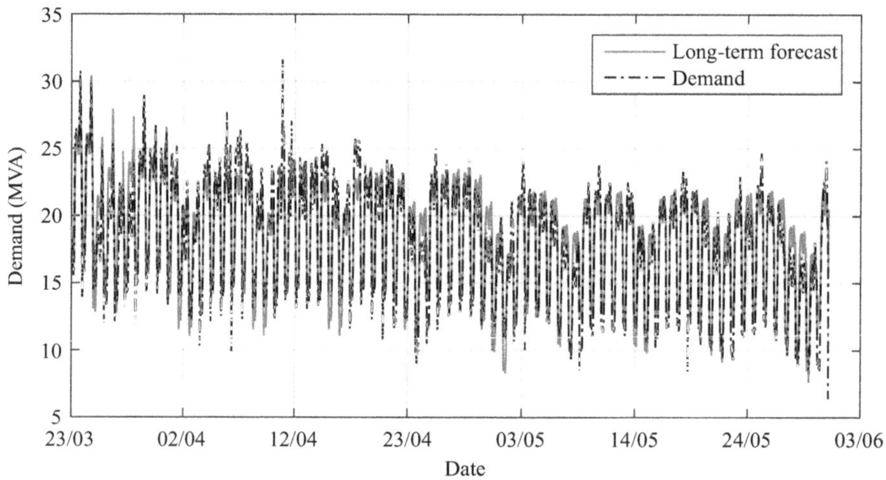

Figure 11.11 The demand forecast and actual demand for April and May 2016

The explanatory variables considered in the SNS forecasting system are given in Table 11.9. Two models were required for SNS: a long-term model to give an indication of which days are likely to have high demand and a short-term model to finalise schedules and reserve power and energy resources. Both models account for long-term trends, but the short-term model also has an auto-regressive element and accounts for the relationship between the forecast temperature and the demand.

11.5.1.3 Forecasting performance

Long-term forecast error

The long-term forecast for April and May 2016 is shown in Figure 11.11.

There are a number of high-demand spikes in April which were not predicted by the forecast. The demand was more stable in May, though the forecast

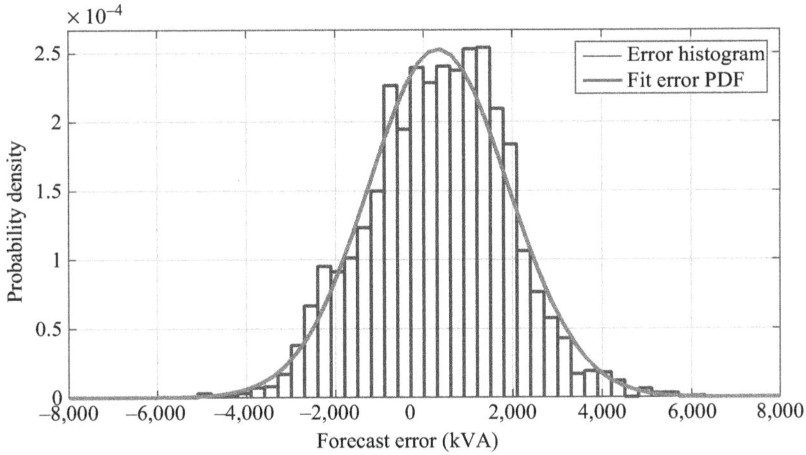

Figure 11.12 A histogram and probability density function of the long-term forecast error

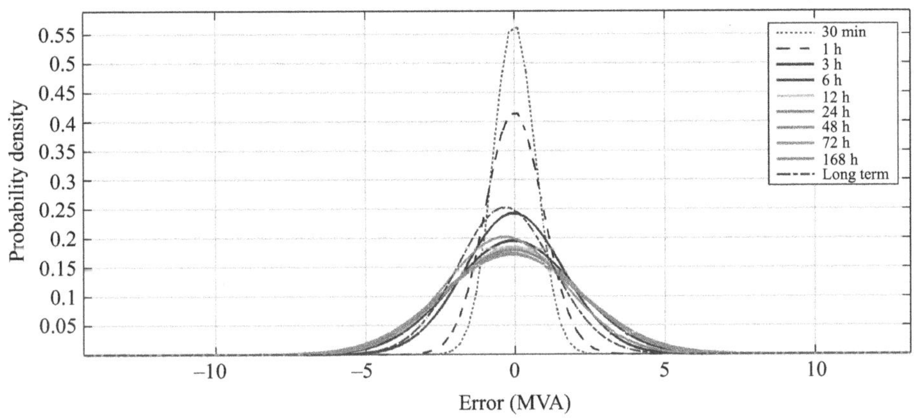

Figure 11.13 Probability density functions of forecast error for different advance periods, and for the long-term forecast

consistently over-predicted the demand. The probability density function for the forecasting errors is shown in Figure 11.12. The error is a reasonable fit to a normal distribution, but the mean of the errors is −328.5 kVA, indicating systematic under prediction. The mean absolute error (MAE) was 1.288 MVA.

Short-term forecasting error

The short-term forecast was run for the same period as the long term, with advance periods of up to 168 h (1 week). Figure 11.13 shows the probability distributions of error at different advance periods, ranging from the next HH to 1 week ahead, and Figure 11.14 shows the mean error, MAE, standard deviation and

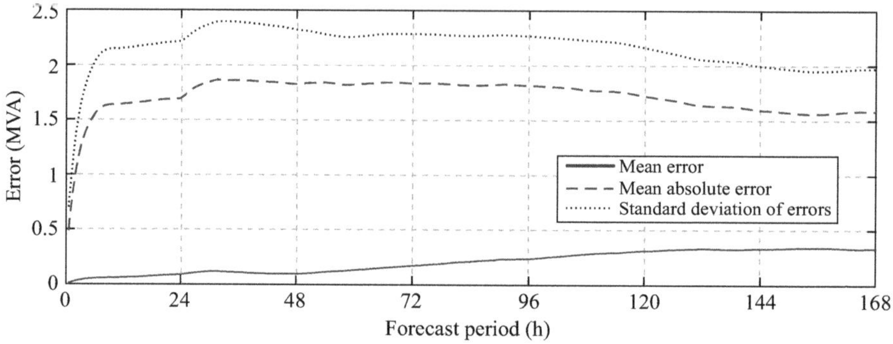

Figure 11.14 Error metrics for the short-term demand forecast

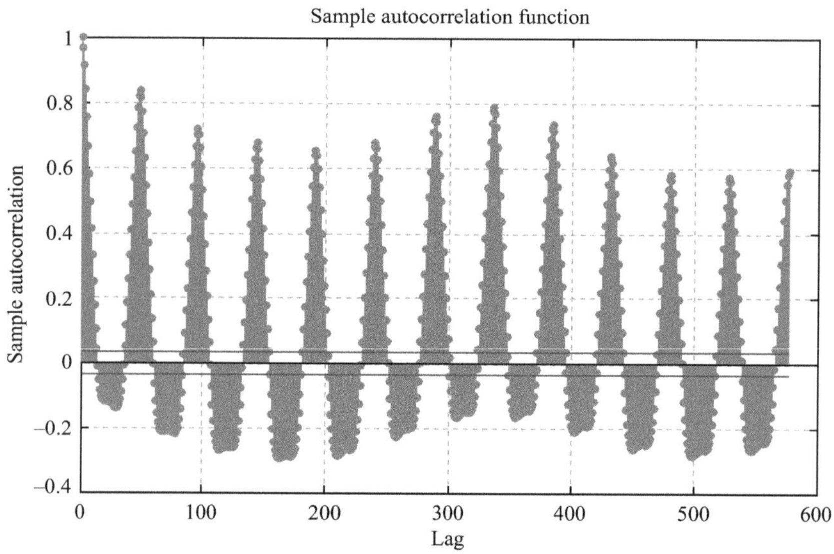

Figure 11.15 Autocorrelation results for the short-term demand forecast

standard deviation of average error for each advance period. The forecast is very accurate over small advance periods, but the accuracy rapidly deteriorates. The error reaches a peak at around 27 h, and then actually reduces as the advance period approaches 1 week (168 h).

In Figure 11.14, we can also see the mean error gradually increasing with the advance period. This slight drift is likely because the baseline demand has increased since the forecast terms were calculated.

Figure 11.15 shows an autocorrelation of the demand for lags of up to 2 weeks. There are both daily and weekly periodicities visible in the autocorrelation, which explains why the mean absolute forecast error decreases as the advance period reaches 1 week (when the same day of the week as was originally observed is being forecast). To take advantage of this effect and improve forecast accuracy, different models could be made for each advance period, using recent data from the same day (i.e. using the previous Tuesday to predict the next Tuesday). This would deliver an improved forecast, but at the expense of being complex to implement.

11.5.2 Principles of optimisation and use in SNS

The approach to optimisation to produce a schedule is characterised by a series of decisions to be taken at discrete points in time. The information available at each stage may be limited, often non-definite and changes as time evolves. What actually occurs at the point at which the schedule is implemented in many cases will not be known until the HH at which the schedule is realised. The scheduling process has been designed to handle the complexity of this situation in a manageable, robust and configurable manner.

A path through the available decisions ('a decision tree') must be constructed so that the maximum benefit is accrued while ensuring power and energy capacity is available to meet all the commitments that have been made and maintain site demands within constraints.

The number of paths that can be taken in a tree problem rapidly becomes intractable as the number of variables and stages in the problem grows. For example, if every service currently in scope for consideration was reassessed for every HH period in a year, then there would be $9^{17,520}$ permutations. Clearly a more efficient way to assess the paths that might be taken is required.

Dynamic programming (DP) is a methodology that has grown around the need to make sequences of decisions [20] and is pertinent to the scheduling problem faced by SOCS. DP is employed to find the minimum cost of executing a process across a number of stages, where at each stage, a decision must be made on the control to be executed to transition to the next state. This is formulated as a recursive problem where the costs of the possible last parts of the path are evaluated, the paths leading to these are added to those costs and this is repeated until the initial state is reached. The lowest cost overall path is then known to be optimal. Uncertainty in the data that defines the decisions can be included. Mathematically this is expressed as follows, the transition from the current system state, x_k to the next, x_{k+1} is written

$$x_{k+1} = f_k(x_k, u_k, w_k) \quad k = 0, 1, \ldots, N-1, \tag{11.2}$$

where k is a discrete time index, u_k is a control action, w_k is a random (disturbance) variable and N is the number of stages. Policies, π (control laws) are defined for each stage

$$\pi = \{\mu_0, \ldots, \mu_{N-1}\}, \tag{11.3}$$

so that the system states can be rewritten

$$x_{k+1} = f_k(x_k, \mu_0(x_k), w_k) \quad k = 0, 1, \ldots, N-1. \tag{11.4}$$

For given functions g_k, $k = 0, 1, \ldots, N$, the expected cost J of using policy π starting at x_0 can be written as:

$$J_\pi(x_0) = E\left\{ g_N(x_N) + \sum_{k=0}^{N-1} g_k(x_k, \mu_0(x_k), w_k) \right\}. \tag{11.5}$$

An optimal policy is one that minimises the expected cost,

$$J_\pi(x_0) = \min_{\pi \in \Pi} J_\pi(x_0), \tag{11.6}$$

where Π is a set of all admissible policies.

Given a suitable structure of an optimisation problem and timely availability and (statistically) well-defined information, it is possible to reduce the solution to a closed-form analytical expression. It is recognised that this is the exception and in most cases, a numerical solution has to be used to complete the task. If applied in an unconsidered way, this means evaluating a huge number of expressions, so it is preferable to understand the nature of the particular problem and apply an approximate technique to achieve a near-optimal result (which could be the optimum). Approximate techniques such as rollout algorithms, certainty equivalent controllers or limited-look-ahead policies use insight from the full DP algorithm to reduce the search space that is necessary to be covered by the numerical solution [21].

The details of the specific scheduling characteristics of the storage system led to the adoption of a limited-look-ahead policy in which scheduling decisions are broken into several independent stages. This takes advantage of the changes to information availability and the ability to act that occur as time progresses. Some examples of these are:

- Contracts with the Ancillary Services Provider (ASP) to supply TSO-related services can only be formed within a tendering process with associated delivery windows and lead times;
- Demand forecasting will yield increasingly accurate results as the input data (principally meteorological) improves with shortening lead time;
- Wholesale market data will become increasingly resolved in time from peak/off-peak pricing to hourly pricing as lead time shortens.

The limited-look-ahead policy views the overall benefits that can be attained within the scheduling window under consideration at a given time and thus, for a tendered service, takes account of the likelihood of delivery across the whole period.

This approach limits the numerical evaluation of possible options to a tractable number of permutations of interleaved service combinations. While not overburdening the system, it prevents a myopic view that would only maximise net worth for a single stage, which could prevent a higher net worth combination of services from being identified.

Rules are defined at each stage to ensure the validity of each service or service combination being considered. Expressions that take the normalised service portfolio parameters evaluate the net worth of each service at the given stage. Once all stages are passed, the most valuable service combination is selected as the schedule for the duration of the optimisation window.

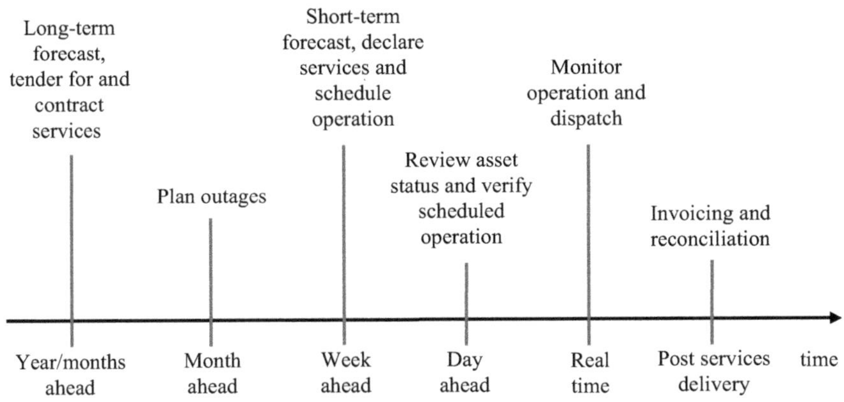

Figure 11.16 High-level timeline for the scheduling and operation of the storage system

Different scheduling time-scales proceed in a similar manner with the difference that the services included in the process will be depending on both the availability of information to usefully allow their inclusion, and the ability to act on a decision (i.e. enter a tendering process, incorporate hourly wholesale pricing). This timeline is summarised in Figure 11.16. The population of a forward schedule is illustrated in Figure 11.17. Long-term schedules are shown to be made available every 12 weeks, medium-term (update) every 4 weeks and short-term (update) every 1 week. An immediate schedule (update) for the following 48-h is produced each day, but this is too short to be visible in Figure 11.17. At the bottom of Figure 11.17, the constituent parts of the full 'live' timetable at week 18 (dotted line) for a 12-month period and their relative ages are shown as an example.

This process requires some initial configuration by the Battery Manager to initially set up services, and also when introducing a new service. However, a degree of diligence is also required to ensure that any individual service can be introduced into SOCS. It is thus appropriate to require the detailed interaction of alternative services with the scheduling process to be considered when being included in SOCS.

11.6 Use of energy storage to deliver grid services

11.6.1 Forecasting, scheduling and optimisation algorithms

The forecasting, scheduling and optimisation algorithms that comprise FOSS are designed to make best use of the limited power and energy resources of the SNS battery system, whilst ensuring a reliable supply to customers connected to the Leighton Buzzard substation, and that any commercial contracts can be fulfilled. The scheduling process is made up of two sets of algorithms: a service layering process which allocates the battery's power and energy resources to a combination of services and an optimisation which evaluates the expected value (EV) of a given branch of services, and selects the service portfolio which will maximise the

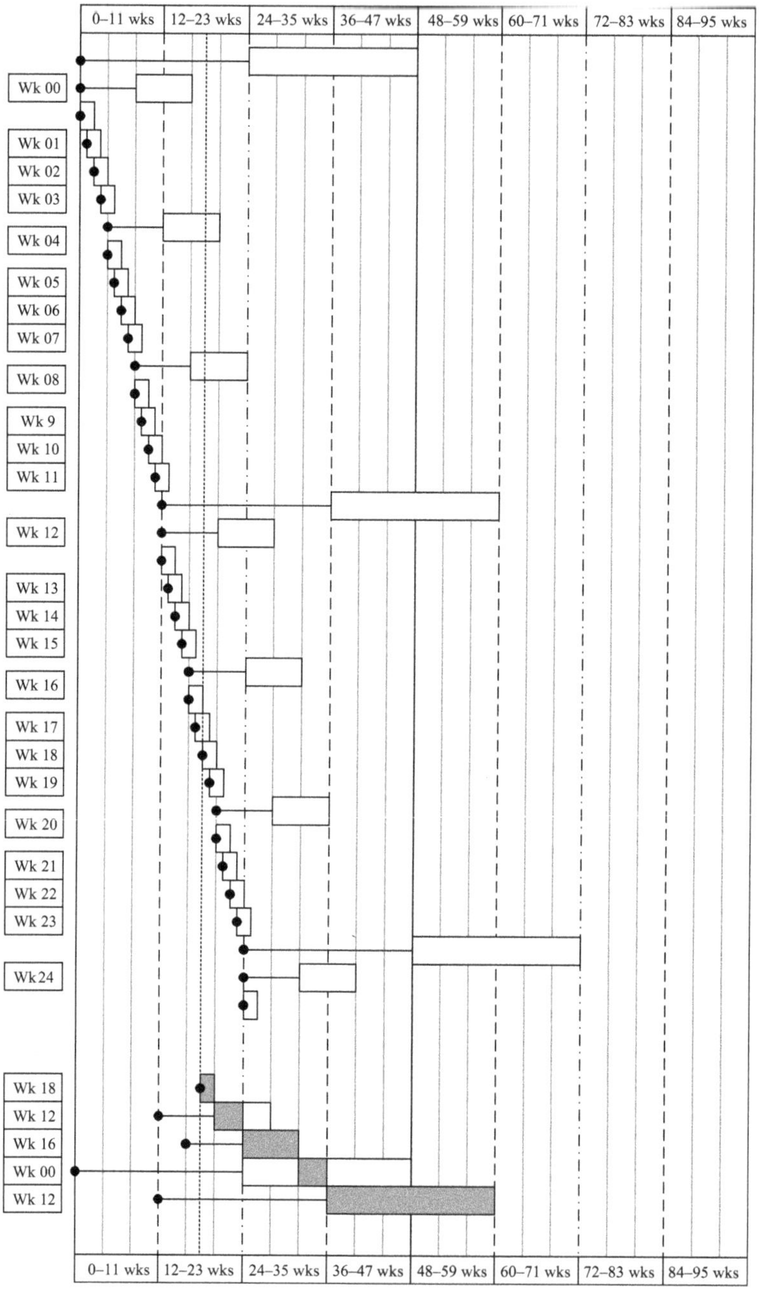

Figure 11.17 Long-term service scheduling

income of the system over the long term. These algorithms are described here, and a more detailed explanation is available in [22].

11.6.1.1 Allocation of power and energy to network services

From the perspective of the storage system, the services described in Section 11.0 can be simplified into three categories: PtG, PfG and bidirectional services, which can be delivered as either PtG or PfG. Each service is committed as a set window, during which the ESS must be available to provide the service if required, but there is no guarantee as to precisely when, or even if, the delivery will take place. Therefore, the scheduling system needs to ensure sufficient power and energy are available for the duration of the service window. The scheduling algorithm calculates the amount of energy, headroom or both that is required, and then infers the last point at which the ESS could adjust its SoC in order to be ready to deliver the service during the scheduled service window. The algorithm can also account for the efficiency of the ESS by reserving more energy for PtG services and less headroom for PfG services. An example of the resource allocation for each service is shown in Figure 11.18.

Once a service window has been added to the schedule, additional services can be added to make use of the battery's remaining resources. If a new service window is added to the schedule, the algorithm calculates the power and energy that are

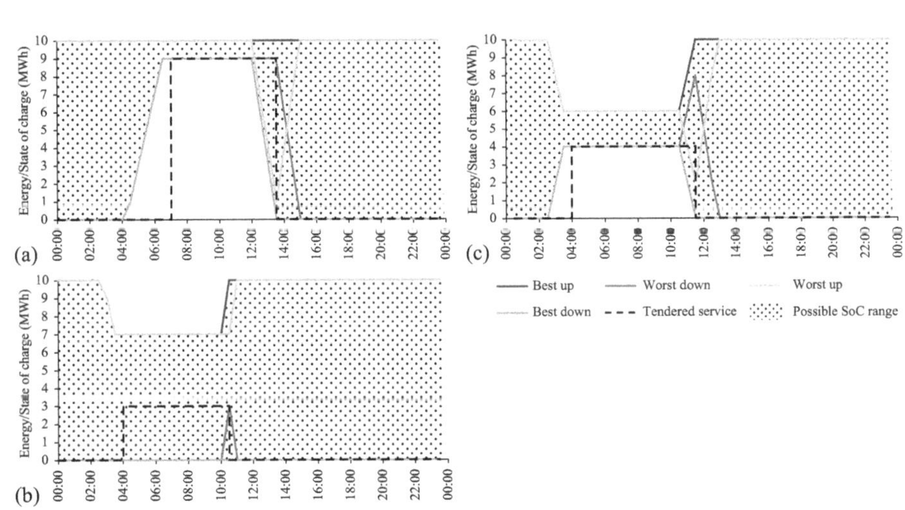

Figure 11.18 Illustrations of the SoC range allocation [1]: (a) A PtG service requires a minimum of 9 MWh to be reserved. The up pathways stay at maximum SoC until the start of the service, and only the worst up and best down pathways show the service delivery. (b) A PfG service requires a maximum of 7 MWh available at the start of the service. Only the best up and worst down pathways show the service delivery. (c) A bidirectional service of the ESS to sit between 4 and 6 MWh. The service is delivered as a PtG service by the worst up and best down pathways, and as a PfG service by the best up and worst down pathways

available at the start and end of that window, and commits them to the service. SoC adjustments are updated to ensure that the ESS's resources are being allocated as fully as possible while ensuring the full-, partial- or non-delivery of each scheduled service. This sequential service-layering process is illustrated in Figure 11.19 [22].

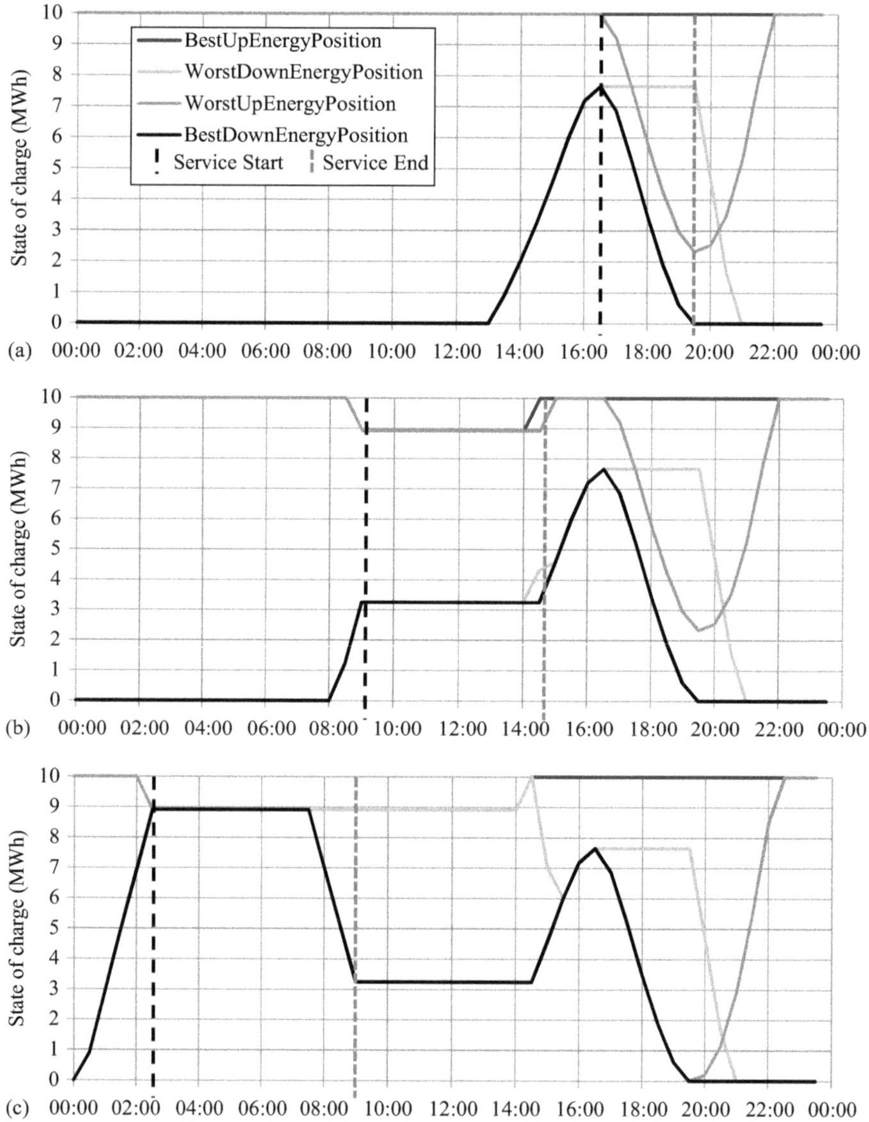

Figure 11.19 Layers of services are added to the service schedule, with the SoC positions, as well as the amount of power that is being tendered, being shown: (a) shows a PS service; (b) shows a PfG service being added; (c) shows a PtG service, immediately prior to the PfG service, being added [22]

11.6.1.2 Optimisation based on expected value of service combinations

Once all the battery's resources have been allocated to a given set of services, the process is reset and a different combination of services is added to the schedule. This can include the same service windows, but in a different order, because the sequence in which the services are layered into the schedule will affect the resources that can be allocated to each service. Once the services have been combined in every plausible sequence, their expected value (EV) is evaluated as a function of (1) the service's availability and delivery fees, (2) the expected cost of energy and (3) the likelihood of each service being delivered. These EVs are then enumerated, and the most valuable service combination is taken forward for delivery by the system.

11.6.2 Practical issues

The operation of SOCS in a business environment poses a number of challenges. These can be broken down in commercial, technical and regulatory.

Commercial challenges involve the processes and decisions required to maximise the value of the storage asset from the provision of PS and its participation in electricity markets. A multi-contract framework is required to enable SOCS to perform its functions:

- Product contracts: These contracts comprise agreements for the operation and maintenance of the hardware systems and include:
    - Battery system
    - Electrical protection and switching systems
    - Power conversion systems
    - Transformers and cabling
    - Fire suppression systems
    - Auxiliary equipment, including lighting, heating, air-conditioning, ventilation, uninterruptible supplies and security systems
    - Communications and control systems
    - Monitoring and metering systems
- Services contracts: These contracts comprise agreements for the provision of services and operation of assets and include:
    - Warranty agreements
    - Communications systems and information technology contracts
    - Facilities contracts
    - Electricity market contracts that include energy supply agreements and access to ancillary services for the TSO
    - Software operation and maintenance contracts
    - Day-to-day operation of the storage system

This complex framework requires continual stakeholder engagement and its challenges are interlinked with the technical practicalities that might arise from the operation of the storage system. These include:

- Uncertainty of load demand at the substation which requires PS can be addressed with careful short-term load forecasting, nevertheless masked load (that includes unmetered intermittent generation), human behaviour and weather and social events affect the accuracy of the forecasts and hence the level of capacity reserved for the PS service and the amount of capacity declared to other services.

- Uncertainty of medium- to long-term load growth at the area which is served by the substation that requires PS (e.g. new connections of generation and/or load) imposes challenges in ensuring security of supply given that the timelines for upgrading the storage system or reinforcing the local distribution network might not align with the rate of load growth.
- Integration and operation of multi-supplier systems reveal incompatibilities or defects that are unforeseen in the specifications/design stage. The lack of a clear 'single-stop' supplier in this niche area of battery storage systems entails multiple failure points that may affect the technical and commercial operation of the asset.
- The demanding response of battery storage systems (i.e. subsecond response markets) requires high-quality equipment and communication systems with appropriate synchronicity features and fail-safe mechanisms. The high number of monitoring points increases the data storage and information flow requirements and subsequently demands increased levels of real-time monitoring, analysis and post-fault response.
- Asset degradation experience is relatively low, given that grid-scale battery storage systems that perform demanding functions are still in their infancy and the impact on the lifetime of the asset is not clearly understood. This impacts the operational characteristics of the asset and in turn the amount and performance of effective storage used in long- to short-term planning.

Regulatory challenges exist when integrating battery storage in electricity networks and affect not only the business model under which an asset is operated, but also the architecture and scheduling processes associated with the operation of the SOCS.

- Network charging: In the United Kingdom and in the European Union, storage is considered as a subset of generation and as such it is not recognised as a separate asset class. In turn, when connected at the distribution network, a storage asset is assigned a network tariff depending on its contribution to security of supply and its utilisation profile. In the United Kingdom, non-intermittent technologies are assigned more favourable tariffs due to their controllability; this affects the level of cost or revenue anticipated from the operation of the asset and impacts the scheduling of its operation.
- Consumption levies: The lack of a separate asset class leaves storage operation exposed to levies both at import and export of electricity.
- Use of storage for network operation: With storage being currently undefined, it is rather unclear whether network operators may own it to contribute to network operation and how this contribution can be quantified given the lack of storage integration into security of supply standards. DNOs in the United Kingdom are currently designing commercial PS/flexibility services but the lack of a common framework poses uncertainties for the long-term planning and operation of storage systems in such markets.

11.7 Operational experience

11.7.1 Introduction

The SNS trials commenced in November 2014 following the commissioning of the battery plant and concluded in December 2016. During the two-year trial period, the provision of numerous services to the DNO and TSO was investigated and unique insight was obtained in the commercial opportunities and technical performance of ESS. The provision of multiple services simultaneously was also trialled during that period, to provide evidence that battery storage can add value to different stakeholders concurrently and improve its business case.

11.7.2 Single service scheduling

11.7.2.1 Peak shaving

PS at a primary 33/11 kV substation typically refers to the reduction of the apparent power load during times that the demand exceeds the rating of the transformers or the overhead lines feeding the substation. The PS requirements can be met with demand reduction or increase of the generation at a local level.

The two-year trial period of the SNS project provided invaluable experience regarding the application of PS for maintaining security of supply. Some of the key learnings from the trials are presented below:

- **Availability requirements**
  An ESS is capable of delivering PS requirements through the provision of active power, reactive power or both, utilising the four-quadrant capabilities of the power conversion system. When compared with other technologies that can deliver a PS service, such as demand response, an ESS is able to provide the service with a fast response and efficiency due to the real-time monitoring and matching of the excess demand with both active and reactive power terms. Notwithstanding the advantages of an ESS, when compared with other technologies, the differing characteristics of an ESS necessitate a different approach to scheduling and dispatching the plant for PS. The plant must be scheduled to charge hours before the expected peak to avoid any breaches of the substation capacity and following the delivery of the service, there is a time period that the plant cannot perform a SoC correction for the same reasons. This increases the total time that plant must be unavailable for other services and has an impact on an ESS business case.
- **Seasonal line loading limits**
  A peak-shaving service may refer to the loading of a circuit or transformer. Both assets may withstand a particular amount of loading during different season, depending on the ambient temperature and the temperature of the hottest point of the asset. For example, a conductor's limit is affected by the weather conditions as different temperatures and wind speeds can increase/decrease its heat dissipation rate.

  In the United Kingdom, the Engineering Recommendation (ER) P27 provides seasonal ratings of the cables and lines that are based on statistical studies by the

Consortium of European Research Libraries (CERL). All ratings in ER P27 refer to each conductor's maximum operating temperature during three periods (winter, spring/autumn and summer), for which an average ambient temperature is assumed. The seasonal ratings for overhead lines take also into account the distance from the earth as greater heights may entail higher withstanding temperatures.

Our operational experience with the SNS highlighted that the pre-defined seasonal limits of the OHL are aligned for the majority of time with the average weather conditions, however, discrepancies were identified during periods of seasonal transition (November and March). During these months, the weather conditions were usually similar to the ones observed in December and February, while the difference between the autumn/spring and winter rating was 6 MVA. This leads to higher utilisation of the plant for PS requirements and reduced significantly its availability for the provision of other services.

- **PS provision with active power, reactive power or both**
  The loading limit of a distribution overhead line is defined by its MVA capacity. This allows a power electronics-interfaced storage system to utilise its four-quadrant power capabilities and deliver PS, through the provision of active power and reactive power or both. The delivery of active power aims to reduce active power flows in the line, while reactive power aims to minimise the reactive power flows in the line and brings the power factor as close to unity as possible.

  Operational trials identified the adequacy of a 7.5 MVA conversion system to address PS while it was proven that reactive power prioritised PS can deliver significant benefits to the ESS owner. The utilisation of reactive power can reduce operational expenditure by minimising the utilisation of the batteries and therefore their degradation during the service delivery, reducing energy related costs and charges, time requirements for SoC correction between different services and enabling the plant to deliver other services during PS, such as Triad avoidance [23]. A PS delivery example utilising both reactive and active power is illustrated in Figure 11.20.
- **Synergies with both local network relief and reducing congestion at grid supply point levels**
  The Leighton Buzzard substation is residential load dominated and its demand peaks during winter evening periods concurrently with the peaks observed at a national level. While Triads have been typically occurring during winter evenings over the last decade, it is expected that due to the integration of distributed energy resources during the last years, and hence the increase of resources chasing these Triad periods, the prediction of the three high demand periods will become harder. Energy suppliers and aggregators offer Triad warning services and the industry has observed that the Triad warning periods have increased due to the increase of this unpredictability. It is anticipated that this current synergy between local PS and national demand reduction will become more complex and more elaborated forecasting capabilities will be needed to address potential conflicts and synergies.

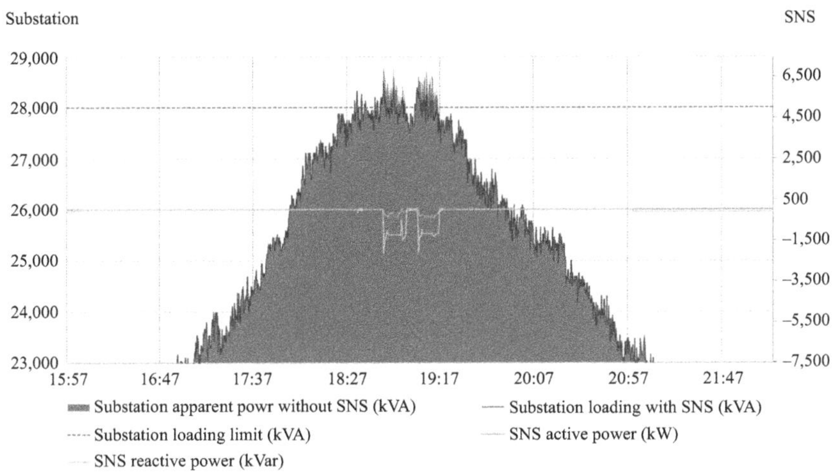

Figure 11.20 Smarter network storage delivering a PS service to the Leighton Buzzard Primary substation by utilising both active and reactive power, with reactive power prioritisation, during the evening hours of 6th March 2015

11.7.2.2 Reserve and capacity market

NG, the TSO in GB, procures two reserve-related commercial products to be able to respond to unforeseen demand increase and/or generation unavailability: STOR and fast reserve. STOR requires a minimum 3 MW power delivery for at least 2 h, this can generally be met by storage systems sized for PS at a primary substation. Fast reserve requires a minimum of 50 MW to be delivered for at least 15 min, a requirement that can be satisfied by only large-scale ESS. Some of the important learnings from the SNS trials are discussed below:

- **Services' structure and economics**
  The current structure and economics of reserve services do not favour ESS in the United Kingdom, as a higher portion of the service value lies in the utilisation payments rather than the availability. The uneven value share brings high uncertainty when developing business cases for ESS, as the number of STOR dispatch signals per year cannot be predicted accurately and the sharp growth of distributed energy resources (DER) further influences the market dynamics, thus making dispatch signals predictions harder. The delivery characteristics of the services require ESSs to retain significant energy and remain idle for prolonged periods. Particular lithium battery technologies require low charge to prolong their lifetime and as such this has a negative impact on the battery degradation.
- **STOR requirements**
  The provision of the STOR service from an ESS requires the export of the contracted power for a period of maximum 2 h. As such, a 2 h full-power declared capacity is required by any resource to respond effectively to a

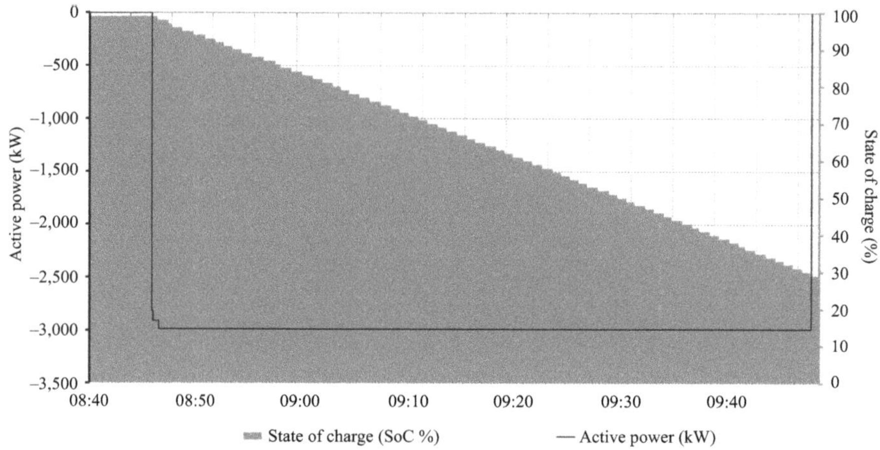

Figure 11.21 During the STOR trial period, a single utilisation event occurred in the morning of 27th August 2015. The storage was called for 1.1 h to deliver the contracted power of 3 MW and was successfully dispatched

dispatch signal. For ESS that have been sized to primarily provide other services, this means that the declared power output will have to be lower than the maximum rating of the plant to satisfy this requirement. An example of the SNS responding to a dispatch signal from the NG is shown in Figure 11.21.

- **Synergies and conflicts with other services**
The active power dispatch for provision of the STOR service can be delivered in parallel to the delivery of reactive power to improve the power factor and/or assist with voltage control to relieve the local network loading. STOR windows are typically during morning (07:30–15:00) and evening periods (16:00–21:30), thus the reactive power delivery can also be utilised for the provision of the PS service, reducing the active power requirements of the service.

  While the two services can work in synergy during STOR evening windows, the two services can be conflicting during morning windows. In case that an ESS is declared for full STOR availability during morning hours, a potential dispatch at 12:00 could lead into risking the delivery of PS and the STOR evening window. This probability is linked with the available headroom that the local substation that could be utilised for a SOC adjustment which requires importing energy to increase the charge of the ESS. As a result, during weeks when the available headroom is low, careful consideration is required in declaring availability for STOR.

The **CM** is a mechanism that ensures that electricity supply continues to meet demand as more volatile and unpredictable renewable generation plants is integration in the generation system and more controllable but emission intensive plants (such as coal) are decommissioned. The minimum capacity required from an asset for the provision of the service is 2 MW, while a typical system stress event lasts for 1 to 2 h (with a maximum of 4 h), a capacity that can be typically met by grid-scale storage applications sized for PS service.

- **CM requirements**
  The participation in the CM offers an ESS a higher value than STOR, as the only requirement of the service is the delivery of the contracted capacity during periods that the TSO issues an Electricity Capacity Market Notice. During the rest of the time, the plant is allowed to deliver other services, increasing its utilisation and maximising its value. In case that a CM Notice is issued while an ESS has been declared to provide balancing services (STOR, FR, Firm frequency response (FFR), FCDM and CMS), according to the CM rules, the plant is not to be penalised for not delivering the balancing service if the contracted capacity is delivered during the event.
- **Synergies with other services**
  Similar to Triads, Electricity Capacity Market Notices are expected to be issued during periods of peak system demand. NG issued two CM notices in autumn winter 2016/2017, both requiring a capacity delivery between 16:30 and 19:00 and one of them being cancelled before the delivery time. Thus, the delivery of capacity during CM notices can work in synergy with export for Triad avoidance and PS services and maximise the profitability of an ESS.

11.7.2.3 Firm frequency response

FFR is a commercial means of providing response services to the TSO. It can be provided via an aggregator to transfer the risk and process of tendering and can be thus declared in a more granular fashion such as on a weekly basis. The operation of an ESS in the FFR market reveals a number of learnings.

- Service requirements
  Firm FR declarations occur in a weekly basis, thus understanding the operational capabilities of the plant (available power and capacity), the requirements for PS services during winter and the droop curve characteristics can lead to minimising the SOC adjustments and maximising the asset value.
  The second-by-second delivery of active power, linked with the frequency measurements through a droop curve, requires a complex architecture for the control system that includes SCADA equipment, power measurement equipment and frequency measurement equipment. Due to the increased complexity, the system has multiple failure points and it is essential to include consideration of adequate levels of equipment redundancy in the design of an ESS.
- Operating costs and commercial variables
  FFR when compared to other ancillary services, such as STOR, has a higher utilisation, as the output setpoint of the plant may be changing every second. The high utilisation increases the overall auxiliary consumption of the plant leading in higher OPEX and degrades the batteries faster. Added to the auxiliary consumption costs, other charges such as DUoS and energy/settlement

charges/PPA are uncertain as they depend on grid frequency and should also be taken into account when developing ESS business case studies [24].

- Synergies with other services
  Similar to previous services that are delivered through the provision of active power, during FFR delivery, reactive power can be utilised for adjusting the power factor and supporting the voltage levels of the local network. However, due to the unpredictability of the frequency and the output power of the plant, it must be ensured that at all times, the total delivered power of the plant is below the set limits to avoid high voltage fluctuations that can lead into tripping circuit breakers. An example of the SNS delivering both dynamic firm FR and voltage control is illustrated in Figure 11.22.
- Contractual variables
  The FFR market in the United Kingdom is an evolving market and due to the integration of a large amount of ESS and other DER that are able to provide this service, it is also a highly dynamic market. Its requirements depend on the amount of synchronous generation that exists on the GB system and the amount of resources that may offer the service. The longest period that a single provider may tender for providing the service is two years, nevertheless contracts through aggregators may be more flexible to ensure that the value of an asset is maximised.

  Changes in the service requirements, such as droop curves, or the technology used to provide the service, such as the increase of declared capacity or changes in control systems, require pre-qualification of the total unit. This is a lengthy process that may reduce the declared availability of the plant and in turn its projected revenues.

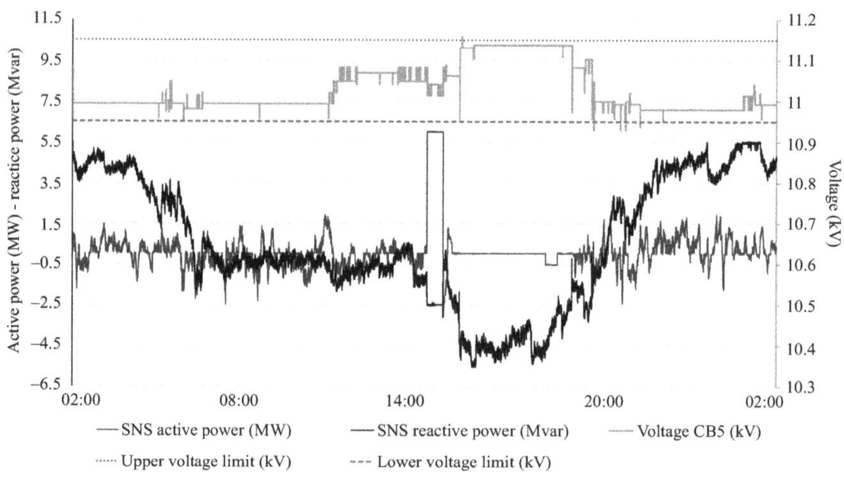

Figure 11.22 SNS delivering dynamic frequency response utilising active power and voltage control utilising reactive power simultaneously during the time of day 9–10 October 2016

11.7.2.4 Energy market

A tolling service is profitable when the differential in energy prices is significant to cover the assets utilisation and operational expenditure while also provides profits for the owner and operator of the asset and the service. In a tolling service, an electricity supplier may identify a high price differential in the short term and may request the usage of a particular operating profile from the ESS to capture the benefit of the price volatility and arbitrage on price.

The current price differential is not significant to justify the sole use of a battery ESS for that service provision. It is expected that the differential will increase in the forthcoming years with the increase of the cash-out price cap, the increased intermittency of generation and the access to multiple markets, including the balancing market and spot markets. An example of an ESS delivering a tolling service is shown in Figure 11.23.

A tolling service however requires a particular amount of energy to be used every HH/settlement period. As such there exists slack in optimising the operation of a battery ESS in:

- Optimising the health of the battery system by balancing the SOC amongst cells to prolong their lifetime
- Utilising plant capacity for arbitraging on price while providing other services such as:
  - FR while the grid frequency remains within a pre-defined deadband
  - Utilising reactive power to improve the power factor or assist with voltage control
  - Maintaining security of supply locally providing PS and optimising use of the plant's reactive power capabilities

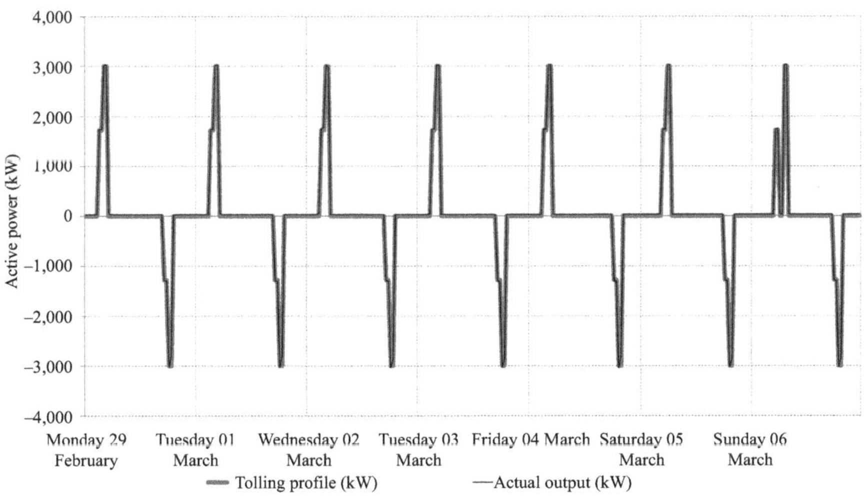

Figure 11.23 Tolling profile and actual output of the SNS during the tolling trials that were conducted during the week commencing 29th February

11.7.3 Multiple service scheduling

ESSs are a multi-purpose technology i.e. storage can do different things at the same time. This capability provides several benefits listed as follows:

- Technical benefits: There are synergies across the whole electricity system between applications and services that ESSs can provide.
  (i) When ESSs export active power during winter evening periods in the delivery of a PS service, they are relieving congested local networks from peak loading. At the same time they contribute to regional peak relief at grid supply points and reduce the peak load at a national level (i.e. provide Triad avoidance services).
  (ii) When ESSs utilise their reactive power capability for PS, the power factor at the point of the ESS connection with the distribution network is also improved. This leads to reduction in losses and improved efficiency of the distribution network.
- Commercial benefits: Revenues from multiple ESS applications can be stacked to improve the business case for storage. In the United Kingdom, a number of applications can be combined to increase the value of the asset and these range from reinforcement deferral, through to provision of CM, Triad avoidance and balancing services. The stacking of services and the development of a business case however needs to be assessed on a per-site basis and varies for each business model i.e. a DNO-owned storage model differs from a large commercial or industrial user's (behind the meter) model and a standalone storage business model.

11.8 Beyond SNS

11.8.1 Managing uncertainty

In SNS, the calculations to determine the value of the services rely on several parameters which are uncertain; the probability of delivery, cost of energy and exact service fees are generally not known in advance, so any values used here are an estimate. This implies that uncertainty should be considered in this evaluation. An imprecise probability approach, where a range of probability values are used for service delivery, could allow for more rigorous evaluations of the EV. In addition, using probability distributions or probability mass functions (PMFs) to represent the energy and fee values would allow this uncertainty to be represented in the calculations. A Bayesian approach could be used, with prior values of the costs and probabilities obtained using the operator's judgement, and posterior values produced after some services have been delivered.

The optimisation algorithm is likely to become infeasible if many of the suggested improvements are implemented. Given that the service scheduling represents a tree structure, a tree search algorithm could be implemented.

11.8.2 Combining smart grid technologies

ESS have the potential to be installed alongside other network interventions, including demand side response, active network management and Real-time thermal ratings (RTTR). In this section, we will discuss the benefits of co-deploying ESS and RTTR, but many of the advantages of this approach are also true for other combinations of technologies.

Both ESS and RTTR can increase the effective capacity of an electrical network – RTTR does this by releasing latent thermal capacity [25,26], while ESS achieves it by supplying demand locally during peak consumption via a PS service [27]. Analysis carried out during the SNS project indicates that a combination of both technologies can defer conventional network reinforcement for longer [28] than either technology on its own.

Figure 11.24 shows data analysis from SNS. Assuming an expected energy not supplied (EENS) of around 40 MWh/year was sufficient to require reinforcement, the EENS of the network with the ESS, RTTR, both and neither, were calculated via Monte Carlo simulation. Deploying either technology was found to bring the EENS below the reinforcement threshold. However, further simulations, under demand growth scenarios from 1% to 10%, found that if only one technology is deployed, the reinforcement threshold will be reached with demand growth of around 9.5%; if both are deployed, the EENS reaches 25 MWh/year with 10% demand growth, and there is potential to further increase the demand while maintaining the initial level of security of supply.

By deploying RTTR, the number of instances of the demand exceeding the network rating – and therefore requiring the operation of the ESS for peak having – will be reduced. This leads to a number of additional hours in a given year in which the ESS can offer balancing services to the SO, rather than being required to provide additional capacity on the local distribution network.

Figure 11.25 shows the lifetime value of the SNS ESS in four scenarios. In all cases, the ESS provides security of supply to a distribution network substation for 10 years, during which time the demand grows by 1% per year. If the system is installed using static ratings, the system makes a loss of around £5M if it only carries out PS,

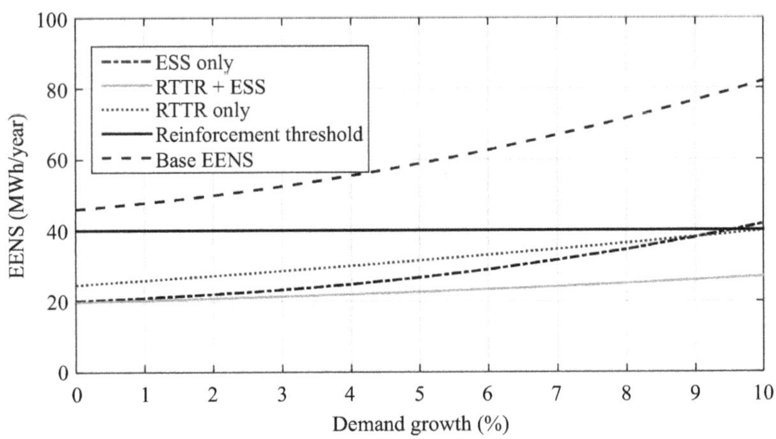

Figure 11.24 Example of network investment deferral for an ESS, RTTR and the combination of both technologies [28]

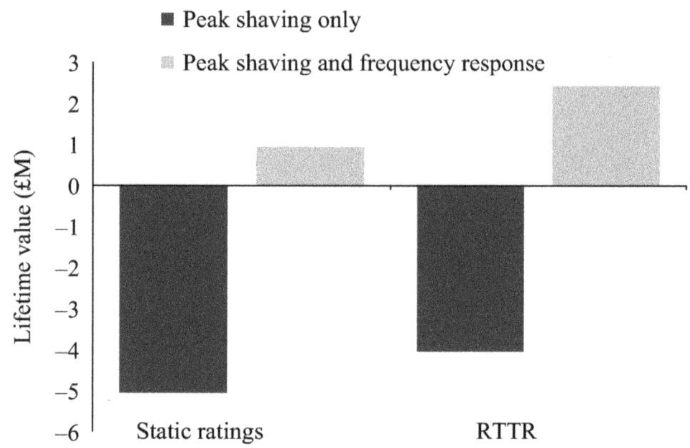

Figure 11.25 Bar charts showing the lifetime value of the SNS system, using static and real-time thermal ratings, and offering either peak shaving or PS and frequency response services

and a profit of around £0.9M if it offers FR as an additional service. With RTTR, the loss is reduced to £4M in the PS only scenario, and the profit is increased to £2.4M when FR is offered – the additional value provided by RTTR in this case is from £1-1.5M. This analysis was carried out as part of [28], and includes an estimate of the cost of installing the RTTR system. This analysis only considered one commercial service – in a real implementation, many service options are available, and the value of the ESS could be increased by selecting the most appropriate combination of services. The value from reinforcement deferral is also excluded, so that the figure for 'loss' could be considered as the cost of reinforcement that can be offset.

11.9 Summary

This chapter has looked at several aspects of the SNS project undertaken by UK Power Networks in order to explore the practical operation of grid-connected energy storage in delivering services to multiple stakeholders in the electricity supply industry. The forecasting, optimisation and scheduling elements of the management software have been outlined. Practical issues arising from commercial, technical and regulatory domains are discussed. Operational experience of the storage system has been given in which a substation capacity constraint is managed, ancillary services are provided to the TSO and the energy market is accessed through a tolling contract. These services are delivered both individually and in combination. Some possible next steps are outlined.

In addition to the technical operation of a storage system, the market and regulatory environment are also crucial to the viability of any project [3]. The SNS project explored alternative business models and the effect of the regulatory environment on the adoption of energy storage in electricity networks as well as the technical requirements.

References

[1] UK Power Networks. (2014). Demonstrating the benefits of short-term discharge energy storage on an 11kV distribution network. Available: http://innovation.ukpowernetworks.co.uk/innovation/en/Projects/tier-1-projects/demonstrating-the-benefits-of-short-term-discharge-energy-storage/

[2] UK Power Networks. (2016, 25 May). Smarter network storage. Available: http://innovation.ukpowernetworks.co.uk/innovation/en/Projects/tier-2-projects/Smarter-Network-Storage-(SNS)/

[3] O. H. Anuta, P. Taylor, D. Jones, T. McEntee, and N. Wade, "An international review of the implications of regulatory and electricity market structures on the emergence of grid scale electricity storage," Renewable and Sustainable Energy Reviews, vol. 38, pp. 489–508, Oct 2014.

[4] SSEN (2016, 26/01/2017). SSEN opens constraint managed zone procurement process. Available: http://news.ssen.co.uk/news/all-articles/2016/12/ssen-opens-constraint-managed-zone/

[5] G. J. Tsekouras, N. D. Hatziargyriou, and E. N. Dialynas, "Two-stage pattern recognition of load curves for classification of electricity customers," IEEE Transactions on Power Systems, vol. 22, pp. 1120–1128, 2007.

[6] M. Clark and L. Lampe, "Electrical grid peak reduction with efficient and flexible automated demand response scheduling," in Electrical and Computer Engineering (CCECE), 2015 IEEE 28th Canadian Conference on, 2015, pp. 818–823.

[7] D. Greenwood, N. Wade, P. Papadopoulos, N. Heyward, P. Mehtah, and P. Taylor, "Scheduling power and energy resources on the smarter network storage project," presented at the 23rd International Conference and Exhibition on Electricity Distribution, Lyon, France, 2015.

[8] National Grid. (2017, 17/02). Balancing services. Available: http://www2.nationalgrid.com/uk/services/balancing-services/

[9] D. M. Greenwood, K. Y. Lim, C. Patsios, P. F. Lyons, Y. S. Lim, and P. C. Taylor, "Frequency response services designed for energy storage," Applied Energy, vol. 203, pp. 115–127, 2017/10/01/2017.

[10] National Grid. (2016, 07/07/2016). Enhanced frequency response. Available: http://www2.nationalgrid.com/Enhanced-Frequency-Response.aspx

[11] C. Mullen, P. C. Taylor, V. Thornley, and N. S. Wade, "Use of standby generation for reduction of transmission network charges for half-hourly metered customers," in Power Engineering Conference (UPEC), 2014 49th International Universities, 2014, pp. 1–6.

[12] National Grid, "Provisional Auction Results T-4 Capacity Market Auction 2020/21," 2016.

[13] K. Bradbury, L. Pratson, and D. Patiño-Echeverri, "Economic viability of energy storage systems based on price arbitrage potential in real-time U.S. electricity markets," Applied Energy, vol. 114, pp. 512–519, 2// 2014.

[14] UK Power Networks (2013). SNS2.15 Logical architecture design document. Available: http://innovation.ukpowernetworks.co.uk/innovation/en/Projects/

tier-2-projects/Smarter-Network-Storage-(SNS)/Project-Documents/SNS2.15-Logical-Architecture-Design-Dec+2013.pdf

[15] S. Fan and R. J. Hyndman, "Short-term load forecasting based on a semi-parametric additive model," IEEE Transactions on Power Systems, vol. 27, pp. 134–141, Feb 2012.

[16] H. S. Hippert, C. E. Pedreira, and R. C. Souza, "Neural networks for short-term load forecasting: A review and evaluation," IEEE Transactions on Power Systems, vol. 16, pp. 44–55, Feb 2001.

[17] B. J. Chen, M. W. Chang, and C. J. Lin, "Load forecasting using support vector machines: A study on EUNITE competition 2001," IEEE Transactions on Power Systems, vol. 19, pp. 1821–1830, Nov 2004.

[18] R. J. Hyndman and S. Fan, "Density forecasting for long-term peak electricity demand," IEEE Transactions on Power Systems, vol. 25, pp. 1142–1153, May 2010.

[19] T. Hong, M. Gui, M. Baran, and H. Willis, "Modeling and forecasting hourly electric load by multiple linear regression with interactions," presented at the IEEE Power and Energy Society General Meeting, 2010.

[20] J. Bather, Decision theory. West Sussex, UK: Wiley, 2000.

[21] D. P. Bertsekas, Dynamic programming and optimal control. Belmont, MA: Athena Scientific, 2005.

[22] D. M. Greenwood, N. S. Wade, P. C. Taylor, P. Papadopoulos, N. Heyward, and S. Alilat, "A forecasting, optimization and scheduling system for energy storage systems in distribution networks," in IEEE Power and Energy Society General Meeting. Boston, MA: IEEE, 2016.

[23] R. Moreira, G. Strbac, P. Papadopoulos, and A. Laguna, "Business case in support for reactive power services from distributed energy storage," presented at the 24th International Conference on Electricity Distribution, Glasgow, UK, 2017.

[24] National Grid, "Applicable balancing services volume data methodology statement", Warwick, UK, 2017.

[25] D. M. Greenwood and P. C. Taylor, "Investigating the impact of real-time thermal ratings on power network reliability," IEEE Transactions on Power Systems, vol. 29, pp. 2460–2468, Sep 2014.

[26] D. M. Greenwood and P. C. Taylor, "Unlocking the benefits of real-time thermal ratings through probabilistic power network planning," IET Generation Transmission and Distribution, vol. 8, pp. 2055–2064, Dec 2014.

[27] D. M. Greenwood, N. S. Wade, P. C. Taylor, P. Panagiotis, N. Heyward, and P. Mehtah, "Scheduling power and energy resources on the smarter network storage project," in 23rd International Conference and Exhibition on Electricity Distribution, Lyon, France, 2015.

[28] D. M. Greenwood, N. S. Wade, P. C. Taylor, P. Papadopoulos, and N. Heyward, "A probabilistic method combining electrical energy storage and real-time thermal ratings to defer network reinforcement," IEEE Transactions on Sustainable Energy, vol. 8, pp. 374–384, Jan 2017.

Chapter 12

Optimal coordination between generation and storage under uncertainties

Dinghuan Zhu[1]

12.1 Background

With the increase of renewable energy sources such as wind and solar, the task of maintaining a good balance between power generation and load consumption in power systems becomes more and more challenging. As a countermeasure, the grid-level energy storage technology has been recognized as a very promising technology for helping integrate a significant amount of green but variable renewable generators into the grid [1–3]. With the advances in electrochemistry and power electronics [4], energy storage devices have been attracting more and more attentions for grid-scale applications. Energy storage such as batteries and supercapacitors has the advantage of fast response capabilities over the conventional generators, making them suitable for providing power-balancing support for the grid.

The current industry practice for power balancing consists of primary, secondary, and tertiary frequency controls. Primary frequency control is a proportional control that is used locally by generators to stabilize the frequency. Secondary control is a centrally organized integral control for a control area to regulate the frequency and tie-line flows. Tertiary control is an optimization-based generation resource scheduling and dispatching task involving economic dispatch.

However, this current practice may not be sufficient in the future grid environment with increased power variations from the generation side for four main reasons: (1) the primary and secondary controls are not designed to quickly respond to power imbalance; (2) rapidly varying power output is usually harmful to conventional generators; (3) it does not take into account the increased uncertainty caused by RESs at the tertiary level, very likely resulting in economically inefficient dispatch of generation resources; (4) more importantly, the existing power-balancing framework does not take into consideration the special operational requirement on state of charge (SoC) for the storage type assets, which may adversely affect the safe operation of the entire system.

[1]Dinghuan Zhu is with the Control Design Automation department at The MathWorks, Inc., MA, USA. He received his Ph.D. degree in Electrical and Computer Engineering from Carnegie Mellon University, PA, USA, in 2014.

12.2 A new two-level approach

The traditional three-level framework needs to be restructured and redesigned to address all of the four aforementioned issues. The redesigned approach consists of two levels, including the advanced frequency control (AFC) level and the stochastic optimal dispatch level. AFC incorporates primary control and the frequency regulation task in secondary control. The level of stochastic optimal dispatch which is specifically based on the stochastic model predictive control (SMPC) method takes into consideration not only the uncertainty introduced by RESs and demand but also multiple time steps in a look-ahead horizon. The term "stochastic model predictive control-based optimal dispatch" (short for "SMPC-based optimal dispatch") and the term "stochastic optimal dispatch" are used interchangeably in this chapter to refer to this level. In addition, the power and SoC constraints associated with energy storage devices are considered in both levels.

The relation between the two control levels shown in Figure 12.1 follows the basic paradigm of hierarchical control. The stochastic optimal dispatch level collects all the relevant data from the physical power system to perform the stochastic optimization every 5 to 15 min, yielding the optimal generator and storage settings to be implemented by the AFC level for the current time interval. The AFC level adjusts the generator and storage power output set points once it is instructed by the stochastic optimal dispatch level and continuously takes the responsibility of maintaining frequency within a tight band by imposing physical control actions. Overall, the two-level control keeps the physical power system operating in a safe and most economical manner.

In the new two-level approach, the coordination between conventional generation and energy storage is achieved by assigning the power balancing responsibilities to these resources in line with their capabilities. The regular generators take care of the low-frequency region in the spectrum of power fluctuations while the energy storage devices with their fast response capability are responsible to

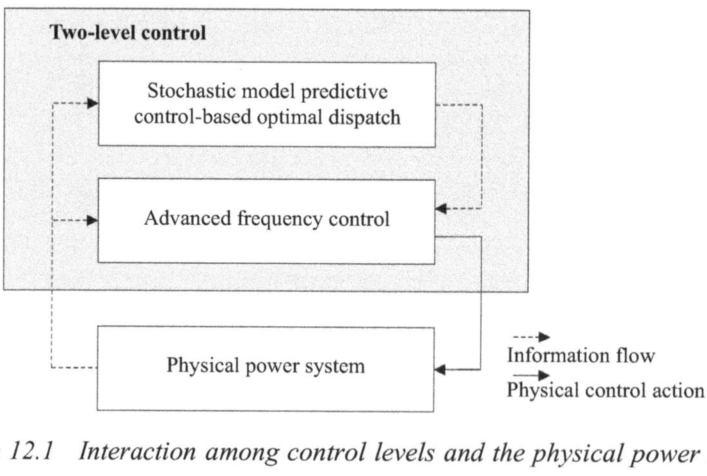

Figure 12.1 Interaction among control levels and the physical power system

mitigate the high-frequency spectrum region. This is the underlying design philosophy of the so-called *time-scale matching*. With the help of fast and high ramping capability provided by energy storage devices, the wear-and-tear effect of conventional generators can therefore be reduced [5].

12.3 AFC level

The AFC level is designed to stabilize and restore system frequency to its nominal value, which addresses the first, second, and fourth issues mentioned in Section 12.1, i.e., the current frequency control scheme fails to quickly respond to power imbalance and does not take into consideration the special operational requirement on SoC for the storage type assets. The design of AFC aims to make full use of both the generator and energy storage assets according to their different capabilities in counteracting real-power mismatches of the system.

In essence, AFC is a unified control that synthesizes both the primary and secondary frequency controls. The two most desirable properties—the paradigm of decentralized control in the traditional primary frequency control and the functionality of frequency restoration in the secondary frequency control—are preserved in the proposed AFC. This level of control operates continuously in real time. The robust control theory together with the decentralized static output feedback design is adopted in order to achieve the robustness and stability of the considered power system as well as the simplicity of the controllers. Another big advantage of robust control is that it allows the synthesis of frequency-dependent weighting functions into the objective function facilitating the implementation of the time-scale matching objective. Thus, the main objective of this control level is still to tightly regulate frequency on all generator buses. The new control design additionally allows us to minimize the required ramping of conventional generators and at the same time to minimize the SoC deviations for energy storage devices.

12.3.1 Feasibility of decentralized PI frequency control

As mentioned in Section 12.1, decentralized control merely exists in the primary level of the current hierarchical frequency control framework in terms of each control area. In other words, there only exist decentralized proportional controllers (P controllers) in current power systems. The combination of the primary control and secondary control actually exhibits the behavior of proportional-integral controllers (PI controllers). The question of whether it is viable to integrate the primary and secondary control levels and design a unified decentralized PI frequency control scheme needs to be answered first. Developing decentralized PI frequency control for generators is a critical first step in optimally integrating energy storage into power systems, as the decentralized PI control makes generators more responsive to power disturbance but only with respect to low-frequency components. The high-frequency components in power disturbance can therefore be compensated by energy storage, which achieves the time-scale matching principle.

The two main technical barriers that currently hinder the application of decentralized PI control in power system frequency regulation are that (1) local frequency controllers are mostly designed based on the sole knowledge of the corresponding local generating unit model and (2) power sharing among generators in responding frequency deviations is not clear when multiple PI controllers are placed in the system even if the system stability is ensured. Due to the first barrier, multiple PI controllers designed in that way would very likely fight against each other and each tries to gain the dominance in regulating system frequency when they are interconnected, causing instability of the system. With respect to the second barrier, the underlying reason is because there is no precise time synchronization scheme implemented among different frequency controllers of different generators in a control area. The integration in all the PI controllers must start at the same time in order for the decentralized PI control to function properly. Otherwise, the power output of each generator depends on not only the power mismatches and the controller settings but also the time when the controller starts its integration process. The generators with the integral control part kicking in early might take more responsibility in balancing power mismatches.

In order to deal with the first barrier, a structure-preserving approach needs to be conceived where the interactions among generators via power transmission networks should be explicitly taken into account. Based on such a system-wide model, it is of more confidence that the yielding PI controllers are able to cooperate with each other to stabilize the system and bring the frequency back to its nominal value. On the other hand, the second barrier can be easily solved by employing the mature technology of phasor measurement units (PMUs) because of their embedded function of Global Positioning System (GPS)-based time synchronization. Voltage angle information that is indeed the integral of the local frequency is directly measured with a precise GPS time stamp by PMUs so that a common time reference among generators that might be located far away from each other can be established.

Definition 1 (Synchronization for PI Controllers): *Multiple PI controllers in a network are said to be synchronized if and only if the integral control part of each controller starts at the same time.*

In terms of the voltage angle which is the integral of the frequency, the time synchronization of PI controllers ensures that the reference values for voltage angles at each bus are measured at the same time. For example, consider two voltage angles at two buses at time t denoted by θ_1^t and θ_2^t. If PI controllers are placed at these two buses, the control inputs at time t in terms of deviations from their nominal values are given by

$$\begin{aligned} \Delta P_1^t &= -k_1 \Delta\omega_1^t - k_2\left(\theta_1^t - \theta_1^{t_1}\right) \\ \Delta P_2^t &= -k_3 \Delta\omega_2^t - k_4\left(\theta_2^t - \theta_2^{t_2}\right) \end{aligned} \tag{12.1}$$

where superscripts t_1 and t_2 are the time instances when the two corresponding angle reference values were measured, respectively; $\Delta\omega_1^t$ and $\Delta\omega_2^t$ are the frequency deviations at time t from the nominal value for Bus 1 and Bus 2, respectively; k_1, k_2, k_3, and k_4 are positive control gains. In such a setting, the two controllers are said to be synchronized if and only if the condition $t_1 = t_2$ holds.

Before we develop the structure-preserving approach for system modeling, we first provide a theorem in the following to establish the uniqueness of the generator power sharing with synchronized multiple PI frequency controllers in an electric power system modeled by the DC power flow model.

Definition 2 (Uniqueness of Generator Power Sharing): *The steady-state generator power sharing for frequency control in a given power network is said to be unique if and only if the power output of each generator merely depends on the power mismatches of the system and the parameters of frequency controllers.*

Theorem 1 (Uniqueness Theorem of Generator Power Sharing Under Decentralized PI Frequency Control): *In an electric power system modeled by the DC power flow model, the steady-state power sharing among generators in responding frequency deviations is unique under negative feedback via synchronized decentralized PI frequency controllers given that the closed loop system is stable.*

Proof. Without loss of generality, we assume that there are $(n+m)$ buses in the network and the first n buses are generator buses and the remaining m buses are load buses.

The DC power flow model, which is widely adopted in the power system steady-state analysis, yields the following relation between bus power injections and bus voltage angles:

$$\begin{bmatrix} \Delta P_G^1 \\ \vdots \\ \Delta P_G^n \\ -\Delta P_L^1 \\ \vdots \\ \Delta P_L^m \end{bmatrix} = H \cdot \begin{bmatrix} \Delta\theta_1 \\ \vdots \\ \Delta\theta_n \\ \Delta\theta_{n+1} \\ \vdots \\ \Delta\theta_{n+m} \end{bmatrix} = \begin{bmatrix} H_{1,1} & \cdots & \cdots & H_{1,n+m} \\ \vdots & & & \vdots \\ \vdots & \ddots & & \vdots \\ \vdots & & \ddots & \vdots \\ \vdots & & & \vdots \\ H_{n+m,1} & & & H_{n+m,n+m} \end{bmatrix} \begin{bmatrix} \Delta\theta_1 \\ \vdots \\ \Delta\theta_n \\ \Delta\theta_{n+1} \\ \vdots \\ \Delta\theta_{n+m} \end{bmatrix} \tag{12.2}$$

where ΔP_G^i, ΔP_L^j are the generator power output deviation at Bus i and the load power deviation at Bus $(n+j)$, respectively; $\Delta\theta_k$ is the voltage angle deviation at Bus k with respect to its reference value, i.e., $\Delta\theta_k = \theta_k - \theta_k^{t0}$ with $t0$ being the initial time when the controller synchronization was done; the matrix H is known as the admittance matrix. Three important properties associated with H are that (1) it is structurally singular; (2) it is diagonally dominant with positive diagonal elements; and (3) its first minors associated with the diagonal elements $M_{i,i}$ are positive.

Since the closed-loop system is stable, the power output of the generator with negative feedback PI control at Bus i in steady state is proportional to the local voltage angle deviation, i.e., $\Delta P_G^i = -k_i \cdot \Delta\theta_i$ with $k_i > 0$. We therefore prove the theorem by proving the uniqueness of the vector for voltage angle deviations given deterministic load changes. The following two cases cover all the possibilities.

1. One single PI controller is installed.
   Without loss of generality, we assume that the first generator adopts the PI control. Thus, rearranging (12.2) gives us

$$\begin{bmatrix} 0 \\ \vdots \\ 0 \\ -\Delta P_L^1 \\ \vdots \\ -\Delta P_L^m \end{bmatrix} = \begin{bmatrix} H_{1,1}+k_1 & \cdots & \cdots & H_{1,n+m} \\ \vdots & & & \vdots \\ \vdots & \ddots & & \vdots \\ \vdots & & \ddots & \vdots \\ \vdots & & & \vdots \\ H_{n+m,1} & & & H_{n+m,n+m} \end{bmatrix} \cdot \begin{bmatrix} \Delta\theta_1 \\ \vdots \\ \Delta\theta_n \\ \Delta\theta_{n+1} \\ \vdots \\ \Delta\theta_{n+m} \end{bmatrix} \tag{12.3}$$

We denote the modified admittance matrix as $\bar{H}$. Next, we are going to prove that $\bar{H}$ is nonsingular. On the one hand, the Laplace expansion along the first row of the original admittance matrix H yields

$$0 = \det(H) = \sum_{j=1}^{n+m} (-1)^{1+j} H_{1,j} M_{1,j} = H_{1,1}M_{1,1} + \sum_{j=1, j\neq 1}^{n+m} (-1)^{1+j} H_{1,j} M_{1,j} \tag{12.4}$$

On the other hand, the determinant of $\bar{H}$ can also be calculated by the Laplace expansion along its first row:

$$\begin{aligned} \det(\bar{H}) &= \sum_{j=1}^{n+m} (-1)^{1+j} \bar{H}_{1,j} M_{1,j} = \bar{H}_{1,1}M_{1,1} + \sum_{j=1, j\neq 1}^{n+m} (-1)^{1+j} H_{1,j} M_{1,j} \\ &= k_1 M_{1,1} + \det(H) = k_1 M_{1,1} > 0 \end{aligned} \tag{12.5}$$

From the third property of the admittance matrix and the condition of $k_i > 0$, $\bar{H}$ is indeed nonsingular and more precisely positive definite so that the vector of voltage angle deviations is uniquely determined given deterministic load changes.

2. Multiple PI controllers are installed.
   We prove this case following the same idea in Case 1. Without loss of generality, we further assume that PI control is applied to the i-th generator. Now, rearranging (12.2) gives us

$$\begin{bmatrix} 0 \\ \vdots \\ 0 \\ -\Delta P_L^1 \\ \vdots \\ -\Delta P_L^m \end{bmatrix} = \begin{bmatrix} H_{1,1}+k_1 & & & H_{1,n+m} \\ \vdots & \ddots & & \vdots \\ \vdots & & H_{i,i}+k_i & \vdots \\ \vdots & & \ddots & \vdots \\ \vdots & & & \vdots \\ H_{n+m,1} & \cdots & \cdots & \cdots \quad H_{n+m,n+m} \end{bmatrix} \cdot \begin{bmatrix} \Delta\theta_1 \\ \vdots \\ \Delta\theta_n \\ \Delta\theta_{n+1} \\ \vdots \\ \Delta\theta_{n+m} \end{bmatrix} \quad (12.6)$$

Denote the newly modified admittance matrix as $\tilde{H}$. On the one hand, the Laplace expansion along the i-th row of $\bar{H}$ in (12.3) gives:

$$0 < \det(\bar{H}) = \sum_{j=1}^{n+m} (-1)^{i+j} \bar{H}_{i,j} \bar{M}_{i,j} = \bar{H}_{i,i} \bar{M}_{i,i} + \sum_{j=1, j\neq i}^{n+m} (-1)^{i+j} \bar{H}_{i,j} \bar{M}_{i,j} \quad (12.7)$$

On the other hand, the determinant of $\tilde{H}$ can be calculated as

$$\begin{aligned} \det(\tilde{H}) &= \sum_{j-1}^{n+m} (-1)^{i+j} \tilde{H}_{i,j} \bar{M}_{i,j} = \tilde{H}_{i,i} \bar{M}_{i,i} + \sum_{j-1, j\neq i}^{n+m} (-1)^{i+j} \bar{H}_{i,j} \bar{M}_{i,j} \\ &= k_i M_{i,i} + \det(\bar{H}) > 0 \end{aligned} \quad (12.8)$$

The positiveness of $\bar{M}_{i,i}$ can be established from the positiveness of $M_{i,i}$ and k_1. Therefore, $\tilde{H}$ is indeed nonsingular and more precisely positive definite so that the vector of voltage angle deviations is uniquely determined given deterministic load changes.

Based on the foregoing proof, it is trivial to see that the theorem holds for the general case of multiple PI controllers by using mathematical induction.

For both the two cases, it can be shown that the vector of voltage angle deviations is uniquely determined by the power mismatches of the system and the parameters of PI frequency controllers. Therefore, the steady-state generator power sharing is unique as the steady-state power outputs of generators are proportional to local voltage angle deviations. This concludes the complete proof of the theorem.

As long as the two aforementioned barriers are overcome and additionally the stability of the system is guaranteed, there should be no problems in implementing decentralized PI frequency controllers in power systems.

12.3.2 Structure-preserving system modeling

Instead of modeling the whole control area as a single aggregated generator like in many existing studies such as [6], the proposed AFC control utilizes a structure preserving modeling approach to faithfully reflect the physical world situation as the frequencies at different locations in a network can be different. As an example, the authors of [6] assume that the frequency of each control area is a global quantity and is uniform across the entire control area.

The structure-preserving model is in fact a set of differential algebraic equations (DAEs), which captures the generator dynamics and the dynamics associated with the energy storage devices with inverters. The electric power system modeled by the DC power flow model serves as the algebraic constraints. Load and renewable generation are modeled as external disturbances. In this setting, the dynamics of each individual device are synthesized together in a way that reflects the physical world situation. The structure-preserving state-space model of the entire system can be written in the following compact form. Interested readers may refer to [7] for the detailed derivation of the structure-preserving model with detailed dynamic models of conventional generators and energy storage devices.

$$\dot{x} = Ax + B_1 w + B_2 u \tag{12.9}$$

where x is the state vector including the generator rotor speed, generator voltage angle, generator mechanical power, generator turbine valve position, energy storage SoC, and energy storage power injection; w is the external disturbance vector including load consumption and renewable generation; u is the control input vector containing generator control input reference and energy storage control input reference.

12.3.3 Basics of robust control

Figure 12.2 shows a typical system setup of an H_∞ control problem. The physical system under investigation is captured by the transfer function $G(s)$; $K(s)$ is the transfer function of the controller to be solved; w is the external input vector which can include disturbances and reference signals; u represents the control input vector; y includes all the measurements from the physical system model; z is the so-called performance vector reflecting our control objectives, which is the key part in H_∞ control design [8,9].

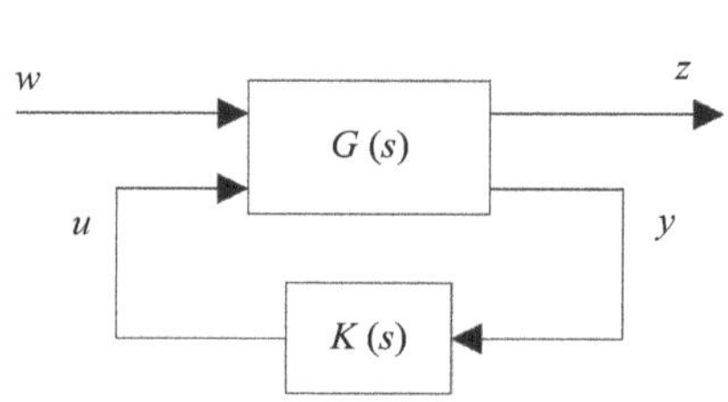

Figure 12.2 H_∞ problem setup

The task in H_∞ control design is to find a stabilizing controller $K(s)$ that minimizes the H_∞ norm of the closed loop transfer function from w to z. The H_∞ norm of this transfer function is defined as

$$\|T_{zw}(s)\|_\infty \triangleq \sup_\omega \bar{\sigma}(T_{zw}(j\omega)) \tag{12.10}$$

where $\bar{\sigma}(\cdot)$ operator stands for the operation of finding the maximum value among all the singular values of a matrix. In this particular case, the operand of $\bar{\sigma}(\cdot)$ is a complex matrix. It can also be proven that $\|T_{zw}(s)\|_\infty$ has the following equivalence in the time domain [9]:

$$\sup_\omega \bar{\sigma}(T_{zw}(j\omega)) = \sup_{\|w(t)\|_2 \neq 0} \frac{\|z(t)\|_2}{\|w(t)\|_2} \tag{12.11}$$

It should now be clear to readers that the H_∞ control minimizes the "amplification" effect between the performance vector and the external input vector. In other words, if the 2-norm of the exogenous input vector is bounded, the resulting 2-norm of the performance vector that includes our control objectives is guaranteed to be bounded as well by a factor of the H_∞ norm of $T_{zw}(s)$.

The simplest form of H_∞ control shown in Figure 12.2 equally treats all the frequency components in both the exogenous input vector w and the control performance vector z. Based on the specific control design objectives, H_∞ control additionally allows us to specify weighting functions for w and z to only focus on the interested frequency spectrum. The augmented H_∞ control problem with weighting functions is therefore written as

$$\min_{K(s)} \|W_z(s)T_{zw}(s)W_w(s)\|_\infty \tag{12.12}$$

where $W_z(s)$, $W_w(s)$ are matrix-valued weighting functions associated with z and w, respectively. The importance of each control objective is encoded in these matrices.

12.3.4 AFC controller design with energy storage

To recap, the main responsibility of the AFC level is to tightly regulate both the frequencies at generator buses and the SoC of the energy storage devices. On top of that, the time-scale matching objective needs to be achieved so that the two types of assets can share the responsibilities of power balancing according to their capabilities. Hence, the performance vector is chosen as

$$z = \left[\Delta\omega_1, \cdots, \Delta\omega_{N_G}, \Delta SoC_1, \cdots, \Delta SoC_{N_S}, \Delta P_{m,1}, \cdots, \Delta P_{m,N_G}, \Delta P_{S,1}, \cdots, \Delta P_{S,N_S}\right]^T \tag{12.13}$$

where $\Delta\omega_i$ and $\Delta P_{m,i}$ are the rotor angular velocity (i.e., the local frequency) and the mechanical power input for the i-th generator, respectively; ΔSoC_i and $\Delta P_{S,i}$ are the SoC and the output power reference for the i-th storage device, respectively.

The key of the AFC level design is to choose appropriate frequency-dependent weighting functions for $\Delta P_{m,i}$ and $\Delta P_{S,i}$ so that the conventional generators take care of relatively low-frequency components in external power disturbances and the energy storage devices handle the high-frequency spectrum region. For conventional generators, more penalties are placed on the high-frequency spectrum region than the low-frequency region. When minimizing the weighted transfer function from w to z, the frequency spectrum of the generator power output behaves like a low-pass filter. The weighting function settings for energy storage are exactly opposite. More penalties are placed onto the low-frequency region. The type of weighting functions adopted in the AFC design can be written as

$$W_{G,i}(s) = \frac{(10s + 20\pi f_c)^n}{(s + 20\pi f_c)^n} \tag{12.14}$$

$$W_{S,i}(s) = \frac{(s + 2\pi f_c)^n}{(s + 0.2\pi f_c)^n} \tag{12.15}$$

where the subscript i stands for the i-th individual generator or storage device; f_c is the separation frequency in time-scale matching (a.k.a. the cutoff frequency in signal-processing domain). Equations (12.14)–(12.15) represent a pair of n-th-order weighting functions.

There exists one disadvantage about H_∞ control, i.e., the calculated controller $K(s)$ is of the same order as that of the physical system model and the control input is dependent on all the states of the system. In AFC, decentralized low-order controllers are desired for practical implementation purposes. To overcome this issue, the decentralized static output feedback technique [10] is employed to incorporate the specific requirement on controller patterns. Decentralized controllers under AFC are designed to be in the form of

$$\begin{aligned} \Delta P_{Gi}^{ref} &= -k_{1i}\Delta\omega_{Gi} - k_{2i}\Delta\theta_{Gi} \\ \Delta P_{Si}^{ref} &= -k_{3i}\Delta\omega_{Si} + k_{4i}\Delta SoC_{Si} \end{aligned} \tag{12.16}$$

where the generator side controller is of PI type and the storage controller has a negative feedback with respect to the frequency deviations and a positive feedback with respect to its SoC.

A custom solver based on iterative linear matrix inequality is developed, as there are no available commercial solvers that are capable of solving the specific AFC H_∞ control problem. For details about the derivation of the solver algorithm, interested readers may refer to [7].

12.3.5 Case study

As an extension to the previous work in [7], this chapter demonstrates the effectiveness of AFC via the modified IEEE New England 39-bus system shown in Figure 12.3. The energy storage is placed at Bus 27 and the wind generator is installed at Bus 26. The system power base is 100 MVA. The maximum input/output power rating of the energy storage is 100 MW (1 p.u.) and its capacity is

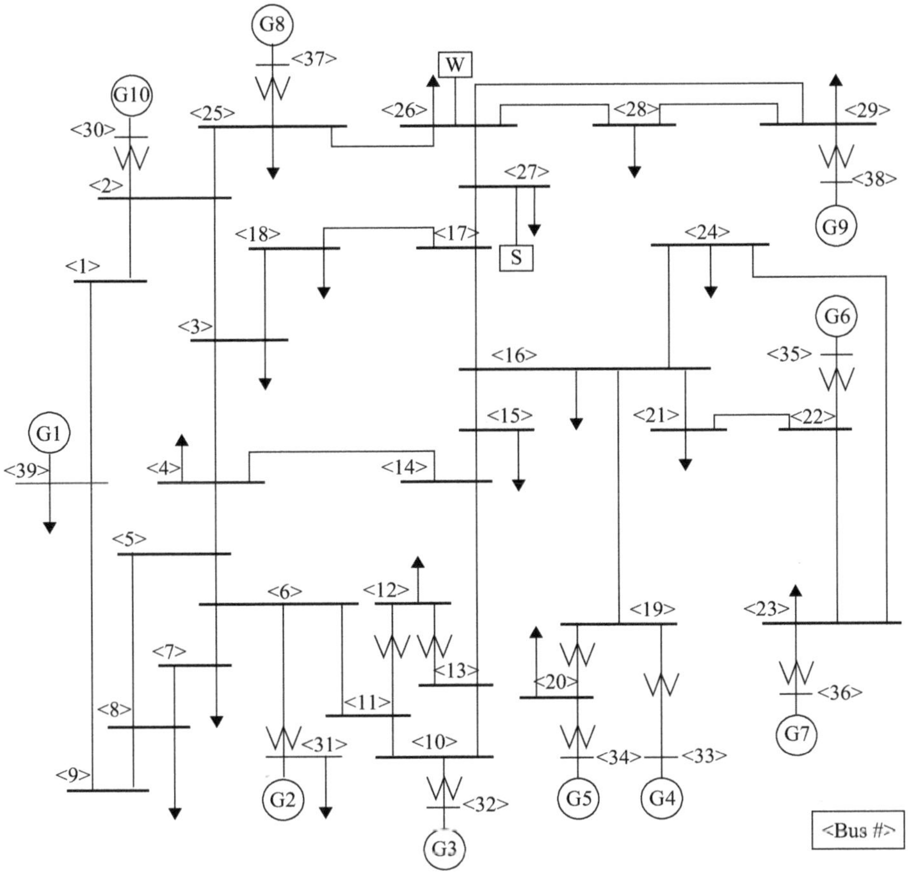

Figure 12.3 Modified IEEE New England 39-bus system [11]

5 MWh (0.05 p.u.h.). The wind power outputs are assumed to be within about ±15% variations around the mean value of 800 MW shown in Figure 12.4.

The magnitude plot of the frequency-dependent weighting functions is shown in Figure 12.5. It can be seen from the figure that the high-frequency region in the spectrum of the conventional generator power output has ten times more weights than the low-frequency region.

In order to illustrate the effectiveness of the AFC control, the traditional primary and secondary controls are used as a baseline, which is denoted by CFC (i.e., conventional frequency control). For the CFC case, all the generators use a typical 5% droop governor setting for primary control and the centrally organized secondary control parameters are determined by experience as the way it is decided in practice. In CFC, the energy storage device is treated as if it is a generator type asset.

The effect of AFC on frequency regulation performance is shown in Figure 12.6. It is clear that the AFC is able to regulate frequencies in a much tighter window compared to the CFC case.

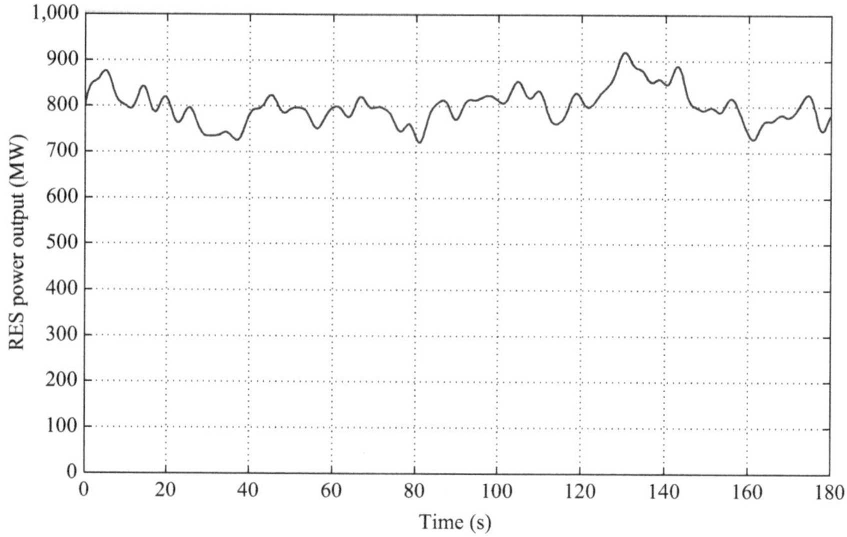

Figure 12.4 Wind power variations

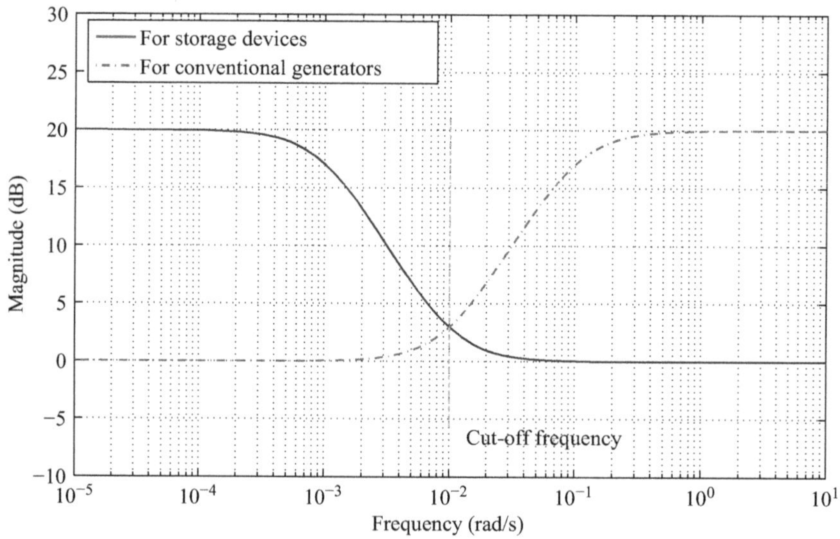

Figure 12.5 Frequency-dependent weighting functions ($f_c = 0.0016$ Hz, $n = 1$)

Figure 12.7 shows the comparison of generator responses between AFC and CFC. The generator power output under CFC fluctuates much more heavily and frequently than in AFC. To better understand the reduced burden on generator ramping, real-power output from Generator 8 in the frequency domain is shown in Figure 12.8. It is evident that the generator under AFC is mainly in charge of wind power variations below 0.01 rad/s compared to its responsibility of up to 0.2 rad/s under CFC.

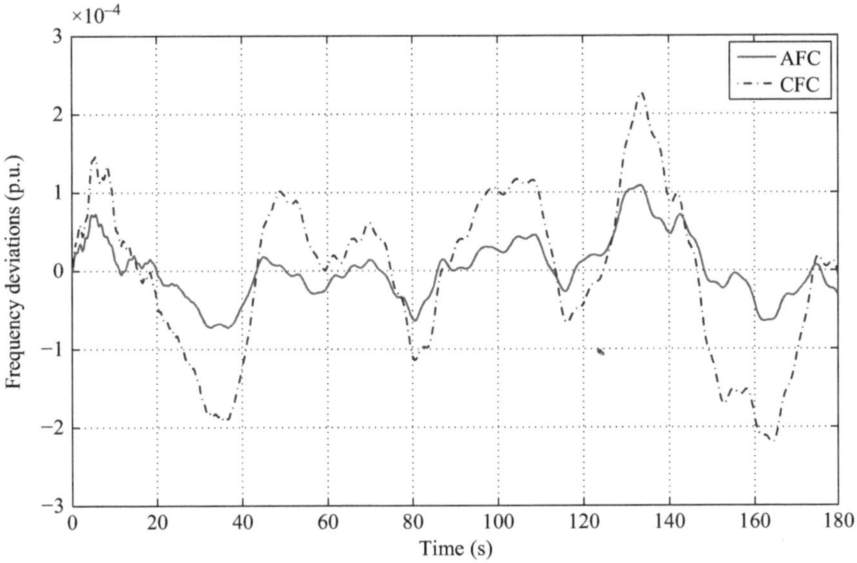

Figure 12.6 Frequency deviations at Bus 37

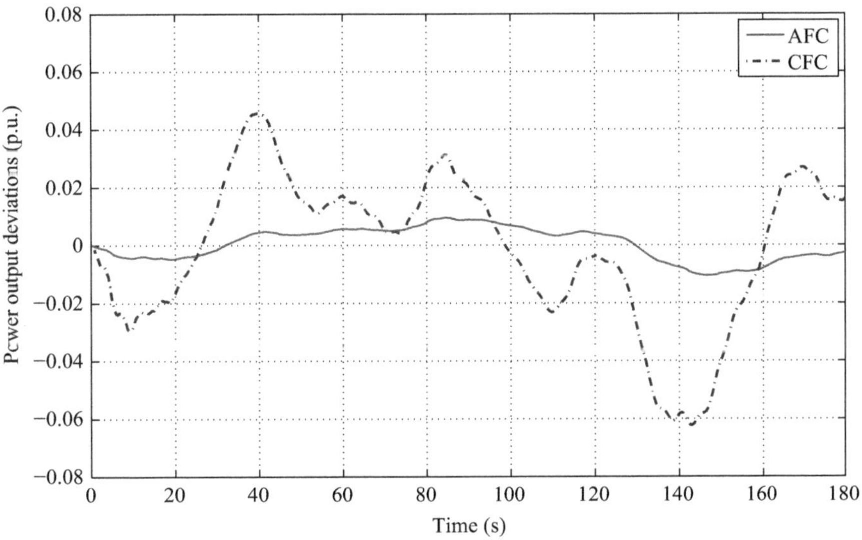

Figure 12.7 Generator response in power output at Bus 37

On the other hand, the power output of the energy storage together with its SoC in time domain is shown in Figure 12.9. Although the power output in storage varies rapidly, its SoC stays within a relatively tight range of $\pm 5\%$. To see more clearly about the performance of AFC in regulating SoC, the comparison of

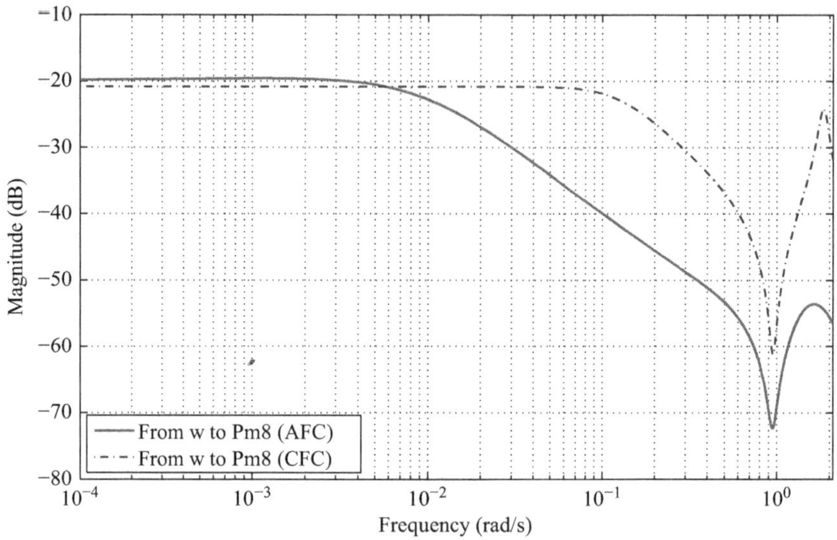

Figure 12.8 Generator response at Bus 37 in frequency domain

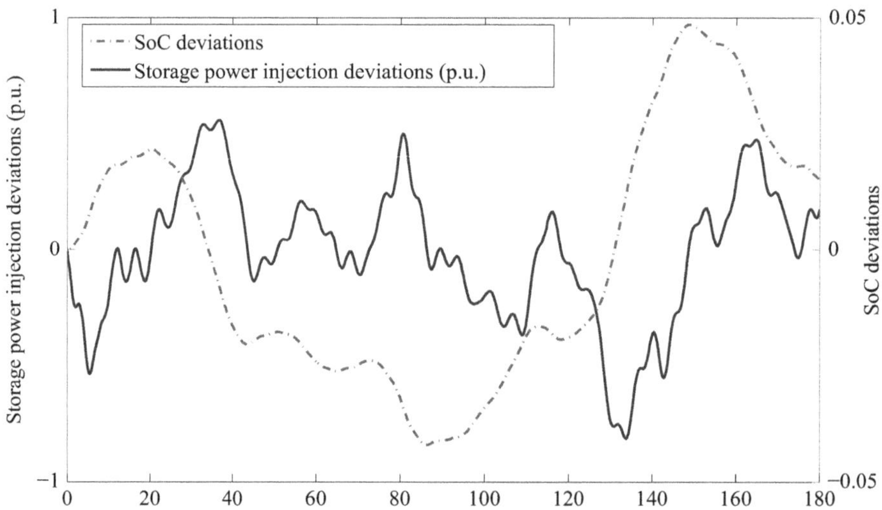

Figure 12.9 Energy storage power output and SoC under AFC in time domain

magnitude plots for the transfer function from external disturbances to the SoC between AFC and CFC is plotted in Figure 12.10. The curve corresponding to AFC is bounded, which backs up the time domain observation in Figure 12.9 that the SoC variations are bounded given that the disturbance input is bounded. In contrast,

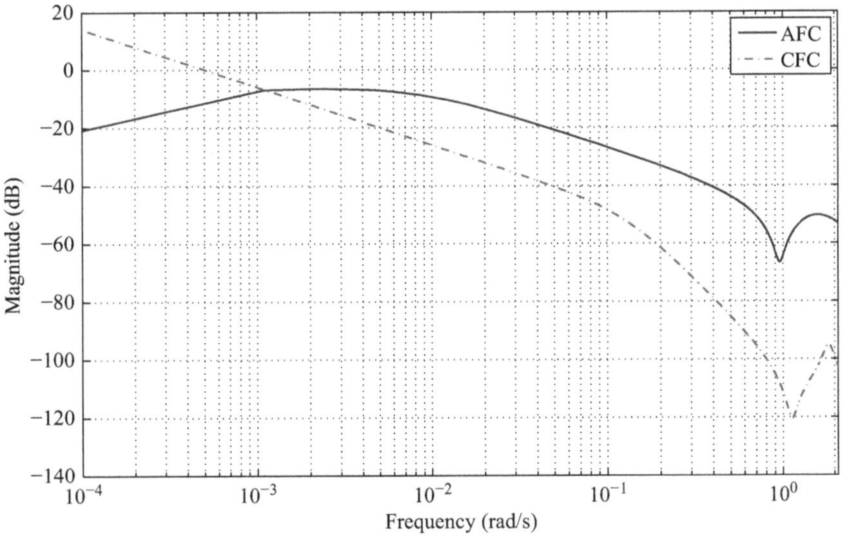

Figure 12.10 Energy storage response w.r.t. SoC in frequency domain

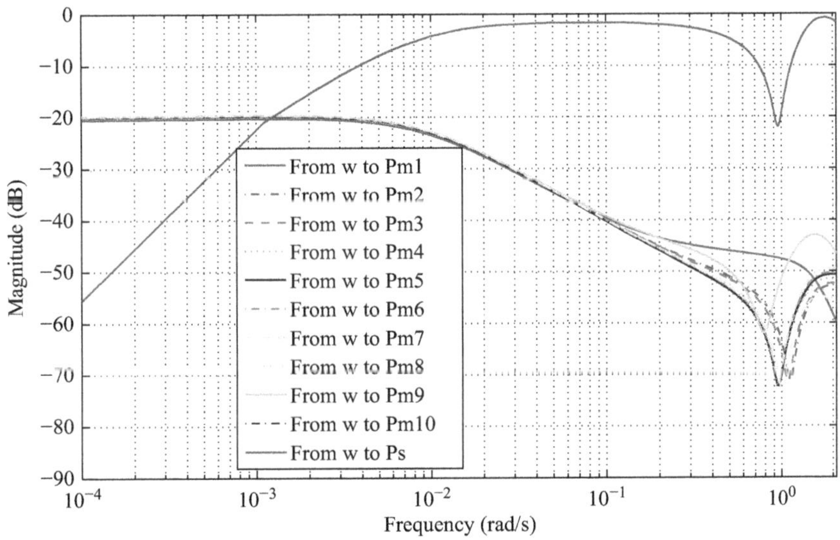

Figure 12.11 Generator and storage power output under AFC in frequency domain

it is very likely that the storage device under CFC will hit its SoC limit if the disturbance input contains a lot of low-frequency components.

The magnitude plot of the transfer function from disturbance to power output of the ten generators and the energy storage device under AFC is shown in Figure 12.11,

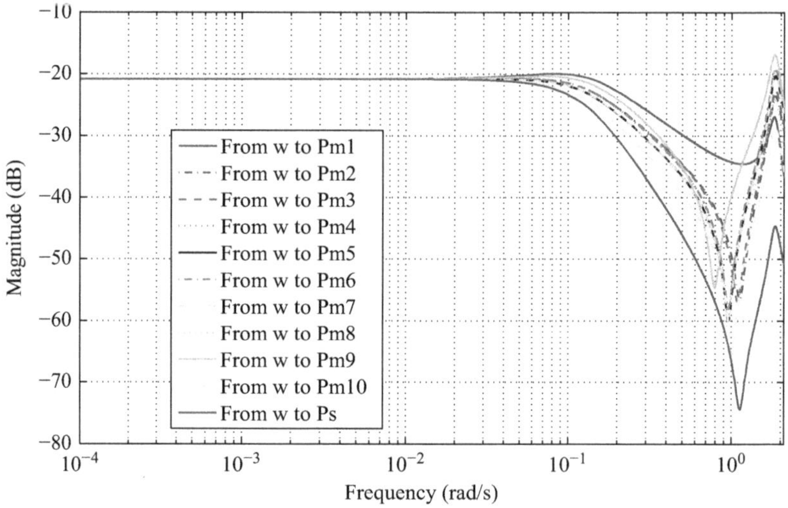

Figure 12.12 Generator and storage power output under CFC in frequency domain

which verifies that the time-scale matching objective is achieved with a separation frequency of 0.01 rad/s (~0.0016 Hz). In contrast, the counterpart magnitude plot under CFC is plotted in Figure 12.12. The behavior of the energy storage device under CFC is almost identical to the generators.

12.4 Stochastic optimal dispatch level

This level aims to deal with the second, third, and fourth issues mentioned in Section 12.1, i.e., the current frequency control scheme does not take into account the increased uncertainty caused by RESs at the tertiary level and in addition, the current control scheme does not consider the special operational requirement on SoC for the storage type assets. Based on the theory of SMPC, the level of stochastic optimal dispatch solves a two-stage stochastic version of the traditional security-constrained economic dispatch (SCED) problem in power systems. The main motivation of reformulating the SCED problem as a stochastic optimization problem is that stochastic optimization is better equipped to handle real-world problems with uncertainties. Dealing with uncertainties is one of the most critical issues that we are facing for renewable generation integration. In addition, the stochastic optimal dispatch level takes into account the limit on SoC for energy storage devices to ensure their safe operation. The time-scale matching objective can be achieved in two ways. One way is to include the ramping costs of conventional generators into the objective function while the other way is to directly specify the limits on the generator ramping rates as optimization constraints. However, the caveat of conducting a stochastic optimization is that the resulting problem is usually very huge in size and is difficult and time-consuming to solve. It is even worse when stochastic optimization is combined with the model predictive control technique. Therefore, optimization decomposition techniques need to be utilized to decompose the overall problem into many subproblems so that the subproblems can be solved in parallel to speed up the solution process.

12.4.1 SMPC basics and problem formulation

SMPC is an advanced control technology that integrates the advantage of explicit inclusion of uncertainties in stochastic programming and the capability of anticipating the future behavior of the target system when making control decisions in model predictive control. A two-stage problem is typically considered in practical applications. The SMPC controller merely implements the first-stage control action of a two-stage problem at each time step. The second-stage decision in theory of stochastic programming is a collection of recourse actions that need to be taken in response to each random outcome for the considered uncertainties. However, these recourse decisions are never implemented under the SMPC setup because a new two-stage SMPC problem incorporating the new information regarding uncertainties and system states will be formulated and solved at the next time step according to the spirit of model predictive control. The reason for this is that there are always errors in predictions and mathematical models, e.g., the solar power output is inaccurately predicted, or losses of the storage devices are not exactly modeled.

A two-stage SMPC problem at time step t has the following general form:

$$\min_{u^s} \sum_{s \in \mathcal{N}} \pi^s \left[\sum_{k \in \mathcal{T}} l_k(x^s(t+k), u^s(t+k)) + l_K(x^s(t+K)) \right] \tag{12.17}$$

$$\begin{aligned} \text{s.t.} \quad & x^s(t+k+1) = Ax^s(t+k) + Bu^s(t+k) && \forall s \in \mathcal{N}, k \in \mathcal{T} \\ & x^{\min} \le x^s(t+k+1) \le x^{\max} && \forall s \in \mathcal{N}, k \in \mathcal{T} \\ & u^{\min} \le u^s(t+k) \le u^{\max} && \forall s \in \mathcal{N}, k \in \mathcal{T} \\ & h(x^s(t+k), u^s(t+k), d^s(t+k)) < 0 && \forall s \in \mathcal{N}, k \in \mathcal{T} \\ & u^s(t) = u^{s+1}(t) && \forall s \in \{1, \cdots, N-1\} \end{aligned} \tag{12.18}$$

where the parameters and variables are given as:

$\mathcal{N}$: the scenario set, $\mathcal{N} \triangleq \{1, \cdots, N\}$,
$\mathcal{T}$: the receding horizon set, $\mathcal{T} \triangleq \{0, \cdots, K-1\}$,
K: look-ahead horizon length,
π^s: probability for scenario s,
$x^s(t+k)$: states at time step $t+k$ in scenario s,
$u^s(t+k)$: control inputs at time step $t+k$ in scenario s,
$d^s(t+k)$: disturbances at time step $t+k$ in scenario s,
$l_k(\cdot)$: cost function at time step $t+k$ in scenario s,
$l_K(\cdot)$: cost function at final step $t+K$ in scenario s.

The first three constraints of (12.18) are the state-space representation of the system under investigation. The forth constraint of (12.18) captures intra-temporal constraints such as the network constraints in our case. The last constraint of (12.18) is unique for stochastic programming problems, which requires that the first-stage decision variables should be identical for all possible scenarios allowing a nonambiguous solution for implementation.

By following the aforementioned general form of the two-stage SMPC problem, the specific optimization formulation for the stochastic optimal dispatch level is given below. Without loss of generality, the current time step t is assumed to be 0.

$$\min_{\Delta P^s_G, P^s_{Si}, P^s_{So}} \sum_{s\in\mathcal{N}} \pi^s \left[\sum_{k\in\mathcal{T}} \sum_{i\in\Omega_G} C_{G_i}(P^s_{G_i}(k), \Delta P^s_{G_i}(k)) + \sum_{k\in\mathcal{T}} \sum_{i\in\Omega_S} C_{S_i}(P^s_{Si_i}(k), P^s_{So_i}(k)) + \sum_{i\in\Omega_G} C_{G_i}(P^s_{G_i}(K)) \right] \tag{12.19}$$

$$\begin{aligned}
s.t.\quad & \forall s \in \mathcal{N}, k \in \mathcal{T}: \\
& E^s_{S_i}(k+1) = \eta_i E^s_{S_i}(k) + \alpha_i T \cdot P^s_{Si_i}(k) - \frac{1}{\alpha_i} T \cdot P^s_{So_i}(k) && \forall i \in \Omega_S \\
& P^s_{G_i}(k+1) = P^s_{G_i}(k) + \Delta P^s_{G_i}(k) && \forall i \in \Omega_G \\
& E^{\min}_{S_i} \le E^s_{S_i}(k+1) \le E^{\max}_{S_i} && \forall i \in \Omega_S \\
& 0 \le P^s_{Si_i}(k) \le P^{\max}_{S_i} && \forall i \in \Omega_S \\
& 0 \le P^s_{So_i}(k) \le P^{\max}_{S_i} && \forall i \in \Omega_S \\
& P^{\min}_{G_i} \le P^s_{G_i}(k+1) \le P^{\max}_{G_i} && \forall i \in \Omega_G \\
& \Delta P^{\min}_{G_i} \le \Delta P^s_{G_i}(k) \le \Delta P^{\max}_{G_i} && \forall i \in \Omega_G \\
& \sum_{i\in\Omega_G} (P^s_{G_i}(k) + \Delta P^s_{G_i}(k)) + \sum_{i\in\Omega_R} P^s_{R_i}(k) \\
& \quad - \sum_{i\in\Omega_D} P^s_{D_i}(k) - \sum_{i\in\Omega_S} (P^s_{Si_i}(k) - P^s_{So_i}(k)) = 0 \\
& -P^{\max}_{ij} \le DF_{ij} \cdot P^s(k) \le P^{\max}_{ij} && \forall ij \in \Omega_L \\
& \forall s \in \{2, \cdots, N\}: \\
& \Delta P^1_{G_i}(0) = \Delta P^s_{G_i}(0) \quad \forall i \in \Omega_G \\
& P^1_{Si_i}(0) = P^s_{Si_i}(0) \quad \forall i \in \Omega_S \\
& P^1_{So_i}(0) = P^s_{So_i}(0) \quad \forall i \in \Omega_S
\end{aligned} \tag{12.20}$$

where the parameters and variables are given as:

$\mathcal{N}$: the scenario set, $\mathcal{N} \triangleq \{1, \cdots, N\}$,
$\mathcal{T}$: the receding horizon set, $\mathcal{T} \triangleq \{0, \cdots, K-1\}$,
K: look-ahead horizon length,
π^s: probability for scenario s,
Ω_G: the bus set for generators,
Ω_S: the bus set for storage devices,
Ω_R: the bus set for renewable generators,
Ω_D: the bus set for demands,
Ω_L: the set of transmission lines,
$E^s_{S_i}$: stored energy for storage device at bus i in scenario s,
$P^s_{Si_i}$: power input into storage device at bus i in scenario s,
$P^s_{So_i}$: power output from storage device at bus i in scenario s,
η_i: self-discharge rate for storage device at bus i,
α_i: power conversion loss of storage device at bus i,
T: the time step size,
$P^s_{G_i}$: power output from generator at bus i in scenario s,
$\Delta P^s_{G_i}$: power output change in a step from generator at bus i in scenario s,
$P^s_{R_i}$: power output from renewable generator at bus i in scenario s,
$P^s_{D_i}$: load consumption at bus i in scenario s,
$C_{G_i}(\cdot)$: generation and ramping costs associated with generator at bus i,
$C_{S_i}(\cdot)$: energy conversion cost for storage device at bus i,
DF_{ij}: the row in distribution factor matrix corresponding to line $i-j$,
P^s: power injection vector in scenario s.

The stochasticity in (12.19) mainly comes from the uncertainties of the renewable power generation over the look-ahead horizon, which is captured by a set of scenarios with corresponding probabilities. The cost function for generators $C_{G_i}(\cdot)$ is quadratic so that the changes in generator power output between two consecutive time steps are minimized and the time-scale matching can be achieved. The solution to (12.19) serves as the set point for the AFC level during the 5 to 15 min time interval. The system dynamics including energy storage and conventional generation together with the limits on states and control input variables are captured in the first seven constraints of (12.20). The eighth and ninth constraints of (12.20) represent the DC power flow model with consideration of transmission line limits. The unique constraint of the two-stage SMPC problem for ensuring identical solution for the first stage decisions is included in the last three constrains of (12.20). It can be seen that the optimization problem here has both cross scenario and cross time step dependencies, making solving the problem difficult.

12.4.2 Solution approach

In order to speed up the solution process for the SMPC problem, optimization decomposition techniques need to be applied. Based on a detailed study of optimization decomposition methods for SMPC problems in [12], the optimality condition decomposition (OCD) method which was first introduced in [13] is identified as an efficient method to facilitate decomposition and parallelization for (12.19). Although OCD is classified into the category of dual decomposition, no parameter tuning is needed for the update of Lagrangian multipliers. In the simplest case, convergence can be reached by just exchanging coupling variable values among all subproblems in each iteration.

A simple two-subproblem example is used to illustrate the basic idea of OCD. Assume we have the following overall optimization problem to solve with decision vectors x_1 and x_2, individual constraints $g_i(\cdot)$, and complicating constraints $h_i(\cdot)$. λ_i and μ_i are the corresponding Lagrangian multipliers associated with each constraint.

$$\begin{aligned} \min_{x_1,x_2} \quad & f_1(x_1) + f_2(x_2) \\ s.t. \quad & g_i(x_i) = 0 \cdots\cdots\cdots \lambda_i, i = 1,2 \\ & h_i(x_1,x_2) = 0 \cdots\cdots\cdots \mu_i, i = 1,2 \end{aligned} \tag{12.21}$$

The first subproblem after OCD decomposition has the following form. Note that the decision vector of the second subproblem x_2 is fixed in the first subproblem and denoted by the bar above the variable. The second subproblem has a symmetric formulation. After each iteration, the two subproblems just need to exchange information about their own decision vectors. The iterative solution process continues until convergence.

$$\begin{aligned} \min_{x_1} \quad & f_1(x_1) \\ s.t. \quad & g_1(x_1) = 0 \cdots\cdots\cdots \lambda_1 \\ & h_1(x_1,\bar{x}_2) = 0 \cdots\cdots\cdots \mu_1 \end{aligned} \tag{12.22}$$

To help understand OCD, we first write out the Lagrangian function for the overall problem (12.21).

$$L(x_1,\lambda_1,\mu_1,x_2,\lambda_2,\mu_2) = \sum_{i=1}^{2}(f_i(x_i) + \lambda_i^T g_i(x_i) + \mu_i^T h_i(x_1,x_2)) \tag{12.23}$$

In order to solve the overall problem, we just need to solve the well-known Karush–Kuhn–Tucker (KKT) system based on the Lagrangian function. Any solvers for nonlinear equation systems such as the Newton–Raphson (NR) method can be applied. The basic idea of OCD is to make the off diagonal blocks in the KKT system zero, which effectively removes any couplings among the subproblems. Figure 12.13 illustrates this fundamental idea in the OCD method.

The flow chart of the complete iterative solution process for the proposed SMPC problem using OCD is shown in Figure 12.14. NR steps are carried out first. If needed, the generalized minimal residual (GMRES) algorithm and the line search algorithm will be employed to ensure convergence. The overall SMPC problem in (12.19) can now be decomposed into smaller subproblems based on different scenarios.

$$\begin{bmatrix} \nabla^2_{x_1x_1}L & \nabla^T_{x_1}g_1 & \nabla^T_{x_1}h_1 & \nabla^2_{x_2x_1}L & 0 & \nabla^T_{x_1}h_2 \\ \nabla_{x_1}g_1 & 0 & 0 & 0 & 0 & 0 \\ \nabla_{x_1}h_1 & 0 & 0 & \nabla_{x_2}h_1 & 0 & 0 \\ \nabla^2_{x_1x_2}L & 0 & \nabla^T_{x_2}h_1 & \nabla^2_{x_2x_2}L & \nabla^T_{x_2}g_2 & \nabla^T_{x_2}h_2 \\ 0 & 0 & 0 & \nabla_{x_2}g_2 & 0 & 0 \\ \nabla_{x_1}h_2 & 0 & 0 & \nabla_{x_2}h_2 & 0 & 0 \end{bmatrix} \begin{bmatrix} \Delta x_1 \\ \Delta\lambda_1 \\ \Delta\mu_1 \\ \Delta x_2 \\ \Delta\lambda_2 \\ \Delta\mu_2 \end{bmatrix} = - \begin{bmatrix} \nabla_{x_1}L \\ g_1 \\ h_1 \\ \nabla_{x_2}L \\ g_2 \\ h_2 \end{bmatrix}$$

Figure 12.13 OCD illustration via a two-subproblem example

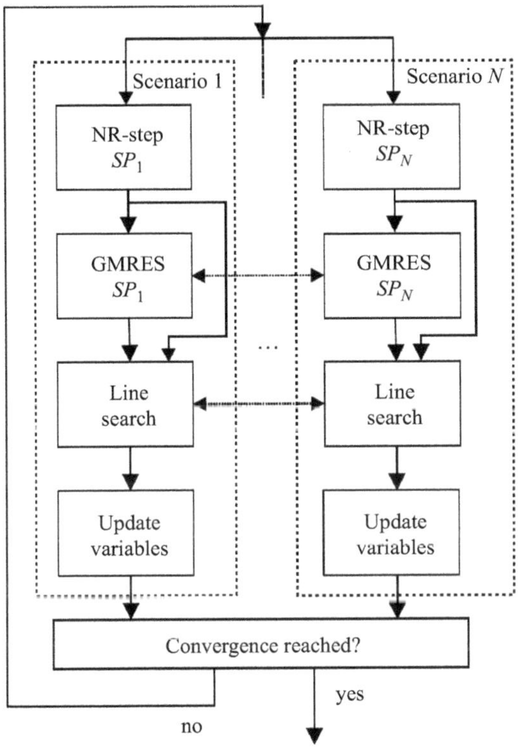

Figure 12.14 Flow chart of the iterative update using OCD

12.4.3 Case study

Since how to efficiently solve the large-scale SMPC problem is the core part of the stochastic optimal dispatch level, the case study section focuses on a single SMPC problem for one time step. The same 39-bus test system shown in Figure 12.3 is used again to demonstrate the effectiveness of the decomposition method.

Table 12.1 Numerical results

Cases	Major iterations	GMRES iterations	Wall clock time (s)	Estimated time in parallel (s)
Direct method	262	N/A	1,864.32	1,864.32
OCD method	234	1780	1,901.33	38.03

For the test case, the power output from the renewable generator at Bus 26 over a 4-hour look-ahead horizon is represented by 50 possible scenarios, each of which is associated with a probability. Each time step is of 5-min length. With this setup, the resulting SMPC problem is very large in size, which has more than 55,000 decision variables and over 353, 000 constraints.

The comparison between the direct method and the OCD method in computing time is shown in Table 12.1. The direct method is to directly solve the overall problem without decomposition by iteratively solving the corresponding KKT system. One cycle from "NR-step" to "Update variables" in the flow chart of Figure 12.14 forms one major iteration. The wall clock time is the time that is spent when a computer program executes in series. The estimated time in parallel is then approximated by the wall clock time divided by the number of subproblems.

Although the estimated time in parallel is an ideal-world estimation, it is sufficient to demonstrate the effectiveness of the OCD method. The OCD method is about 49 times faster than the direct method without any decomposition with respect to the estimated time in parallel.

12.5 Summary

This chapter aims at providing some ideas on how to fit energy storage devices into the constantly evolving grid environment with more and more renewable generation. The core idea of this chapter is to assign tasks according to the capabilities of the resource assets. In the context of real-power balancing, this capability-based task assignment is exactly the so-called time-scale-matching principle. The proposed two-level optimal control approach makes full use of the fast response capability of energy storage to tackle the real-power-balancing problem under the new environment. With the coordinated control of energy storage, better performance in frequency control is seen from the simulation results even that the conventional generators become less responsive to high-frequency portions in the spectrum of external power variations.

References

[1] M. Beaudin, H. Zareipour, A. Schellenberglabe, and W. Rosehart, "Energy storage for mitigating the variability of renewable electricity sources: An updated review", *Energy for Sustainable Development*, vol. 14, no. 4, pp. 302–314, 2010.

[2] R. Fioravanti, Vu Khoi, and W. Stadlin, "Large-scale solutions", *IEEE Power Energy Magazine*, vol. 7, no. 4, pp. 48–57, 2009.
[3] S. Chu and A. Majumdar, "Opportunities and challenges for a sustainable energy future", *Nature*, vol. 488, no. 7411, pp. 294–303, 2012, 10.1038/nature11475.
[4] J. M. Carrasco, L. G. Franquelo, J. T. Bialasiewicz, *et al.*, "Power electronic systems for the grid integration of renewable energy sources: A survey", *IEEE Transactions on Industrial Electronics*, vol. 53, no. 4, pp. 1002–1016, 2006.
[5] D. Lew, G. Brinkman, N. Kumar, P. Besuner, D. Agan, and S. Lefton, "Impacts of wind and solar on emissions and wear and tear of fossil-fueled generators", in Proceedings of the IEEE Power and Energy Society General Meeting, San Diego, CA, 2012. https://ieeexplore.ieee.org/document/6343967/.
[6] H. Bevrani, *Robust power System frequency control*, Power Electronics and Power Systems. New York: Springer, 2nd edition, 2014.
[7] D. Zhu and G. Hug, "Coordination of storage and generation in power system frequency control using an H infinity approach", *IET Generation, Transmission & Distribution*, vol. 7, no. 11, pp. 1263–1271, 2013.
[8] K. Zhou and J.C. Doyle, *Essentials of robust control*. Upper Saddle River, NJ: Prentice Hall, 1998.
[9] S. Skogestad and I. Postlethwaite, *Multivariable feedback control: Analysis and design*, New York: John Wiley & Sons, 2nd edition, 2005.
[10] Y. Y. Cao, J. Lam, and Y. X. Sun, "Static output feedback stabilization: An ILMI approach", *Automatica*, vol. 34, no. 12, pp. 1641–1645, 1998.
[11] M. A. Pai, *Energy function analysis for power system stability*, Boston, MA: Springer, 1989.
[12] D. Zhu and G. Hug, "Decomposed stochastic model predictive control for optimal dispatch of storage and generation", *IEEE Transactions on Smart Grid*, vol. 5, no. 4, pp. 2044–2053, July 2014.
[13] A. J. Conejo, F. J. Nogales, and F. J. Prieto, "A decomposition procedure based on approximate Newton directions", *Mathematical Programming*, vol. 93, no. 3, pp. 495–515, 2002.

Index

  Xtreme Power on Kodiak Island, Alaska 231

www.ingramcontent.com/pod-product-compliance
Lightning Source LLC
LaVergne TN
LVHW051046210926
848486LV00001B/41

* 9 7 8 1 7 8 5 6 1 3 4 9 4 *